I0130910

Human Resources in the Urban Economy

Due to the urbanisation of American society and the economic problems that accompanied it; a series of conferences was held to explore the economics of human resources. Originally published in 1963, this study draws together papers from the first conference dealing mainly with the under-utilisation and misallocation of human resources, as well as wage rates, migration patterns and education in urban societies and the impact they have on the American labour force. This title will be of interest to students of Environmental Studies and Economics.

Human Resources in the Urban Economy

Edited by
Mark Perlman

RFF PRESS
RESOURCES FOR THE FUTURE

First published in 1963
by Resources for the Future, Inc.

This edition first published in 2016 by Routledge
2 Park Square, Milton Park, Abingdon, Oxon, OX14 4RN
and by Routledge
711 Third Avenue, New York, NY 10017

Routledge is an imprint of the Taylor & Francis Group, an informa business

© 1963 Resources for the Future, Inc.

Publisher's Note
The publisher has gone to great lengths to ensure the quality of this reprint but points out that some imperfections in the original copies may be apparent.

The publishers would like to make it clear that the views and opinions expressed, and language used in the book are the author's own and a reflection of the times in which it was published. No offence is intended in this edition

Disclaimer
The publisher has made every effort to trace copyright holders and welcomes correspondence from those they have been unable to contact.

A Library of Congress record exists under LC control number: 63022775

ISBN 13: 978-1-138-96320-7 (hbk)
ISBN 13: 978-1-315-65891-9 (ebk)
ISBN 13: 978-1-138-96321-4 (pbk)

Human Resources in the Urban Economy

Papers presented at a conference of the Human Resources Sub-committee of the Committee on Urban Economics, Resources for the Future, November 16 and 17, 1962, plus an earlier working paper

EDITED BY MARK PERLMAN

RESOURCES FOR THE FUTURE, INC.
1755 Massachusetts Avenue, N.W., Washington 36, D. C.

Distributed by
THE JOHNS HOPKINS PRESS
Baltimore, Maryland 21218

RESOURCES FOR THE FUTURE, INC.

1755 Massachusetts Avenue, N. W., Washington 36, D. C.

Resources for the Future is a nonprofit corporation for research and education in the development, conservation, and use of natural resources. It was established in 1952 with the co-operation of The Ford Foundation and its activities since then have been financed by grants from that Foundation. Part of the work of Resources for the Future is carried out by its resident staff; part supported by grants to universities and other nonprofit organizations. Unless otherwise stated, interpretations and conclusions in RFF publications are those of the authors; the organization takes responsibility for the selection of significant subjects for study, the competence of the researchers, and their freedom of inquiry.

The conference at which these papers were presented was part of the program of RFF's Committee on Urban Economics. Members of the Committee are: Harvey S. Perloff, of RFF (chairman); Harold J. Barnett, Washington University; Joseph L. Fisher, of RFF; Lyle C. Fitch, Institute of Public Administration; Alvin H. Hansen, emeritus professor, Harvard University; Walter W. Heller, Council of Economic Advisers; Werner Z. Hirsch, University of California (Los Angeles); Edgar M. Hoover, University of Pittsburgh; Howard G. Schaller, Tulane University; Leo F. Schnore, University of Wisconsin; Arthur M. Weimer, Indiana University; and Robert C. Wood, Massachusetts Institute of Technology.

RFF publications staff: Henry Jarrett, editor; Vera W. Dodds, associate editor; Nora E. Roots, assistant editor.

Preface

Difficult problems are emerging in the wake of the rapid urbanization of American society and many of these are largely economic in nature. Some years ago, Resources for the Future, with the support of a special grant from The Ford Foundation, organized a Committee on Urban Economics to spur the development of a broader understanding of the urban economy and to explore application of the tools of economic analysis to critical urban problems.

One of the major fields of concern of the Committee has been that of human resources, especially households in their twin roles of producing labor services for the urban economy and as consumers of goods and services. It was evident to the Committee that there was much that could be drawn in substance and method from the well established fields of labor economics and consumer economics. The initial problem was to define the issues in a way which would attract scholars in these fields to turn their attention to urban configurations. Questions of under-utilization and misallocation of human resources in the city particularly seemed to deserve attention, but questions of differential wage rates, migration patterns, and education, as they influenced the character of the supply in the urban labor market, also called for deeper understanding. On the consumption side, greater knowledge concerning the role of income distribution, discrimination, social patterns, and similar crucial variables could be expected to suggest new approaches to solving persistent community problems.

In the autumn of 1960, the Committee asked Mark Perlman, then of The Johns Hopkins University, to review the work done in the fields of labor and consumer economics, to consult with scholars in these fields, and to recommend a conference and research program. The main focus of his effort was a working paper to suggest lines of research endeavor. This paper was reviewed in detail at a seminar held in May of 1961 by a group of scholars and practitioners who had for some time been concerned with the problems of human resources in the city. Two additional papers were discussed at this seminar as well, one by Herbert Klarman (of the Hospital Council of Greater New York) and one by Burton Weisbrod (of Washington University).

Perlman's recommendation for a series of conferences on the economics of human resources that would permit interested scholars to exchange ideas and help advance the field was well received and subsequently a Committee on the Economics of Urban Human Resources was formed to sponsor the conferences. This committee included: Robert Aronson (Cornell University), Henry Cohen (Deputy City Administrator of New York City), F. Ray Marshall (University of Texas), Richard Muth (University of Chicago), Jacob Mincer (Columbia University), Harvey Perloff

(Resources for the Future, _ex officio_), Mark Perlman, Leo Schnore (University of Wisconsin), Martin Segal (Dartmouth College), and Lowdon Wingo, Jr. (Resources for the Future). Perlman served as chairman until after the first conference; Marshall thereafter replaced him.

The papers presented in this volume are the product of the first conference, held in November of 1962. The Committee sponsoring the Conference had decided, in the main, to draw on research already under way which touched on themes significant to urban economics, rather than to request papers on special topics involving entirely new research. This seemed a useful means of bridging the gap between the established work under way in labor and consumer economics and the quite new approaches suggested by the urban focus. Thus, this volume, at one and the same time, reveals what might be borrowed -- to a great advantage -- from the more traditional ways of looking at production and consumption and highlights the need for new and specifically-focused urban research.

Mark Perlman undertook to edit these papers and organized this volume. I want to acknowledge our gratitude to him: his efforts highlighted the main themes and converted these papers into a book.

Harvey S. Perloff

Editor's Introduction

The use of human resources has always been recognized as an economic problem. At various times concern centers on "do we have enough?" At other times unemployment comes to the fore and the question is then raised, "do we have too much?"

At this time in the United States (and likely in the future during the 1960's), we are confronted with both questions. In briefest terms, the situation is this: The rate of technological development with widespread substitution of capital for labor is greater now than ever before. Industrial demand for large quantities of some types of highly skilled personnel at attractive unit labor costs is great and can be presumed to be growing. As a result, shortages of some skills will exist at current wage levels. In short, our advanced technology requires that someone, be it individuals, firms, or society-at large, prepare a flow of skills for the future. By way of contrast, the demand for unskilled personnel at any hourly rate equal to or above the legal minimum has fallen away most sharply. Thus we face the prospect of having little or no gainful employment for great numbers of unskilled individuals. What is true of the unskilled generally is true of some workers with particular outmoded skills, as well.

This situation is made more crucial by the fast increase in our population: the "baby-boom" of the forties and fifties is just coming into the labor force. Also, social reform has encouraged once tightly-held-down minority groups to migrate from their subsistence farms to urban labor markets. Thus we are faced not only by the press of what to do with our current stock of unemployed, but also what to do with the vast flow of young adults, not a few of whom incorrectly anticipated employment as unskilled, or at best as semiskilled, operatives.

The key point is disaggregation. We must know more about the stock and flow of critical skills. We must know how many employed workers there are, what are their age and earnings distributions, where are they located geographically, and how mobile are they. Also, we must know much more about the unemployed: Where are they geographically, who are they ethnically, what is their age distribution, and, most of all, what skills do they possess?

This book is devoted to the questions of conceptualizing and analyzing human resources allocations in American cities. The city, as a subsequent essay points out, is becoming not only the "consumer" of human skills; it is also becoming the "producer" of these skilled humans. It is the theater where traditions (both good and bad) are inculcated or changed as the labor force operates and grows.

The book consists of four major parts. Briefly, they are (1) my original working paper setting out my views of the dimensions of the field to be explored; (2) three papers on the general topic of the effects of discrimination in the labor market; (3) three papers on consumer economics; and (4) four papers on the economics of labor force migration.

PART I

My paper starts with a plea that labor economists turn from the more special problems of industrial employment to the consideration of what urban living means to individuals in their inter-related roles of producing and consuming. I argue that the modern American city is a complex organism whose unfolding economic patterns are to a large degree shaped by its own past decisions and experience. Its citizens (used in the social rather than the political sense) are affected by the costs of industrial obsolescence and ultimate relocation, by the impact of social and ethnic prejudice not only in the labor market but also in the money market, and by the burden of having to pay for investment in the training of new and "raw" populations. I conclude by returning to the original point, namely, how consumer decisions shape the quality of the producing labor force; in particular, the problem of assessing and forecasting the use of public funds to subsidize households' consumption.

PART II

The general theme of Part Two is the labor force participation and performance of "minority" groups. *

F. Ray Marshall (University of Texas) reviews the material on the problems of entry of Negroes into the skilled trades. His interest is principally (and properly) positivistic. He examines where Negroes suffer discrimination and argues that inadequate training is only part of the story. Unions, cultural traditions, economic vagaries, and managements' attitudes all play roles in restricting opportunities for Negroes to become apprenticed. Once Negroes do have training, however, other factors become operative; these include seniority rules, job-security clauses in collective agreements, and Fair Employment Practice legislation.

Harry Gilman (Rutgers University) discusses one aspect of the frequently heard proposition that Negroes are "the last to be hired and the first to be fired." He notes that Negroes are principally in jobs which are business-cycle sensitive. Why they are there is not part of his analysis. But his data suggest that if one standardizes layoffs for this type of sensitivity, it is not at all certain that Negroes suffer disproportionately from layoffs.

John Korbel (University of Wisconsin) considers the determinants of women's entry into and departure from the labor force. His essay is focused on the use of simulation analysis for studying female labor force participation rates. Therefore, much of his paper is oriented to method or technique. His substantive conclusions include an observation that less than half of the women accounted for the bulk of the movement to and fro, and that this relationship remains so over time. Reasons for the swing -in and -out of the labor force are offered; reasons for changing jobs

---

* One paper on this theme, by Paul H. Norgren and Samuel E. Hill (Princeton University Research Project), "The Employment Effects of Fair Employment Practice Laws," is not reproduced here. It will appear as a chapter in their forthcoming book.

(while perforce remaining in the labor force) are similarly presented.

Some of the remarks presented by three of the discussants conclude the section. The discussants were Henry Cohen (Deputy City Administrator of New York City), Herbert R. Northrup (University of Pennsylvania), Jacob Mincer (Columbia University) and Albert Rees (University of Chicago).

PART III

The papers in this section stress the influence of decisions by consumers on performance by producers, with an unusually wide variation in treatment of substantive questions. Harold W. Guthrie (University of Kentucky), using data from the Michigan Survey Research Center, surveys the difference between a variety of certain centers with regard to a host of expenditure decisions and attitudes. He divides the country into six types of urban and non-urban units. He then considers four basic socio-economic classifications -- income, age, education, and family size. Thereafter, Mr. Guthrie turns to differences between the areas in their housing and land patterns, consumer debts and confidence regarding personal job security, and the use of tax revenues for public investment purposes.

John Richards (Texas Western College) considers the factors affecting residential location, in general, and in Baton Rouge, Louisiana, in particular. He explains his dissatisfaction with earlier works in locational economics because they fail to explain population scatter and they offer little or no predictive insights. He presents his own model to explain residential location choice based on the theory of consumer choice rather than on conventional location theory.

Richard L. Pfister (Dartmouth College), drawing on data collected not from tourists, themselves, but from the receipts of businesses catering to tourists, concludes that New Hampshire had about four times its "expected" share of population, income, or retail sales. The usefulness of his study is apparent when one recalls how little is known about recreation as a publicly sponsored industry.

Discussants at this session were J. B. Lansing (University of Michigan), Richard F. Muth (University of Chicago), and Jack L. Knetsch (Resources for the Future, Inc.).

PART IV

The concluding part contains four essays presented at the third session of the conference, the remarks made by three discussants, and an appended statement written by Martin Segal, whose earlier work was vigorously criticized in one of the essays. The major theme of the session was the significance of wage differentials for human resource allocation.

Belton Fleisher (University of Chicago) considers the depressant impact of Puerto Rican migration to New York on labor's earnings but adds that he believes that non-labor cost factors have caused industry to leave that area, nonetheless. Early in his paper, he tries to determine the effects of this migration upon migration flows internal to the United States. Later he attempts the same thing for wage movements. In general he notes the similarity of the Puerto Rican immigration patterns to earlier European immigrant patterns. He feels that it was the Puerto Ricans who depressed wages in the industries where they have located.

Martin Segal, whose views he rejects, explained the situation by pointing to other regions' competitive pressures. Mr. Segal's rejoinder is included among the comments of the three discussants.

Martin Segal (Dartmouth College) writes on changes in occupational wage differentials among female office workers and male workers in manufacturing plants in sixteen cities and assesses explanations provided by Reder, Douty, and others. His paper, focusing on changes in occupational wage differentials as between major cities during the 1950's, is an interim report. It consists of several parts: (1) He uses Bureau of Labor Statistics data to consider whether there is a consistent regional pattern associated with changes in one direction or another: there is not. (2) As regards short-run fluctuations, again he finds no regional explanation. (3) He analyzes salaries of women workers and speculates whether the behavior of occupational wage differences among women office workers was sui generis. (4) Finally, he assesses Reder's and Douty's explanations and concludes that both are unsatisfactory.

Robert L. Bunting (Cornell College, Mt. Vernon, Iowa) treats labor mobility and wage improvement by analyzing data from the Bureau of Old Age and Survivors' Insurance continuous work history sample. Tapping this generally neglected sample, he is impressed with the responsiveness of mobile workers to wage differentials. He questions this conclusion at some length, but in the end remains satisfied with it.

And last, William Goldner (University of California, Berkeley) employing "factor analysis," discusses interurban wage differentials, making allowance for differences in skill. He devotes a great deal of effort to explaining and defending factor analysis as a method for this type of research. Among the substantive findings is the noting of a general wage-rate stability among most occupations between areas.

The discussants were Robert Aronson (Cornell University), Richard A. Easterlin (University of Pennsylvania), and W. Lee Hansen (University of California, Los Angeles).

This is an initial effort in a complex field. It is the hope of the Committee on Urban Human Resources that this volume will stimulate others to take hold of some of the intriguing research problems posed by the fact that most of our population now works and lives in cities.

Mark Perlman

Center for Regional Studies
University of Pittsburgh
Pittsburgh, Pennsylvania

August, 1963

Contents

PART IV

AREA WAGE DIFFERENTIALS AND MIGRATION

PART I. A STATEMENT OF RESEARCH SCOPE

The Economics of Human Resources in the American Urban Setting: Some Concepts and Problems

by Mark Perlman

Urbanization an old phenomenon but with new facets - Why concentrate on Standard
Metropolitan Statistical Areas? - Problems of urban misuses of human resources are
associated with decay of once modern and thriving industrial plants, the impact on
firms in one industry of economies of scale in another, the range of the impact of
prejudice in all the factor markets, and the burdens of caring for dependent, needy
populations - Problems of urban subsidies to households, including to whom and when
to give public aid - Migration is a very limited answer to the production problem;
it poses great questions on the supply side.

The purpose of this paper is to focus interest on what seem to me to
be the most pertinent questions relating to the economic aspects of human
resources in the contemporary American urban setting. The conventional
concept of resources relates to the production of goods and services. By
extension, therefore, it is logical to conceive of human resources in terms
of the characteristics of the labor force. However, in this paper I propose also to consider human resources from another, less conventional
side -- namely, consuming households. The reasons for doing so are
three. First, as Perloff and others[1] have pointed out, the location of production activities is usually placed near its markets. Thus, the study of
the location of industry, and in these terms labor resources, must take
account of the facts of consumption -- the nature and location of effective
demand. Second, as the means of private transportation have proliferated,
if not exactly improved, individual laborer's decisions regarding residential location have come more and more to involve consumption questions.
That is, they live where the housing, schooling, recreational, and some
other household consumption factors are best suited to their needs. Just
as the growth of cities once represented essentially the relocation of rural
labor, so recently the mass decisions to migrate to suburbs reflect decisions taken principally as consumers.

Author's note. Harvey S. Perloff and Lowden Wingo, Jr., both of Resources for the Future,
Inc., have given me so many ideas that one of my roles as the author has been to select the most
pertinent ones. Responsibility for the selection and interpretation of the ideas is, of course, my
own.

Residential location has been traditionally viewed as a junction of proximity to job opportunities and ease of transportation. In the past this was generally true. However, the growing importance of the consumption factors in choosing one's homesite is modifying the traditional rationale of urban location principally as production centers. Consequently, the simple association of human resources and job opportunities, although still useful for some analytical purposes, appears to be growing less so. Instead, increasing emphasis will have to be put on urban centers as places to attract consuming households both in the sense of encouraging the development of the appropriate skills needed for future industry, and in the sense that the community will be able to hold its best workers because it is a pleasant place for them to live.

As the title indicates, this paper concentrates upon problems in the urban setting. This setting, as we conceive it, is the site of services provided to households in order that most of them can contribute to the labor force. The quality of the services (inputs, if you will), directly affects the quality of the labor outputs. Thirdly, therefore, we elect to stress the consumption side both because increased technological demands on the working population (the output) have required and will continue to require increased attention to the quality of the "consumption" inputs absorbed by the household. But my point goes further; there is good reason to believe that the community rather than the single household to an increasing extent plays the determining voice in specifying the inputs which particularly affect the quality of the labor force. Education has long been largely the problem of the community. Health and recreational facilities are becoming increasingly the community's responsibility. Welfare services to the impecunious, the handicapped, and to the aged are also each decade becoming more and more social rather than happenstance individual household responsibilities, not only because minimum standards, themselves, are being raised, but also because as a result of improved medical knowledge each decade sees the old becoming a larger share of the population.

Thus, three generations ago, it might have been adequate for virtually all analytical purposes to conceive of the city as a huge manufacturing environment into which were fed raw materials, machines, labor, and entrepreneurial talent. True, cities at that time did provide many services, but because so much of the labor was unskilled and because households consumed relatively so little in the way of either public goods or services, the quantity and quality of public inputs into the household could largely be ignored except by a few socially-conscious reformers. Now, I suggest the city must be conceived differently. Besides being the site of the manufacturing of goods and saleable market services, it has become the conscious molder of men, not only in general terms of their social values (which social economists have long understood), but also in terms of their physical and intellectual competence to participate in traditional industrial activities. Once, rebellious or unteachable students could look forward to earning their own living as unskilled manual laborers. Now, so direct a "solution" seems increasingly less feasible. If it cannot teach them in their youth, the community bears the burden of their dependency in the years when they should be most productive. For these reasons, therefore, I urge the inclusion of the household consumption side in this discussion of human resources, even though this may be an innovation to the usual understanding of the term.

One other note deserves mention in this introduction. What should we consider as coming within the scope of "urban setting"? The Bureau of the Census has a convention of using an incorporated area with a population of 2,500 or more. Similarly, social economists have preferred to employ a concept of a community of individuals which is aware and utilizes the principles of economic specialization in manufacturing and consuming. I seek no quarrel with either definition. I submit, however, that for our purposes it is sufficient to talk of the problem in terms of the larger urbanized units. Thus, most of my data concern what currently is termed Standard Metropolitan Statistical Areas (SMSA). Certainly there is no reason to argue that smaller cities do not have the same problems that large ones do; and perhaps they are no less pressed by a need to solve these problems. Yet, because I am concerned with raising conceptual issues in this paper, I think we can live safely with the arbitrary decision I have suggested.

The substantive body of my observations is found in the ensuing four sections of this paper. The first section is a summary of some pertinent background of information; second, a general discussion of the city as a focal point of labor force analysis; the third presents the companion analysis of the city as a site of household inputs; and, finally, I attempt to put some tentative lines of research activity into priority perspective.

SOME BACKGROUND THOUGHTS

Commonly, the period of city dominance in America is dated, depending on the region, some time after 1878,[2] although in fact it was not until 1920[3] that it was apparent that over 50 per cent of the American population lived in urban areas.[4] By 1950, almost two-thirds of the population lived in urban areas.[5]

Yet, it is not the simple fact of the urbanization of America which is most interesting to me. Urbanization is an old process; after all, imperial Rome was a city of several million, and Greece was a geographical entity composed of city-states. Rather, what stands out is the fact that there has been a striking paucity of economic information regarding America's urban problems. This country has been statistics-conscious for almost 200 years, and we have been collecting data relating to wages, hours, and prices for almost 100 years.[6] Yet, in spite of these relatively early starts, we have not connected statistical data collection with the peculiar needs of the urbanization process until the very recent past. Instead, we have concentrated virtually all of our thinking upon the industrialization process, identifying the metropolitan city's role only in an offhand, or casual fashion. For example, the Census Bureau announced that in 1957 the ten largest metropolitan areas in the United States were responsible for 40 per cent of the value added production, that the next fifty-two areas were similarly responsible for an incremental 25 per cent. The remaining 35 per cent was divided unequally between about 130 smaller metropolitan areas, a vast number of cities, towns, and villages, and the countryside.[7] The import of this study was to stress the locational characteristics of industry; I argue that it might just as well have been used to illustrate the external economies found in and social costs falling upon cities. It is largely a matter of emphasis -- but no less important because of it.

The 1950 Census created a population classification, the Standard Metropolitan Area, to describe metropolitan cities and their surrounding industrial and residential suburbs. That concept was further refined about 1957 and was used in the 1960 population census; it is now called the Standard Metropolitan Statistical Area. There were 212 such areas in the country in 1960. They contained 113 of the 179 million persons in the country.[8] This was 63 per cent of the population. In 1950 these areas contained 59 per cent of the country's population. In the ten-year interval, 84 per cent of the country's population growth was in these 212 centers.

I have already proposed that we concentrate attention on the SMSA. I do so for three reasons. First, they contain the largest part of the nation's people. Second, the range of social problems facing urban areas, these being largely the products of concentration and the aging of physical plants, are for the most part felt by large cities because of the nature of the problems, themselves, and to a lesser extent because of the fact that most (but not all) of the largest areas are fairly old. Third, good data are now becoming available and, as any social research worker knows, we should make maximum use of what is relatively easy to obtain.

The 1960 censuses of population and housing will greatly augment our stock of economic knowledge regarding these 212 areas. Better household income data will be published. Commuting patterns will be presented. But there are other sources, as well. The Continuous Work History Sample of the Bureau of Old-Age and Survivors Insurance of the Department of Health, Education and Welfare has material that can be used far more than it has been.[9] Also surveys like those of the University of Michigan Survey Research Center contain some interesting, if statistically biased, material.[10]

In sum, I propose that we concentrate our study of human resources in the urban setting on the SMSA's, and that at first, at least, we work the data fields which have already been tilled.

THE PRODUCTION SIDE

Malallocation Because of Industrial Decline

The Problem. Were we to enumerate the critical economic problems of the sixties, high on the list would be the relative decline of several of our major manufacturing cities as places for people to work. It is not only that technological changes occur which cause industries to expand and contract, that technological advances within given industries levy new and vastly different demands on cities' labor forces, but mostly that the effects of these two distinct kinds of changes are felt differently in specific cities. Whatever may be the shifts in demands for labor skills on the nation as a whole, the impact on individual cities is frequently of a more severe nature.

Let us exemplify the situation by pointing to a "single, dominant-industry" community like Detroit -- where that "dominant industry" has been declining as a source of employment opportunities. (Admittedly, one could just as well name any number of cities in New England, the Mohawk Valley of New York, Ohio, or Pennsylvania.) What stands out at first glance is the fact that the auto industry reached its production peak in 1955 when 7.9 million cars were produced. The employment peak came in 1953 when employment was 903,800. By 1959 industry employment had declined to 731,600. In addition, Detroit's share of the national production of automobiles has also fallen because Detroit's auto plants are outmoded. They are being replaced by plants in all regions of the country. Detroit has responded to this relocational challenge by attempting to modernize its factories in order to increase productivity. But

from the standpoint of job opportunities for Detroit workers the net result of this response has been unsatisfactory; the savings have not been
transmitted into sufficient price reductions to justify compensatory output increases, and there are fewer jobs for the local labor force. Nearly
130,000 automobile manufacturing jobs, more than one-third of those in
the Detroit auto industry in 1950, had been eliminated by May, 1959. In
the same period the proportion of Detroit's non-agricultural wage and
salary employment associated with that industry declined from 27.8 per
cent to 18.2 per cent. Unemployment has mounted and the supply of skills
embodied in the unemployed worker is wasted.

The fact of technological unemployment often calls forth three "solutions." First, bring new industry to Detroit to absorb surplus labor; second, help the surplus labor to adapt to the situation by acquiring new skills;
and third, encourage the unemployed to migrate to industrially growing
areas. Each of these "solutions" presents very difficult problems.

First, what industry or industries are currently mobile or expanding
and looking for a place to go, and, even if one exists, why should it move
to Detroit? At present we lack a simple systematic way to describe in
sufficient detail Detroit's supply of labor, including such characteristics
as skill, and the basic types of industrial experience the workers have
accumulated, in order to be able to advise entrepreneurs why Detroit is
an area justifying establishment of new firms.

Many communities, suffering decline as Detroit has, have looked to
two types of future employment: the growth of records-handling companies (like insurance offices), or the growth of research and development
firms. These were the industries to which Boston, one of the earliest
outmoded or blighted areas, turned.[11] It can easily be seen, however,
that there is a sensible limit to the number of insurance company head-
quarter cities that can simultaneously exist. Moreover, there are few
metropolitan areas in the country that have a Harvard and an MIT, each
the pre-eminent research institution in its respective fields. And even
areas that can be said to be similar to Boston in possessing such research
talent come to the development of "research and development" industry
some decades after Boston "discovered" it, brought it to realization, and
commenced enjoying real external economies of scale.

The second "solution" is industrial re-education. That is, re-equip
the unemployed, giving them new skills which are in demand. As a "solution," this has advantages, but only if it is possible to develop a sufficiently large number of labor-hungry firms in the area and if it is in fact economically feasible to retrain the unemployed laborer.[12]

The problems in this second "solution" are also found in the third. It
reduces Detroit's welfare costs if its unemployed migrate elsewhere. But
unless the migrants find jobs in some new location, the national community is no better off. Migration, however, poses some problems of its own.
Who will be the migrants? Clarence D. Long[13] argues convincingly that
the hard core of unemployed are the unskilled and particularly the older
unskilled workers. Thus, it is far from clear that these individuals will
have sufficient real economic incentive to leave the depressed metropolitan areas, where the chances of employment might be only slightly worse
than they would be were they to move westward or to the South. It can be
argued that these unemployed workers have good economic, and particularly strong social reasons to see their lives out, even at reduced living
levels, where they presently are, rather than to move to strange, new communities. If this argument has practical merit, the latter can be fortified

with the economic logic that it is cheaper to the nation (if not to a specific community like Detroit) to avoid the third "solution" (i.e., migrate to new job opportunities). Instead, these unemployed must be treated as permanent welfare cases. Perhaps we should call them the disabled veterans of past technological battles.

But this comment does not dispose of the third "solution." There is a migration-prone group, consisting generally of young adults. Over short time periods, their leaving may save specific communities welfare expenditures; losing them, however, may prove to be an expensive way for a community over longer time periods to plan its future. A community which exports a large proportion of its high school graduates not only cuts itself off from the future, but it also wipes out as unrealized the considerable investment it has made in the graduates' educational conditioning, including besides expenditures on schools, no less real expenditures on public health, juvenile correction, and publicly supported recreation. As a "solution," out-migration of the young is drastic -- akin to surgical removal of a leg; it may save the "economic" life of the community in the short run, but the long-run multiplier effect on the cost is heavy.[14]

Whatever conclusions I have suggested about decaying, one-industry areas are also (albeit fortunately less) true of areas with diversified industrial activities, where almost invariably some of the industries show signs of decline. Putting it bluntly, the illness of economic decay is endemic and, in the context of technologically advanced economies (where new processes, if not machinery, alone, can quickly eliminate the need for large groups of workers), it is extremely contagious.

My points are chiefly these. Metropolitan areas with blighted industries should look to out-migration to solve their problems only as a last resort. Rather, they should look to the possibilities of attracting firms in dynamic industries which can absorb the unemployed, without or with additional training. Human resources, however, do not represent the only social or economic factors in plant location. Others, such as the availability of capital and public services, the existence of necessary secondary services and tax advantages, are similarly important. In order to attract the "right" firms, each metropolitan community ought to know what its inventory of human skills are and what it could, within some short period of time, be made to be. If in order to attract some desirable "Firm X" in order to absorb several hundred semi-skilled unemployed operatives, it is necessary to train (or recruit elsewhere) 25 or 100 individuals with special skills, the community ought to at least be cognizant of these possibilities. In other words, the community ought to possess certain demographic data which would help it compare economic choices.

The Nature of Appropriate Data. Fortunately, the collection of the relative data regarding labor force, showing cities as a job-producing unit, need not be started de novo. The decennial synthesis has established several series which reflect just this purpose. Principal among these are series illustrating (1) age and sex distribution of the population, (2) participation by persons fourteen years and older in a labor force, (3) years of formal schooling completed by persons twenty-five years and older, (4) class of worker in major occupation group of employed persons, and (5) industry group of employed workers. These data are readily available on an urban and rural breakdown, and permit at least some comparative analysis of large cities and SMSA's.

Clarence Long[15] and Gertrude Bancroft[16] in separate studies have

taken these materials and produced two particularly useful monographs
showing empirical trends during the past half-century and more. Their
studies have been focused on national changes and are principally useful
for our purposes in their provision of a basis for national comparisons
and in pointing out the method of analytical endeavor. These works, however, were not intended to focus on or to highlight the various aspects of
labor force development in urban areas, but their techniques, as we have
already noted, should be applied to specific local areas. Thus, one
necessary line of research is the study of labor force development in
specific urban areas. The result of these studies will illustrate the effects
of different kinds of industrial specialization and conceivably could indicate
the direction of future industrial development. To be most useful,
however, studies of different labor forces should be presented in terms of
a series of standard dimensions in order to permit comparisons both
between cities at a moment of time and of a single city over time.

Starting from what we now have, we could include in our profile system (or brace of standard comparison) components like (1) size and participation rates of labor force, (2) age structure of the labor force, measured in medians and coefficients of variation and classified according to a
small number of basic patterns, [17] and (3) formal educational structure
of the labor force. In addition, it would be useful to know (4) the quantities of types of skills presently available in the areas. Although some of
these data are not presently collected in sufficient detail to be useful, they
could be assembled, if only on a sample basis, fairly easily. In addition,
(5) it would be useful to know what the principal secondary skill and classifications workers have. Knowledge of profile indicators (4) and (5),
when coupled with the wealth of information regarding job cognates found
in the Department of Labor's Dictionary of Occupational Specifications,
would provide a degree of sophistication about the possibilities of industrial relocation or expansion which individual cities presently lack. A
sixth dimension or indicator could reflect prevailing wages paid each of
the major job or skilled categories.

These profiles could be used for several major purposes. The heart
of each major urban area (SMSA) in the United States is a politically defined city. Generally, each of these cities is suffering from a sort of
angina pectoris. The central city generally bears an inequitable proportion of the non- or less-productive population. Sooner or later we must,
it seems to me, realize that core cities cannot carry disproportionately
heavy burdens supporting the aged and educating the young. That they are
doing so can readily be seen by comparing SMSA and core city profiles.
Moreover, the degree of disproportionality is also easily seen when different core cities and their respective suburban areas are compared. Finally, differences between intra-urban areas (census tracts, for instance)
can be analyzed; and the same tract can (as was true of the city as well)
be studied over time.

Malallocation Because of Industrial Specialization

Why it Develops. The nature and degree of industrial specialization directly affects the nature and the number of jobs of specific types which are
available in any area. Obviously, in a steel manufacturing city like
Pittsburgh, there will be a disproportionately large number of jobs in the
steel industry. Not so obvious, however, is the second point; because of
internal economies of scale, each large steel firm may do its own truck-

ing, provide its own cafeteria service, handle its own accounting, and so forth. Consequently, the development of some tertiary or service industries may be stunted, and although some firms inevitably do get established, their number is smaller than one would expect, using size of the urban area or some measure of manufacturing as a point of reference. The train of consequences is simple to follow. Because the large industrial firms find it adventageous not to contract the service out, there are fewer external economies of scale for firms engaged in the particular service industries. Their unit costs, consequently, remain relatively high and are reflected by their prices The high prices of these services result in making the area less attractive to other manufacturing firms, firms of a size or a type which would normally expect to contract out these service functions. The gross consequences are that there will be fewer jobs not only in the service firms, but that there may well be fewer in all other kinds of business and manufacturing endeavor. In sum, industrial specialization may be an autogenerative condition.

Of course, there are certainly other factors as well which can "distort" the labor market's spectrum of jobs. The level and distribution of household income in the area will directly affect both the number and size of firms providing consumer services. Cities with a relatively high median income and enjoying a small average deviation may, quite reasonably, be expected ceteris paribus to have more restaurants, laundries, private recreation centers, and so forth, than cities with a low median income which is distributed very unevenly. Also, cultural traditions undoubtedly play a role. Certain groups, irrespective of income considerations, are more theatre-minded than others; other ethnic groups favor bowling alleys. The degree to which ethnic groups retain their traditional mores, their numerical strength, and their economic importance will naturally determine in part the qualitative spectrum of jobs available.

One other point should be mentioned under the rubric of industrial specialization. A city which serves as the home office of a very large firm may derive from that relationship some unexpected economic characteristics. It will have a large number of well-paid executives whose families will purchase a large number of household commodities and services. Less obvious, however, is the effect on the capital structure of the community. Wilmington, Delaware, and Tulsa, Oklahoma, for instance, enjoy unexpectedly great importance as financial centers.[18] The explanation is simple: both the large corporations and wealthy executives bank in the local financial institutions in these cities. In turn, these financial institutions have large deposits. These deposits can then be used to make loans not only in the local market but, in wider context, mainly the national and international markets. One effect will be an increase in the number of positions in the financial intermediaries. Ultimately, external economies of scale develop and a new form of industrial specialization then arises.

The Problem. The consequences of an industrial specialization are interesting in themselves when the area is expanding. Yet what happens when the specialized industries cease to grow, or even decline, introduces a state of labor force distress which would only otherwise be hinted at. Wilmington and Tulsa are in the former, more fortunate, category. Detroit and Pittsburgh are in the latter. Boston and Hartford, Connecticut, for historical reasons have pioneered as insurance company centers. This form of specialization historically came about as a result of those cities' pre-eminence as manufacturing centers in an earlier era. Great economies of scale have developed favoring concentration of insurance activity in those cities. However, it does not follow that the economies of scale will continue. Nor is it entirely likely that the "insurance capital" of the nation will move to areas enjoying considerable surplus deposits in their financial intermediaries.[19] From the long-term point of view, it may be worth studying the development of financial intermediaries, particularly their locations, in order to determine future location of labor force needs. I mention this point because so many cities, presently suffering a decline of employment in second industries (manufacturing), look to the development of insurance companies and other financial intermediaries to absorb the slack. It seems to me that the logic of location for financial intermediaries other than insurance companies is associated not only with the presence of a large, cheap, unemployed labor force, but, in most instances, with the propinquity of large investors and sources of capital funds.

Malallocation Because of Racial or Ethnic Prejudice

The Problem. One of the other critical economic problems facing many but far from all American metropolises is the malallocation of human resources because of racial differences. Gary Becker notes that this phenomenon occurs because some parties are willing to pay a price in order to deal with certain kinds of people.[20] I do not disagree with the definition of discrimination; I do, however, suggest that there may be some technical, economic considerations which explain their behavior as well. In particular, I am impressed with the costs of obtaining information regarding credit or entrepreneurial skill when no instrument has been created to collect data or to analyze them. Explicitly put, I want to raise the question whether there are economic reasons for the high cost of capital for Negro-owned business and industrial enterprises.

Every new-coming minority group has, I think we can agree, both a cultural heritage regarding occupational choices and a need to tap the resources of other ethnic groups in order to prosper. For a variety of reasons, many immigrant Jews had a cultural history emphasizing the occupational virtues of retailing and one or another of the major professions. Fortunately, for them, the same heritage put some social premium on capital accumulation. The consequence was that, while Jews may have had some difficulty in attracting non-Jewish customers or gaining necessary admission to professional schools, as a group they had relatively little trouble in getting the capital necessary to finance their enterprises. Their capital market functioned efficiently because their own tradition encouraged capital accumulation and because they had previously had experience with the inter-personal informational system necessary for the wise allocation of capital. It took but a short time for the supply side of the money market for a small enterprise owned by Jewish entrepreneurs to lose its completely Jewish flavor, not because non-Jewish capitalists were eager to assimilate Jews socially, but because the inter-personal information system necessary for the operation of any money market became even more widespread and efficient and non-Jews were interested in benefiting from its success.

The efficiency of money market activity obviously affects the business and industrial life of every urban area. Where there is a tradition of ethnic segregation, its influence is all the greater. In Baltimore, for example, the social life of Negroes is quite distinct from that of the non-Negro population. Although many restaurants and perhaps a few nightclubs are "integrated," the truth is that this integration is more nominal than real. There are, however, relatively few Negro nightclubs in light of the fact that 35 per cent of the city's population is Negro. Why, I ask, are there so few such clubs?

Part of the answer lies in the income and age distribution among the city's Negro population. They are not, by white standards at least, as prosperous as other groups. The proportion of Negroes who are below the "nightclub age" is also disproportionately high. And it might be argued (although I reject the point), that Negroes are not "interested in nightclubing." It also may be argued that it is harder for Negroes to get liquor licenses than it is for non-Negroes -- the evidence in this regard is mixed.

My own guess is that the explanation lies in the relative difficulty Negroes experience in getting venture capital. The usual credit agencies, having ample opportunity to make "satisfactory" if not the maximum rate of return on the funds they presently have available, lack adequate economic information both about the profitability of the Negro nightclub market and the business acumen of would-be Negro entrepreneurs. Understandably they are reluctant to lend money at attractive rates, if at all. Consequently, that industry, one which is job-opportunity rich, does not develop in Baltimore Negroes, who spend their incomes on other, probably less job-opportunity rich, services and commodities. The "fault," I hasten to add, is not entirely the banks' or other lending institutions'. As compared to the Jewish cultural heritage, the Negroes' seems to put little emphasis either on a capital accumulation or in fashioning the instruments which make borrowing more efficient.

What to Study. If my definition of the problem has merit, one interesting study would compare the slow development of the money market for Negro business enterprises in New York City, for instance, with the development of similar institutions among Jews or Puerto Ricans. But my basic point is that we who are interested in job opportunities may well want to consider the part that poorly functioning money markets for ethnically-identified firms and industries play in limiting employment in many major urban areas. If analysis confirms my hunch, the policy implications are clear -- the pay-off for improving the necessary communications would justify the necessary effort in terms of jobs created.

Malallocation Because of Qualitative Deficiency

The Problem. We have already made reference to a fear that our core cities are becoming burdened with a disproportionately large share of welfare dependent citizens. Putting aside the school children and youths, what we leave for consideration are those who are unemployed (possibly all but unemployable) because they are in some sense old or physically handicapped, or lack skills because they were not educated or were once the victims of racial, ethnic, or religious prejudice. If individuals in these groups are to be made productive (that is, incorporated into the employed labor force), the transition can be made in several ways. First, as we noted earlier, the absolute level of job opportunities within a city may be increased by attracting additional firms or by making the capital market operate more effectively in order to make it possible for more firms to engage in business or industry. With an absolutely greater number of jobs, these unemployed may find work.

Second, in some instances migration may be the best, if not the only, solution. Third, particularly in the case of the aged and some of the physically crippled, there is no "solution" except to treat them indefinitely as social charges.

But it is the fourth possibility, retraining, which is most promising. Drawing upon the individual's skill and industrial experience, it may be possible to rehabilitate him economically by giving him additional training or counseling.[21] Economists can, it seems to me, make three contributions in this regard. They can analyze the probable costs and probable results (as measured in terms of employment opportunities) of different lines of training. They can suggest different methods of financing this rehabilitation period and, finally, they can suggest the type of return on public expenditure on this kind of effort as compared with similar returns for providing capital funds advantages to entrepreneurs or even leaving the people on relief or welfare rolls.

What to Study. Adams and Aronson,[22] Shultz and Myers,[23] among others, have touched on the experiences of different cities with counseling, and to a lesser extent with job-retraining. None has set up an analytical matrix to compare the costs and benefits. Were an economist to team up with someone learned in the discipline of industrial education, the products might be quite useful.

Counseling and re-education programs can be financed in four ways. The recipient can be asked to underwrite the cost either by pre-payment or by lending him the cost of his tuition and asking that he repay the amount at a later date. Each firm can be required to contribute to a fund for these purposes; in fact, the experience rating system, by now firmly established in the unemployment and workmen's compensation programs, could be used. These programs can also be financed by taxing groups of firms (or however one defines industry) directly or by working in collaboration with unions' welfare funds. And, finally, the community can be asked to subscribe the amount directly out of tax revenues. Obviously, all of these methods can be explored. I presume that the first, if it can be made to work, is the most compatible with our economic heritage, although, to be sure, the others don't represent any really alien traditions.

The amount which communities have to contribute to viable rehabilitation and counseling programs could then be considered in light of the alternative uses of the funds for relief purposes. It is conceivable that when one considers a likelihood of an unemployed worker being able to solve his own job problems in time and without any community aid, it is economically wiser to leave him on the relief rolls rather than to engage in what might be an expensive effort to retrain him. In any case, some analysis along these lines is worth consideration.

THE CONSUMPTION SIDE

I have already made my case for discussing the effects of household consumption patterns on household output. A major remaining difficulty is that the topic, although previously considered by several different groups, lacks the vigorous intellectual integration which makes for ready analysis. The situation is far from hopeless, however.

The Dimensions of Consumption

Let us start by considering urban areas conceptually as a collection of households. Each area has the purpose (among others) of creating a high-quality population, some of whom will participate actively in the labor force and some of whom will do so indirectly. All consume. The question is simply how to improve the quality of the population. Presumably this can be done by increasing the total quantity of the appropriate consumption items, providing sufficient attention is also paid to the distribution of these items. [24]

In the main, what is consumed by households are goods and services produced either by private firms or supplied by the various levels of government. Our ideological mythology has tended to minimize the part played by the latter. It is not necessary here to go into the finer points of welfare economics; suffice it to say that in practice, the government has played, plays, and in all probability will continue to play an important role in shaping household consumption patterns.

The government plays this role in a variety of time-honored ways. It provides a spectrum of services ranging from police protection to sewage disposal and including such activities as education, welfare services, health services, and recreational opportunities. It also affects the economic life of the community in many ways. It engages in monopolistic practices and brings to the public the advantages of economies of scale (as in the case of the police force). It sets minimum standards and forces unwilling agents of production to re-price their products and services (minimum price and wage laws). It taxes some goods and subsidizes others (cigarettes in contrast to education). It uses its general police powers to discriminate between property holders, providing that some may subdivide their property and that others may not. Most of all it taxes individuals according to their presumed ability to pay and uses the revenue thus acquired to furnish to the deserving certain goods and services at prices bearing no relationship to cost. In this manner, the government causes the rich "to keep" the poor.

Governmental discrimination along the lines of wealth has become, if anything, more widespread in this period of great urban concentration. The government now takes responsibility for subsidizing housing, just as in an earlier century it invaded the field of education. It also appropriates vast sums (largely collected through taxation) for recreational purposes, for public health activities, and for the more obvious welfare or relief programs.

Whether the government has to play a larger or a smaller role regarding the financing of "inputs" into households is admittedly a subject dear to the hearts of social philosophers. As a topic, however, this debate has been for practical purposes resolved as an "illogical" compromise. What emerges is an "agreement" that the government should take steps to help those below the "social minimum"; it should also take responsibility for some wide citizenship-type programs of which schooling, public health, and some kinds of recreation are examples I have already cited. The rationale for this compromise is simply that the cost to the community of neglect is far greater in the long haul than the cost of the programs.

The technical question of determining the nature and degree of public-supported programs thus assumes importance to us. If we want certain standards to be established, we have to know what the facts are at current levels of governmental assistance and what may be the direct results of augmented intervention.

Development of a Profile of Indicators

What we have said about the usefulness of urban profiles regarding each urban area's labor force applies as well to the development of a set of indicators appropriate to the economic facts of household consumption. Our immediate question is what to include in the profiles and why.

At the risk of redundancy, may I note again that the point of focus of all of this paper is the economic significance of urban living -- what are cities as places to work and as places to prepare people to work? For this latter reason we are more interested in family income than either income per employed worker or per capita earnings. Putting this matter another way, it is worth adding that we must not become so enamored of our market system's value index that we overlook the important effects that non-working wives and mothers (to say nothing of built-in baby-sitters) have on the actual performance of the labor force.

The first indicator should reflect the numbers of families with some reference to their size composition: Are there many very large or many two-person families? Secondly, the profile should indicate both median family income and some index pattern of distribution. It seems desirable to include, as the third indicator, the age structure of the population: How many are school children, and what are their ages? How many of the population are past sixty-five years of age, and so forth? In 1950, the figures were as shown in Table 1.

Table 1. Comparison of Percentage Age Structure of Four American Urban Centers, 1950[1]

	Pre-school (0 - 4)	Public school (5 - 19)	Super-annuated (over 65)	Total (dependant ages)
Baltimore city	10.3%	20.1%	6.9%	37.3%
Baltimore suburban	12.4	15.3	2.5	30.2
Buffalo city	9.0	19.6	8.4	36.0
Buffalo suburban	11.1	22.4	7.3	40.8
Houston city	11.4	21.3	4.7	37.4
Houston suburban	13.2	26.7	4.5	44.4
Seattle city	9.2	16.8	10.2	36.2
Seattle suburban	13.6	22.1	7.1	42.8

[1]Source: Census Reports.

In all cases the core cities carried a larger proportion of the "super-annuated" than did the SMSA. In Baltimore alone the core city had a larger percentage of its population in school. And in every case the babies tended to live in the suburbs. (As a Baltimorian I hazard the explanation for Baltimore's sole atypicality. Negroes are welcome in the city; they are not welcome in the suburbs. Negro families tend to be fairly young. Consequently, they thrust a disproportionate load on the city's educational system.)

The implications in the table are interesting. The burden of the dependent ages is (with the exception of Baltimore) carried by both the core city and the suburbs; the young by the suburbs, the old by the core cities. In the short run, the social costs of the young are greater -- they need schools, parks, and police protection. Fortunately, the suburbs are somewhat wealthier. The question, of course, is whether the added wealth is sufficient to cope with the augmented needs for publicly supported services. If not, the suburbs will have to finance their services through state aid. In this sense, the few wealthy in the city through their income tax payments help the suburbs. As a city dweller (already paying substantially a higher tax rate than the suburban citizens), I note my findings with displeasure. But such is social organization. (Fortunately, Baltimore is an exception.)[25]

As a fourth indicator it would be useful to know the relative ethnic characteristics insofar as consumption is concerned. So long as racial and religious differences play parts in American economic activity, it is important to have those kinds of data. Some additional indicators might reflect present patterns of consumption of publicly supported educational, medical, welfare, and recreational services as well as some information about police activity including the rehabilitation of those convicted of felonies.

Some Implications of What the Profile Indicates. Again, as was true in the case of the labor force profile, the profile of households could serve to compare cities at any moment in time and compare a given city over time. In other words, the development of this set of profiles is only the first step. The second and more difficult phase is to attribute meaning to the indicators. What do changes in the indicators suggest? Let us consider the matter on a census tract basis.

An increase in the number of households should directly effect the demand for housing; moreover, information regarding age and income distribution will complement the picture. If the number of "young households" increases and their income tends to be low, the housing pinch will quite naturally be felt in cheap housing areas. Destruction of cheap housing (as in slum clearance) will force impecunious families to seek housing where they can find it. New, so-called low-cost housing projects may not be the answer. First, they may not be available in sufficient quantity, and second, because they may not truly be low unit-cost housing, but low room-cost housing. Impecunious families may not be able to contemplate costs per room in their more pressing need to conserve total outlay for house room. If new, low unit-cost housing is not available, then there will be pressure to convert existing facilities. In most instances, this means downgrading of neighborhoods. It also means that there will be increased pressure on the school facilities, on public health activities including garbage and trash collection, as well as publicly supported medical stations, and on public recreational facilities. Necessary levels of police and rehabilitation services may need to be reconsidered, depending upon the ethnic structure and mores of the new group.

If the increase in the number of households involves a young but higher income group, there will be other, no less serious, demands. Schooling standards may have to be re-examined and different kinds of additional facilities provided. The pattern of publicly financed recreational centers may be different. Parking problems may take a different form, etc.

If the increase in households, on the other hand, involves an older group, schooling needs may not appear to have any importance, but, again, depending upon the distribution of income, public health and recreational needs will have to be thought through.

The types of services furnished to some ethnic "minorities" deserves special consideration. In Baltimore we have observed that some "hillbilly" families' children require types of "formal" education heretofore not expected. Never having lived near central plumbing or having ridden buses before, these children require urban-living orientation courses long before they are exposed to the more usual academic curricula. And, in some cities, it may be desirable to separate school children of different grades geographically in order both to speed the process of their learning English and to undermine juvenile gang structures. Also, planned, publicly supported recreational services may be an economic substitute for increased police and rehabilitation expenditures, too.

Levels of family income will undoubtedly affect the demand for local personal services industries. It seems reasonable to expect that large, young, and impecunious families will not use restaurants as frequently (on a relative basis) as small, old, wealthy families. The reverse conclusion probably applies to drive-in theatres and certain other forms of privately financed (but publicly zoned), recreational facilities. On the whole, the hypothesis that higher levels of family income will generate a larger number of local personal industries seems tenable, as does its companion that personal service industries usually generate more employment than does manufacturing.

The range of possible income, age, and ethnic background relationships is great. I suggest that one of the early lines of research of endeavor is to consider the effects of different combinations on the economics of urban development, particularly as measured in terms of opportunities for groups of individuals. Precisely put, which specific public goods and services affect the quality of the labor force directly or indirectly? Can some measure be set up to consider the "returns" on alternative types of "investment"? Also, which specific goods and services provided by private firms will have greatest beneficial effects and how can the community encourage consumption of these? Will tax relief or outright subsidy be necessary? Finally, one has to take account of time lags. The community must have some basis for making projections of its needs because schools, hospitals, sewers, water systems, and even baseball fields take time to plan and to build. Perhaps the greatest use of the profile system is its "warning effect." Shifts in coefficients of variation of household income will show up dramatically some time before median incomes necessarily change.

Who Pays and What to Subsidize.

In our discussion of rehabilitation of an unemployed labor force, we discussed the choices which could be made about underwriting the cost. That analysis is pertinent to this discussion as well, and it can be extended beyond the topics of education. Presuming that certain types of recreation are more desirable than others, can the community encourage the preferred activity by providing facilities as well as leadership. Are swimming pools "better" than zoos? And, can pools be made operationally self-supporting (as contrasted with the cost of the initial plant)?

The point goes beyond the community's using its revenues to offset directly the prices of different commodities or services. Rather, what can the community do to encourage proper consumption patterns to develop? Education and the provision of accurate information is one obvious method. Its cost may be small or large depending upon the difficulty of obtaining the necessary data. Another way is to provide inducements to individuals by giving tax relief. For example, he who takes care of a dependent university student is given an additional exemption.

One aspect of the financing question that particularly intrigues me is the present anomaly regarding investment in physical plant and human skills. I refer to personal investment in educational improvement as a means to augment the community's supply of human resources. Although we encourage entrepreneurs to improve their physical capital equipment by terming it a necessary business expense (and even devise ingenious tax arrangements to permit fast, and occasionally faster or countercyclical tax write-offs to speed these moves), we do not extend the same encouragement to individuals.[26] The incentive to continue one's education can be an aesthic desire to enrich culturally one's life or a hope to increase one's income (less the rate of taxation at the margin). In brief, I argue that it is certainly socially desirable to consider higher or extended education as an investment process; treating it as such, it makes sense to consider it on a plane with business expenses. It may be asked, what about music lessons; should they be considered socially desirable? The answer obviously is, why not? They increase the quality of consumption and should in the long run have a positive effect on the labor force, to say nothing of the level of effective demand for personal services. Ultimately, of course, the line will have to be arbitrarily drawn somewhere. It seems reasonable to argue that music lessons are desirable, cha-cha lessons although still desirable are probably less so, and lecture series about the occult, perhaps are not socially desirable except for advanced students of psychology.

But, too insensitive an application of this principle causes extremely adverse results. To encourage activity in one direction may divert resources from continuing in others. For instance, about fifteen years ago the Federal Government put large amounts of funds into urban renewal programs. Presumably, the Congress intended to eliminate low-standard dwellings. What it did, however, was to eliminate necessary low-unit cost housing (slum clearance) and cause those whose dwellings were thus demolished to move to once better neighborhoods. The wave of moves continued until the wealthiest households moved out of the city's corporate limits and, from a tax revenue standpoint, the city became handicapped in furnishing its "free" services. Of course, the explanation offered was that the physically rehabilitated part of the city would attract the wealthy and quite probably the older wealthy (whose household needs did not include large demand for extensive public schooling). Instead, the rehabilitation programs also included superhighways into and out of the heart of the city, and whatever locational reasons the sought-after older rich may have to live near the center of things, these reasons were more than compensated for by the resulting ease of fast access to suburban cheap land. In other words, planners did not think through the logical consequences of what first appeared to be a simple program to improve household consumption "inputs." There is little reason to conclude that housing "inputs" were necessarily improved, and it is clear that property evaluations, basis of most urban taxation, were decreased.

One has to study the range of operations of given social mechanisms. With this information, it should be possible to devise techniques which serve a social purpose while enlisting individual self-interest. The steps to be taken are three. First, after examining the inventory of what the city has, decide what "inputs" are needed and would pay off best. Second, plan several programs to facilitate the supply of these desired inputs. And, third, consider the side effects of each program and choose the one with the least harmful consequences and possibly the one which ties self-interest to the public weal.

THE IMPLICATIONS OF MIGRATION

The economic reasons for migration are basically two; improved job opportunities and the chance to improve household "inputs." Job opportunities can be measured in employment rates and real earnings anticipation. This choice involves consideration of how great the probability is of finding steady work, and, possibly irrespective of the amount of work, how great earnings will be over some standard time span like a year, five years, a generation, or even what constitutes the balance of a normal work life. The publication of area unemployment rates (and their classification into the small number of categories) is a direct product of the analysis of employment rates. Dr. Weisbrod of Washington University has recently proposed an index of lifetime real earnings anticipation; in fact, he has gone further and proposed the measurement of real earnings anticipation in terms of their present value.[27] It seems to me that these two measures of job opportunities can be further refined in order to take cognizance of differing skill and age mixture in distressed areas. This observation follows quite logically from my earlier observations regarding the uses to which a proposed profile of indicators can be put.

Here, I suggest another line of consideration. What will migration do to the pattern of household consumption in their sending and receiving areas? If it is the young who migrate, to what degree will their movement affect the necessary "inputs" of schooling, public health and welfare, and publicly supported recreation? What change will their migration put on the existing labor force and prevailing wage rates, and in that sense upon levels of household income? It is both the speed of migration -- that is, the speed by which areas lose or gain households -- and the net changes, themselves, which are most enticing to study. A fast in-migration can be very deceptive; in New York City responsible analysts writing in the New York Times, for instance, estimated the Puerto Rican migrants in Manhattan at half a million at a time when there was scarcely half that number there. Such erroneous guesses, whatever their cause, reflect not only basic ignorance of what was occurring but also relatively little appreciation of what the change meant economically to the city in terms of the necessary household "inputs." Most of all, however, it shows that changes which occur over short time periods seem to be physically greater than they really are.

The problem seems to be to prepare to cushion fast changes. Social adjustments invariably occur, but they take time. Given half a century the Irish and Jewish immigrants became economically acculturated. One wonders whether somewhat more conscious social planning, if based on adequate data, could speed up the process for the more recent in-migrant groups. Once the burden of raising wage levels and through that medium household "inputs" fell on labor unions, the inevitable social price that was levied was union resistance to newcomers. With better social planning, the burden could be handled so as to minimize both unreasonable competitive bitterness over job rights, and the time required to make the situation stable.

The lines of inquiry are clear. From changes in the profiles it should be evident what modification should be made in planning household "inputs." If the necessary modifications are made as they are needed, rather than some time afterwards, friction can be minimized. So, what will be the necessary modifications? Studies of what the desired household "inputs" are should have a high priority. They will vary according to the characteristics of the in-migrant group. They will also vary according to the industrial structure of the given city. One final line of inquiry worth considering is whether it is possible to transfer unused "input" facility from areas undergoing out-migration to those in process of receiving migrants. If it is possible to transfer some of these facilities (and in many instances they may be "human capital," such as teachers, social workers, or public health personnel), it would be worth knowing the incentives that can be most readily used.

SOME TENTATIVE RESEARCH TOPICS

A. Profiles of cities' labor forces, involving measuring (averages, distributions) of skills and educational reservoirs, ethnic and age characteristics, wages, and type of industry experience.

B. The development of standard methods for projections of labor force requirements by cities in the context of labor standards (wages, hours, and working conditions).

C. A study of the criteria used to give training (what training, given to whom) in cities where technological unemployment (for whatever cause) occurs.

D. A standard measure for labor force migration between areas reflecting, besides numbers and classes of skill, labor standards.

E. A standard measure of the economic effects of labor force migration on the labor conditions in the short and long receiving and sending areas.

F. What does industry specialization with large key firms (local or branch) do to the profile of job opportunities in a city? Can a standard profile be developed to illustrate these effects so that comparisons can be made or so that remedial action be planned?

G. Can standards be devised for comparative purposes of measuring the effects of inequities in the money market (because of the race and ethnic origin of the borrower) on the profile of job opportunities?

H. Profiles of cities' households, involving indirect measuring (averages and distribution) of age, income, education, health and welfare, police, recreation characteristics.

I. Profiles of cities' households involving indices of changes in consumption patterns. Relation of this consumption to local (private and public) production.

J. Consideration of the impact of different kinds of migration (age, wealth, ethnic qualities, previous education and skill attainments) on local (private and public) production.

K. The effect of extreme housing density involving the poor, the racially or ethnically discriminated-against on expenditures and resources in the public sector. Can this problem be handled locally?

NOTES

[1] Harvey S. Perloff, et al., Regions, Resources, and Economic Growth (Baltimore: Johns Hopkins Press, 1960).

[2] A. M. Schlesinger, The Rise of the City 1878-98 (New York: Macmillan, 1933).

[3] In 1910, the Census report put 45.7% of the almost 92 million people in the United States living in urban areas -- that is, incorporated areas having 2,500 or more persons. Criticisms of this, the old, definition can be found in Donald J. Bogue, The Population of the United States (Glencoe, Ill.: The Free Press, 1959), pp. 20-24.

[4] Historical Statistics of the United States, Colonial Times to 1957 (Washington: U. S. Government Printing Office, 1960), p. 14.

[5] Using the 1950 definition. Cf. Bogue, op. cit., p. 30.

[6] Mark Perlman, Labor Union Theories in America (Evanston: Row, Peterson, 1958), chapter 1.

[7] U.S., Bureau of the Census, 1957 Annual Survey of Manufacturers, Series MAS-57-5, "General Statistics for Standard Metropolitan Areas with 40,000 or More Manufacturing Employees."

[8] U. S., Bureau of the Census, 1960 Census of Population Supplementary Reports, PC (S1)-1, dated April 10, 1961.

[9] Cf. Vincent F. Gegan and Samuel H. Thompson, "Worker Mobility in a Labor Surplus Area," Monthly Labor Review (Dec., 1957), pp. 1451-56; see also, Bunting's paper, p. 208.

[10] The Michigan Center has recently published a useful description of its work. "Economic Survey Data: Basic data available to academic researchers from the Economic Behavior Program, Survey Research Center, Institute for Social Research," Ann Arbor, Mich.,1960. Processed. Also, cf. Guthrie's paper, p. 137.

[11] These industries were projected in most studies of the future as being of increasing national importance, as well.

[12] On a priori grounds it must be assumed that many of the unemployed are ill-endowed to compete for any jobs, particularly when law, custom, or collective bargaining agreements require that they be paid "respectable" wages.

[13] His views on this topic were presented at the Catholic Economic Association's St. Louis Meetings (December, 1960), "A Theory of Creeping Unemployment and Labor Force Replacement"; presumably they will be published soon.

[14] Of course, it is possible that some communities can't be revived, and in these cases, planned out-migration may reduce their burdens. These changes are the characteristics of economic change, if not always growth.

[15] Clarence D. Long, The Labor Force Under Changing Income and Employment (Princeton: Princeton University Press, 1958).

[16] Gertrude Bancroft, The American Labor Force: Its Growth and Changing Composition (New York: Wiley, 1958).

[17] Each indicator should consist of several components. For example, the second indicator (age structure) might appear as follows: median age; coefficient of variation; and classified according to whether there is a small, roughly typical, or large proportion of (a) women, or (b) non-whites.

[18] Cf. Table 19, "Inter-City Business Loans for Major Cities. . .," in Otis Dudley Duncan, et al., Metropolis and Region (Baltimore: Johns Hopkins Press, 1960), p. 119.

[19] Location of insurance centers need not be dependent upon propinquity of policy holders. Banks, however, have good reasons to be geographically close to their depositors.

[20] Cf. his Economics of Discrimination (Chicago: University of Chicago Press, 1957).

[21] It is important to add that retraining may not help the national picture simply because the skill added in one locale may result in a worker elsewhere becoming unemployed. Also, skill-hungry workers may migrate to areas where new skills are being taught. If these migrants are more able than the local people as students, industry will prefer them. Consequently, a specific city's unemployed may not be directly helped.

[22] Leonard P. Adams and Robert L. Aronson, Workers and Industrial Change: A Case Study of Labor Mobility (Ithaca, N. Y.: New York State School of Industrial and Labor Relations, 1957).

[23] George P. Shultz and Charles A. Myers, The Dynamics of a Labor Market (New York: Prentiss Hall, 1951).

[24] Increased quantity should be so defined to include some reference to improved quality.

[25] I am told that the cities' super-annuated are more liable to need welfare assistance than the suburbs'.

[26] This idea is developed by Dr. Richard Goode in a forthcoming article, "Educational Expenditures and the Income Tax."

[27] Burton A. Weisbrod, "An Expected-Income Measure of Economic Welfare" (processed, 1961).

*PART II.   THE LABOR FORCE PERFORMANCE OF
           MINORITY GROUPS*

Racial Factors Influencing Entry into the Skilled Trades

by F. Ray Marshall

Race question as an urban economic problem - Its budget effect - Limitations on Negro
entry into the trades considered in light of past and present conditions of Negroes -
Present pattern of Negro employment and reasons for it, including inadequate education and training - Also Negroes are probably excluded from apprenticeship by unions,
cultural tradition, economic vagaries, and management - Negroes offered dirty and
servile work, but management's attitude is less important than other factors and
changes relatively more readily than others because of market conditions and legal en-
actment - Role of unions varies from rigid anti-Negroism to accommodation of different forms - Recent desegregation of seniority rosters, union emphasis on job-security,
and overt anti-discrimination programs typifying many current unions' attitudes -
Unions also support Fair Employment Practice legislation - Generally unions support
FEP and similar programs for their own reasons - Problem of membership attitudes -
Influence of moral pressures on AFL-CIO, but latter lacks enforcement power -
Conclusions.

The "race question" has been one of the most persistent domestic
problems this country has faced. Surely no other issue has incited such
passionate debate, and no other question has so tested the nation's moral
fiber or forced such a re-examination of basic principles. This was not,
however, an important urban problem until Negroes started migrating
en masse from southern agriculture to urban areas, especially in the
North. Today less than 10 per cent of the Negro work force remains in
agriculture, and most Negroes live outside the states that made up the
Confederacy (see Appendix A). Thus, Negro employment problems are
now in large measure questions of urban economics.

Negro concentrations confront most cities with serious economic as
well as social problems. To the extent that Negroes have low incomes,[1]
are employed below their abilities, are not able to develop their productive potential, suffer much higher rates of unemployment (unemployment
for non-whites in 1960 was 8.7 per cent as compared with 4.7 per cent
for whites), must live in inferior housing and attend inadequate schools,
they will constitute heavy burdens on cities in terms of welfare, police,
and health costs, but will make relatively small contributions to urban
income. Underemployment of Negroes also constitutes an important
resource waste that the nation can ill afford. Moral and international
political considerations aside, at a time of growing concern over shortages of skilled manpower, it is imperative that Negroes and other minor-

ities be able to develop their skills to the limit of their potential.

While the factors influencing the entry of Negroes into the skilled trades are many, varied, and interrelated, I shall be concerned generally with factors influencing the Negro's ability to prepare himself for and acquire more productive and higher paying jobs. More specifically, I shall focus on the following questions: (1.) What is the occupational status of Negroes relative to whites and how is this status changing? (2.) Why are Negroes concentrated disproportionately in unskilled and semi-skilled jobs? (3.) Finally, can we expect the market, or employers, or unions to take action to provide greater opportunities for Negroes or must "outside" forces intervene to alter racial employment patterns?

In approaching these questions, I shall ignore some of the more important methodological issues confronting those who would analyze this problem. I shall not, for example, concern myself with the extent or magnitude of discrimination as compared with other factors producing differences in Negro and white employment patterns. There is no question in my mind that discrimination (defined as denial of employment opportunities to Negroes solely because of their race) exists, but I seriously doubt that it can be measured precisely. Nor shall I concern myself with the lack of adequate employment statistics or with their commensurability. I likewise avoid the question of establishing criteria as to how many Negroes there "should" be in particular categories; rightly or wrongly, my standard is the distribution of white employment. And I shall not attempt to define "skill," but concentrate on upgrading in terms of income and more desirable jobs. I would also make it clear at the outset that I consider factors that interfere with the individual development of minorities to their fullest potential to be economically wasteful and morally wrong. Moreover, my discussion is admittedly evaluative, and I make no effort to attach other than evaluative weights to the different factors influencing the Negro's ability to improve his occupational position.

In the following sections I shall examine various racial employment patterns and attempt to explain the forces influencing those patterns, particularly: cultural and historical factors; education and training, especially vocational, apprenticeship, and on-the-job training; and the influence of management and unions. I conclude with some implications for public policy.

FACTORS INFLUENCING THE PATTERNS

The overwhelming weight of evidence shows that Negroes are concentrated disproportionately in unskilled, menial jobs, are rarely promoted to supervisory positions over whites, and occupy skilled jobs mainly in Negro communities or occupations traditionally reserved for Negroes. Moreover, while Negroes have improved their occupational positions in the past twenty years, they still have a long way to go to approximate the white patterns. Furthermore, improvements in racial employment patterns through shifts to better jobs have shown remarkable rigidity through time and differ geographically only in degree (see Appendix B).

The general facts concerning the status of Negro employment are scarcely subject to debate, but the complex of interrelated factors producing these patterns makes it less clear why Negroes occupy inferior occupational positions. An important impediment to the Negro's ability to acquire skilled jobs is undoubtedly his easy identification and the image of inferiority stamped upon him by slavery. Not having worked in a variety of skilled, technical operations, Negroes have become stereotyped for certain jobs by employers, white workers, and even themselves. Since the Negro is regarded by many whites as an inferior person, those who would perpetuate the feeling of superiority for their crafts or occupations often seek to bar Negroes. The more superior a job is considered to be by its occupants, the more they have resisted "dilution" by Negroes. Examples are supervisors everywhere, stock wranglers, electricians, clerks and checkers in longshoring, plumbers, locomotive engineers, white-collar workers, physicians and others. Negroes and whites frequently work together in the South, but not in the same jobs for fear this would imply equality. Nor is this factor peculiar to the South. A New York study, for example, discovered that whites suddenly found the jobs they had been doing onerous and degrading when Negroes were hired for them.[2]

Negroes are also restricted in their employment opportunities by a host of cultural and social factors. Since Negroes usually live in segregated neighborhoods, they rarely learn about jobs with few or no Negroes in them, and they apply for the kinds of jobs they know they can get. Since aspirations are conditioned by one's associates, few Negroes are motivated to try for skilled jobs.[3]

Education and Training

Negroes are inadequately prepared through education and training to compete on an equal basis with whites. While the educational level of non-whites is improving, it still lags behind the level for whites by almost three years.[4] Only 15.6 per cent of non-whites had four years of high school or college in 1960 as compared with 32.5 per cent of whites.[5]

Levels of formal education do not tell the whole story, however, because Negroes have usually had inferior education, and actual levels of education generally bear little relationship to ability. For example, a Department of Defense study of inductees for the year ended September 1, 1959, tested Negroes and whites on the basis of mental abilities. Group I was the highest classification and slow learners and mentally slow persons fell into Group IV. The monthly ranges of proportions of Negroes and whites falling into these groups were:

	Negroes	Whites
Group I	0.3%-1.8%	9.8%-20.3%
Group IV	3.0-20.3	6.4-10.9

It was further found that whether a person fell in a particular group bore little relation to formal education.[6] It has also been found that at grade eight, Negroes are two to five years behind whites in achievement.[7]

It should be noted, moreover, that poor training for Negroes is not a problem peculiar to the South. A study of 1961 Negro high school graduates in Newark, for example, found that thirty-three of the boys who graduated failed the Army GCT test and had a reading level of below the fourth grade and that girl graduates who had typing courses could not type thirty words a minute.[8]

Vocational and Apprentice Training

Negroes have also had inadequate vocational and apprenticeship training as compared with whites. Even while they are in vocational schools, moreover, Negroes are usually at a disadvantage because they have relatively less formal education than whites. [9]

In addition, vocational school training frequently perpetuates Negroes in traditional occupations. The usual practice in the South, for instance, has been to have segregated vocational schools where Negroes are trained only for occupations they have traditionally held. There are, however, some excellent vocational training schools in the South associated with Negro colleges, like Tuskegee, which train Negroes for a number of skilled trades. However, the graduates of these schools are usually unable to acquire additional apprenticeship training to become journeymen or they are unable to pass various city and state licensing laws. These college programs also offer limited opportunities to Negroes because they are open only to high school graduates, a qualification which bars most Negroes in the South. Obviously, if Negroes are going to acquire greater skills through vocational training they must be permitted to study the same courses and work in the same occupations as whites. There are indications that the Federal Government will move to deny federal funds and other aid to discriminatory vocational schools, [10] and efforts are also under way to reduce discrimination in hiring.

Minority group organizations have given considerable attention in recent years to apprenticeship training as a means of upgrading Negroes. Attention has been focused on this type of training by the virtual absence of Negroes from such programs and the decline in unskilled and semiskilled jobs. Colored workers can reduce their rate of unemployment relative to whites by moving into skilled categories which have lower rates of unemployment. [11] It should be noted, however, that in the short-run the higher rates of unemployment among Negroes cannot be entirely eliminated by upgrading, because Negroes have higher rates of unemployment at every occupational level (see Appendix B, Table 8).

Negro leaders are also motivated to seek greater apprenticeship training for Negro youths by the realization that this training is probably going to be increasingly important for the expanding skilled trades of the future. Vocational training has rarely given students sufficient practical and theoretical training to equip them to become well-rounded craftsmen. Apprenticeship also has important advantages over other means of acquiring skilled trades (i.e., armed forces training, "picking-it-up," and upgrading within plants) and assumes greater importance as technological innovations increase the need for well-trained craftsmen. At present, however, apprenticeship training furnishes only about 70,000 to 80,000 of the 500,000 skilled craftsmen needed each year. [12] While apprenticeships have been a relatively small source of skilled workers generally, they have been very important to the construction industry. [13] Even so, it is estimated that apprenticeship training will furnish a relatively small proportion of the projected needs of skilled building tradesmen by 1970 unless something is done to increase the number of apprentices. [14]

There is strong evidence that Negroes have been systematically barred from apprentice training. In New York, for example, of 10,111 apprentices in 1950 only 152 were non-white; in 1959, only 2 per cent of 15,000 New York apprentices were non-white. [15] A Connecticut survey of graduates of programs registered with the State Apprenticeship Com-

mittee between September, 1952, and June, 1954, could identify only 8
Negroes among 1,509 graduates. This study concluded, even though a
precise count of Negro apprentices was not possible, partly because of
the unco-operative attitude of the State Apprenticeship Council, that
". . . it seems fairly conclusive the Negroes possess decidedly limited
opportunities for acquiring training in certain of the highly skilled
trades."[16]

How are we to account for the virtual absence of Negroes from
apprenticeship programs? The answer to this question is not at all clear,
but some of the more important factors include:

Union Exclusion. Most of the apprenticeship programs in the building
trades, for example, are jointly administered by unions and employers,
though unions frequently in fact control the programs. While it might be
argued that unions do not limit the total number of apprentices, because
non-union employers indenture no more apprentices than union employers,[17] it cannot be denied that unions limit the total number of Negro
apprentices. The union's motives for restricting apprenticeship to
whites include: racial discrimination and a desire to maintain the "prestige" of the trades; a desire to restrict apprenticeship to sons and rela-
tives; general monopoly instincts, which only incidentally affect Negroes.
There is, however, frequently a conflict between national and local
unions over apprenticeship training and racial discrimination. Many
national unions oppose racial discrimination by their locals, as well as
other restrictions on the numbers of apprentices, because of the unfavorable reputation gained thereby for the unions; and nationals usually are
interested in growth while many locals are interested in restricting the
work to present members.[18] Union racial practices will be discussed at
greater length below.

Cultural Factors. As noted above, few Negroes apply for apprenticeship
training, probably because of a general belief that they will be excluded,
but also because many Negroes who could qualify for this training are
more interested in the professions. After all of these factors are accounted for, however, the evidence from every section of the country
supports the conclusion that Negroes have been denied apprenticeship
training solely because of their race. Recently, a few unions (IBEW
Local 3 in New Yor'., for example) have actively sought qualified Negro
applicants.

General Economic Conditions. Since fixed ratios of apprentices to
journeymen are common, there are limited opportunities for apprentices
when journeymen are unemployed. There is also greater resistance to
Negroes entering the skilled trades when white craftsmen are unemployed. Paradoxically, though, when conditions improve, many Negroes
(and whites) drop out of (or refuse to enter) apprenticeship training because there are more lucrative alternatives that require less rigorous or
lengthy training. Apprentices are characteristically paid 50 per cent of
journeyman wages, an important deterrent to non-whites who have lower
incomes than whites and perhaps cannot receive support from their families; indeed, minority youths frequently must help support their families

Management Attitudes. Before apprentices can be indentured, they must
normally get jobs. Management attitudes therefore condition the extent

to which Negroes will be barred from apprenticeship training. This is particularly true of the majority of apprenticeship training programs in plants, which are typically controlled exclusively by management. Management's motives for exclusion might be fear of reaction from white employees and the public, or might be based on personal prejudice. The influence of management on racial employment patterns will be discussed at greater length below.

In-Plant Training

Many workers also acquire greater skill through on-the-job training programs of varying degrees of formality. Here again, however, the evidence suggests that few Negroes participate in upgrading programs leading to the more highly skilled and supervisory or management positions (see Appendix B). Since it is frequently assumed that Negroes will not be promoted to skilled or managerial positions, management has hired Negroes who are generally unqualified for promotion. This has been an important obstacle to the upgrading of Negroes in those southern plants that have reduced the formal barriers to upgrading in the last ten years.

INFLUENCE OF MANAGEMENT ON NEGRO EMPLOYMENT OPPORTUNITIES

Management attitudes are probably more important determinants of the extent to which Negroes are promoted within plants than they are of Negro participation in apprenticeship programs because management has greater control of promotion; and there is ample evidence that management is prejudiced against using Negroes in skilled and supervisory categories.[19] The prevailing management attitude seems to be that Negroes are not suited for skilled jobs or that customers or white workers will react unfavorably to the upgrading of Negroes. It goes without saying that it is unfair and false to apply these stereotypes to individual Negro workers. While there is some ground for the fear of white worker reaction, the evidence seems conclusive that a firm management position on upgrading will usually prevent overt reaction from white workers, especially where management's equalitarian policies are supported by the local or international union. It is sometimes argued that employers have no status reasons to oppose upgrading because top management's status is not threatened by promoting Negroes. This is only partly true, however, because in the South management representatives frequently fear the loss of status in the community if they take unpopular racial action.

Management has historically preferred Negro workers for certain kinds of jobs or for certain of their attributes. Since there is a tendency for whites to leave undesirable jobs when economic conditions improve, management has found that Negroes are more stable in these occupations because they are "locked in" by limited occupational mobility. Moreover, many managers believe that Negroes are "better suited" than whites for hot, disagreeable work requiring great physical strength. Some employers have also considered that Negroes have other attributes which suit them for certain kinds of work. Some employers in the South, for instance, have expressed the belief that Negroes are better than whites for certain kinds of cargo unloading because they have better "rhythm." Moreover, employers have historically been willing to hire Negroes in preference to whites because of racial wage differentials and because Negroes were safeguards against unionism. The great transformation wrought by the New Deal, the CIO, and World War II eliminated this preference, however, because it virtually abolished the racial wage differential and changed Negro leaders from anti-union Republicans to pro-union Democrats.

It would appear that management attitudes are only marginally sig- nificant in determining whether or not Negroes will be upgraded. There are other forces which will tend to support or oppose basic attitudes and cause action to be taken which may or may not be contrary to manage- ment's prejudices. One such force is general market conditions, espe- cially the supply of labor, as indicated by the fact that Negroes have made their greatest occupational gains during the labor shortages accompanying World Wars I and II. (See Appendix B, Table 3). The general pattern seems to be that when white males are not available, white women will be hired if they can do the work and are available, then Negro men and women will be hired, in that order.[20]

The supply of labor is important because management will be more willing to hire Negroes in particular categories if enough colored crafts- men are available to do the work if whites quit (which they will rarely do in industrial jobs) or if they boycott particular employers (which is rela- tively easy to do in construction work). For example, some of the highest paid construction craftsmen in the South are Negroes in occupations like bricklaying where Negroes have sufficient numbers to do the work if whites refused to work with Negroes. Indeed, strongly prejudiced em- ployers will not discriminate against southern Negro longshoremen for fear they will be boycotted by Negroes and there are not enough whites to do the work; colored longshoremen have an added advantage in that work stoppages or delays can be very costly to the companies. Moreover, the knowledge that there are enough Negroes to do the work will usually deter whites who might otherwise be inclined to leave good jobs.[21] In other cases where the jobs are less desirable, the introduction of Negroes has caused whites to quit and be replaced by Negroes, but employers are willing to risk losing white employees if there are sufficient Negroes to take their places.[22] Thus, if the employer has some other reason to hire or promote Negroes, like wartime labor shortages, pressure from the Negro community or government agencies, he will be inclined to do so if the supply of Negroes is adequate.

The personal preferences of employers have also been counteracted with legal action by the NAACP and other Negro groups. It can, more- over, be expected that Negroes will continue to use their growing politi- cal power to overcome their economic disadvantages. The FEP laws in twenty states and many municipalities, together with the various govern- ment contract committees established since the beginning of World War II, have undoubtedly caused employers to upgrade Negroes where they would not have done so in the absence of these measures. Critics argue that these laws and orders constitute unnecessary interefence with management's prerogatives and introduce inefficiency. However, these measures probably have been used to a great extent by management as excuses to upgrade Negroes when economic and moral grounds were not strong enough to compel such action. Racial discrimination would appear to be the real factor making it difficult for management to exercise ra- tional employment decisions; legal measures, to the extent that they free management from concern over irrational reactions from white employ- ees and customers, actually improve the operation of the market. These governmental actions also have the symbolic effect of assuring minority workers that a majority of the people are opposed to discrimination and probably reduce suspicions of discrimination; the high proportion of complaints found to be without sufficient grounds to credit the allegations also suggests that management is being protected from unfounded charges of discrimination. This is not to deny that FEP commissions can be used to get jobs for Negroes by threatening to take respondents to public hearing where racial discrimination does not exist; there is some evidence that this has been done in a few cases, but most of the commissions appear to have scrupulously avoided such action because they realize their effectiveness would be destroyed thereby. Moreover, most of the commissions seem to carefully investigate charges of discrimination and seek to adjust valid complaints by conciliation.

The belief that the market will eliminate discriminatory hiring practices cannot be supported by the evidence. Racial discrimination in employment is part of a vicious cycle that responds to market conditions only under the most unusual conditions -- such as war -- and then only slowly. Under normal market conditions, racial employment patterns change very slowly in the absence of positive action by minority groups or the government. To be sure, legal action cannot prevent prejudice but it can prevent some discriminatory actions.

The level of employment is an important factor influencing the Negro's ability to move up the occupational ladder, but it has had different effects at different stages. During periods of labor shortage, management appears increasingly willing to overcome its personal prejudices and face the opposition of white employees and customers by upgrading Negroes. On the other hand, the recent activities of the Federal Government contract committees, which threaten to deny government contracts to discriminating firms, suggest that management is probably more vulnerable to economic sanctions during recessions when the consequences of losing federal contracts are greatest. This has been an important factor behind the desegregation of seniority lines in the petroleum refining, rubber, paper, automobile, steel, and other industries in the South since 1954.

THE INFLUENCE OF UNIONS

Unions have had more effect on the ability of Negroes to acquire skills on the railroads and in the construction trades than in industrial jobs. Since unions are normally strong enough in the former occupations to control upgrading, they have blocked the advancement of Negroes.[23]

Where unions have control of the supply of labor and maintain de facto closed shop conditions, they have been able to deny Negroes employment in the skilled trades by barring them from membership. While the number of unions with formal color bars in their constitutions or rituals has declined from at least twenty-two in 1930 to only three today (the Brotherhood of Locomotive Firemen, the Order of Railway Conductors, and the Brotherhood of Locomotive Engineers) a number of unions bar Negroes from membership and from apprenticeship by informal means. It should be emphasized, however, that the locals and not the internationals appear to be mainly responsible for these forms of discrimination. Moreover, most of these are craft organizations, since industrial unions rarely have either the ability to control the supply of labor or the motive to exclude Negroes, though there have been a few cases in the South of exclusion by industrial organizations. Scattered locals of most of the major craft organizations have barred Negroes at various times and places, but locals of the following organizations have had persistent reputations for racial exclusion going back for at least thirty years: Granite Workers, International Brotherhood of Electrical Workers (IBEW), United Association of Journeymen and Apprentices of the Plumbing Industry in the United States and Canada (UA), Flint Glass Workers, Structural Iron Workers, and Asbestos Workers. In addition, there have been frequent charges of discrimination against locals of the following organizations: Bricklayers, Masons and Plasterers; Plasterers and Cement Masons; United Brotherhood of Carpenters and Joiners; International Union of Operating Engineers; Lathers International Union; Painters, Decorators and Paperhangers; International Association of Sheet Metal Workers; and the Elevator Constructors.

It is difficult to generalize about the unions which bar Negroes from membership. They are not restricted to any particular geographical area, because there are actually stronger bars against Negroes in some union locals in the non-South than in their southern counterparts, particularly in such occupations as bricklayers, hodcarriers, common laborers, bartenders, waiters and service employees, where Negroes have long traditions in the South and are excluded in northern or western locals. Nevertheless, while some craft unions have had equalitarian racial policies and some industrial locals have barred Negroes from membership, as a general rule the unions which practice exclusion are craft organizations. Similarly, with the exceptions noted, southern unions are more likely to bar Negroes than those in other areas.

While craft locals not only bar Negroes from membership but have the ability to deny them employment or to restrict employment to certain areas, an examination of the statistics on the proportion of Negroes in various building trades fails to disclose any correlation with union growth, probably because the differences are qualitative and the impact of unions has not been to displace Negroes from occupations so much as it has been to displace them from the better jobs within those occupations.[24] The common practice for plumbers and electricians in the construction industry (as contrasted with industrial plants) has been to deny membership to Negroes and to restrict their employment to Negro neighborhoods. Unions in these occupations have not, to be sure, had the power to restrict Negro employment without the co-operation of employers, other unions, governmental licensing agencies, and sometimes vocational schools where union-controlled apprentices train. Indeed, it frequently becomes difficult to establish blame for discrimination because of "buck passing" between these parties. The basic power of the electricians and plumbers rests in their control of the supplies of skilled manpower, supplemented by agreements with employers to hire only union men and further augmented by the inability of outsiders to come into an area unless they pass a city or state license examination, over which the unions have varying degrees of control. Union control is rendered complete by the refusal of other craftsmen to work with non-union men. Thus, Negro plumbing and electrical contractors throughout the country have been forced to restrict their activities to areas where the unions will permit them to work, which usually means "patch" work and in Negro neighborhoods. Because of their restricted employment opportunities, these Negro workers are unable to acquire sufficiently varied experience to become well-rounded craftsmen.

Even where unions admit Negroes to membership, they might re-

strict their employment opportunities by placing them in segregated
locals, as is common for carpenters' locals in the South (actually, however, a relatively small proportion of union locals in the South are "segregated" in the sense that separate unions are maintained in the same
plant or trade for Negroes and whites) or by discriminating in referrals.
Some unions have perpetuated employment segregation by maintaining
segregated seniority rosters, which restrict Negroes to certain menial
jobs. This has been the chief form of discrimination by industrial unions
in the South, and is more rigid where separate seniority rosters by
crafts are formalized by having different unions in each craft, as is true,
for instance, on the railroads. Separate seniority rosters are not necessarily written into the contracts, but might be based on "informal"
understandings that Negroes and whites will move into different jobs and
that Negroes will never get jobs which will place them in supervisory
positions over whites. Even those few firms following "equal employment" policies in the South have almost never promoted Negroes to positions over whites.

It would be a mistake, however, to conclude that unions caused segregated rosters, because racial divisions of labor existed long before
unions were significant in the South. Indeed, the virtually unorganized
southern textile industry has very few Negroes and rigid racial segregation. Almost without exception, moreover, those firms which have desegregated their lines of progression in the South are unionized. The
impact of the union would appear to be to restrict the variety of work
Negroes have been permitted to perform and to formalize existing lines
of progression. The union can be accused of discrimination, however,
where it fails to apply contracts equally. Indeed, the courts have ruled
that Negroes have a legal cause of action against unions which deny them
equal promotion opportunities, and most national unions formally oppose
racial discrimination; these have been factors making it possible to desegregate some jobs.

Whether or not desegregation of seniority rosters will result in the
upgrading of many Negroes, however, depends upon the sincerity of
unions and management in promoting equal opportunities for Negroes, as
well as the following:

1. The Negroes' willingness to apply for and prepare themselves for
upgrading.

2. The employers' willingness to hire qualified Negroes. (Employers can make a mockery of equal promotion clauses in contracts by
hiring Negroes who are not qualified for promotion. For example, contracts might provide that high school graduates will be promoted on an
equal basis, but employers might hire only Negroes who have not finished
high school.)

3. The pace of employment changes after rosters are desegregated.

4. The relative seniority of Negroes. (As noted earlier, one of the
factors accounting for the differences between Negro and white employment everywhere is the recentness of the Negro's commitment to non-
agricultural employment. There is some evidence, for example, that in
those industries where Negroes have been employed for many years, as
in Detroit automobile factories, Negroes have been able to retain their
proportions during economic recessions because layoffs are based on
seniority, and Negroes had moved into many different job categories. In
this connection, however, plant-wide seniority might benefit Negroes
during upgrading, but it might work to their disadvantage during layoffs

where whites have been in the plants much longer. Similarly, the possibilities of introducing plant-wide seniority are limited by technological
considerations which determine the kinds of jobs that are related.)

    5. The nature of the tests that are to determine promotion and trans-
fer. If these tests turn on subjective factors, such as "attitude" or
"personality," it will be quite possible for management to deny promotion to otherwise qualified Negroes. Moreover, the rules concerning how
many times the tests may be taken, whether or not seniority must be sur-
rendered to move to other categories, whether Negroes must enter the
bottom of new seniority rosters (and thus take reductions in pay), the
willingness of whites to teach Negroes the new jobs, will all determine
the extent to which formal desegregation in lines of promotion will result
in greater opportunities for Negroes.

    It would be a mistake to conclude that the impact of unions on Negro
employment opportunities has been entirely negative. Unions have made
it possible for Negroes to hold their positions in industry by introducing
the principle of seniority. Moreover, the existence of unions gives
Negroes legal rights to equal treatment that they would not have if unions
were not in the plants. For example, when a union is certified as bargaining agent by either the National Labor Relations Board or the National
Mediation Board, it acquires the duty to represent all workers in the bargaining unit fairly, and the employer becomes jointly liable with the union
for this duty.[25] A few unions have also actively sought to get Negroes
hired and upgraded. Outstanding in this connection have been the United
Packinghouse Workers, who have followed more equalitarian policies in
the South than perhaps any other union. The Packinghouse Workers have
not only worked to desegregate jobs and facilities within plants but have
taken other measures, including insistence upon non-discrimination
clauses in contracts, to get Negroes hired into a variety of jobs which
were previously all white. A number of other unions, including the International Union of Electrical Workers, the United Auto Workers, and the
United Rubber Workers have taken measures to protect Negro interests
but have not been as vigorous as the UPWA in promoting equalitarianism.

    Unions have also improved the employment opportunities of minori-
ties by promoting FEP legislation. Labor organizations have provided
major financial and organizational support for these laws; the AFL and
the CIO promoted a federal FEP law and the AFL-CIO is on record in
favor of such legislation -- which AFL-CIO President George Meany
argues is necessary to combat union as well as employer discrimination.
That unions have supported FEP laws and have been among the most
intransigent respondents before the commissions created by those laws
is not entirely paradoxical because there are many different kinds of
unions and union members.

    Unions have also been instrumental in causing the abolition of racial
wage differentials, partly because the basic logic of the union must be the
"common rule," but also because unions have sought to abolish racial
wage differentials for moral reasons. Perhaps the most important agency
for removing racial wage differentials was the War Labor Board, but key
cases were brought by such organizations as the Oil Workers International Union.[26] Moreover, to the extent that unions have promoted wage
leveling they have benefited Negroes who were more heavily concentrated
than whites in the lower paying jobs. Whatever arguments may be raised
concerning the union's ability to influence wages generally, it seems
clear that unions have improved wages and fringe benefits for Negroes.[27]

Prospects for the Future

On balance, however, the evidence seems to support the conclusion that unions have generally done very little to alter the racial employment pattern except where this was incidental to such general objectives as wage equalization and wage leveling. Moreover, unions are unlikely to do much to promote the interests of Negroes except where this is in keeping with their general political and economic objectives. This is not because the national unions and the AFL-CIO favor discrimination, but because they generally either have little power to change the practices of local unions, where the real problems are, or have no compelling reasons to do so.

That is to say, unions will not take steps to abolish racial discrimination within their own ranks or in employment until the importance of the issue is raised in the operational scale of union leaders and members. One of the reasons labor unions declare against discrimination while permitting it to continue is that equalitarian racial practices take lower priority than other objectives facing the unions. When other basic motives (achievement of political goals, getting re-elected, or organizing a bloc of unorganized workers) are threatened by the continuation of discrimination, unions can be expected to take action to change these practices. One function of Negro organizations and government civil rights organizations has been to raise the priority of anti-discrimination policies through publicity, outside pressures, and other tactics. But the effective implementation of public policy requires an understanding of the structure of the labor movement and the vulnerability of its various components to moral, political (either government or intra-union), economic, or physical powers.

It is widely assumed, for example, that unions will respond to moral pressures brought to bear on them through publicity (public hearings, recommendations of fact-finding boards, public statements by prominent officials, etc.) Experience demonstrates, however, that moral pressure is more effective against the AFL-CIO and intermediate federations (and against employers) than against national or local organizations. The federation's vulnerability to charges of immoral conduct is due to the fact that it has different functions than national and local organizations. The federation is mainly a public relations organization; it is the keeper of the labor movement's conscience and seeks to influence legislation and perform other functions in the interest of organized labor. In order to accomplish these objectives, the AFL-CIO must appeal to non-labor groups, including Negroes, and hence must be concerned with its public moral image. Moreover, the federation is the most conspicuous part of the labor movement and there seems to be a direct relationship between conspicuousness and vulnerability to moral pressures.

If this is true, and if the AFL-CIO is really sincere in its public declarations, why then does it not vigorously enforce its policies against offending locals? The reason, of course, is that the AFL-CIO is a purely voluntary organization and actually has little other than moral power over its affiliates, since it seems generally established that, except in unusual cases, national unions lose very little by being expelled from the federation. Moreover, the record seems to show that while some national unions are more vulnerable to moral pressures than others, national unions are generally more sensitive to economic or political pressures in the short run than they are to moral pressures. This is because national

unions are mainly collective bargaining organizations and thus are primarily motivated by economic rather than political objectives. However,
if the union has broad social objectives, it will be more concerned with
its moral image than if its objectives are purely economic. This is one
reason industrial unions are more likely than craft organizations to adopt,
equalitarian racial policies. It could be argued, however, that in the
long run moral power is more important because third parties make their
decisions about an organization largely on moral bases, and, in an inter-
dependent society, third parties will ultimately have an opportunity to
vote or take other action that will damage the organization violating prevailing moral sentiment. Thus, the labor movement has witnessed growing public regulation in the past twenty years partly because the practices
of some of its affiliates have caused it to lose moral power.

The real problem of racial discrimination is usually at the local
level and whether or not locals will take corrective action depends upon
the extent to which they may be compelled to do so by parent internationals or by outside organizations. Generally, local unions seem to be relatively impervious to moral pressures, but more responsive to economic
pressures. There are a number of cases in which local unions have defied the most vigorous public and official denunciations of their practice
in Cleveland, Washington, D. C., Milwaukee, Hartford, San Francisco,
Los Angeles, and other places. The Hartford case, involving an IBEW
local, was resolved, after the local defied the Connecticut Commission
on Civil Rights, when the Superior Court found it guilty of contempt for
refusing to admit Negroes to membership and fined it $2,000, and $500 for
each week it remained in contempt.[28] These, and other experiences,
seem to support the conclusion that recalcitrant local unions will not
ordinarily change their practices unless they are confronted with economic losses or unless their leaders face defeat. But the extent to which
the national union can compel a local to take equalitarian racial action
seems to depend primarily on the power relations between the local and
the national organizations.[29]

CONCLUSIONS

The evidence suggests that unions probably will not solve their racial
problems without outside pressures because of intra-union power considerations and because leaders normally assign low priority to racial
policies. This realization was undoubtedly behind AFL-CIO President
Meany's request for a Federal FEPC to help the labor movement solve
the race problem in its own ranks. As suggested above, it is also doubtful that employers would normally take action to promote equal employment opportunities in the absence of outside pressures.

This does not mean, however, that legislation alone can solve the
problem of equal employment opportunities for Negroes. But legislation
can establish a framework that will make it easier for equalitarian procedures to be enforced. Within this framework, implementation will require the co-operation, and sometimes the conflict, of a variety of groups
including organizations of the aggrieved individuals, outside pressure
groups and fact-finding organizations, employers, and the regular internal union power structure. Any appreciable improvement in Negro
employment levels must also require heroic efforts by the Negro community to prepare Negroes for better jobs if and when the racial barriers
are lowered. Indeed, there is now evidence that in many areas Negroes

could get better jobs if they were qualified for them.   Thus, since this is
a "chicken and egg" problem, significant changes will require action on
a broad front.

NOTES

[1] Median income for white persons over 14 years of age in 1949 was $2,053 as compared with $973 for non-whites; in 1959, the medians were $1,502 and $3,024. Thus, while non-whites gained relative to whites, the non-white median was still only about half that of whites. (U.S., Bureau of the Census, Census of Population, 1960, United States Summary, "General Social and Economic Characteristics," Table 95.)

[2] Jacob Seidenberg, Negroes in the Work Group, New York State School of Industrial and Labor Relations, Cornell University, Bull. No. 6, February, 1950.

[3] Persons concerned with improving the economic status of Negroes frequently stress the Negro's lack of motivation and failure to take advantage of available opportunities. The 1958 report of the Ohio Governor's Advisory Commission on Civil Rights, said, for example: ". . . employers, Negroes, educators and placement personnel repeatedly indicated that the Negro had not taken full advantage of training and educational opportunities, and means should be found to encourage him, to motivate him to do so in the future."

[4] Median years of schooling completed by non-whites in 1960 was 8.2 as compared with 6.9 in 1950; by whites, 9.7 in 1950 and 10.9 in 1960.

[5] The educational distributions for whites and non-whites according to the 1960 Census were:

Years of Schooling	Whites	Non-whites
None	2.0%	6.8%
1-4	5.5	20.9
5-6	7.1	14.6
7	6.6	8.4
8	18.4	12.3
High School, 4 years	22.2	12.1
College, 4 years	10.3	3.5

The median education levels of whites and non-whites for selected Southern and non-Southern states in 1960 were as follows:

Southern states	Whites	Non-whites	Selected non-southern states	Whites	Non-whites
Louisiana	10.5	6.0	Massachusetts	11.6	10.3
Alabama	10.2	6.5	Michigan	11.0	9.1
Georgia	10.3	6.1	Minnesota	10.8	9.9
Florida	11.6	7.0	Iowa	11.3	9.5
Arkansas	9.5	6.5	Indiana	10.9	9.0
Mississippi	11.0	6.0	Connecticut	11.1	9.1
Texas	10.8	8.1	Colorado	12.1	11.2
Kentucky	8.7	8.2	Oregon	11.8	9.9
North Carolina	9.8	7.0	S. Dakota	10.5	8.6
Tennessee	9.0	7.5	Rhode Island	10.0	9.5
South Carolina	10.3	5.9	Pennsylvania	10.3	8.9
Oklahoma	10.7	8.6	Utah	12.2	10.1
Virginia	10.8	7.2	Washington	12.1	10.5
			Wisconsin	10.4	9.0
			Wyoming	12.1	9.3

[6] Cited by New York Times, December 13, 1959.

[7] L. B. Granger, "Community Factors Affecting Motivation and Achievement in a Decade of Decision," Louisiana Education Association Journal, May-June, 1962, p.4.

[8] Naomi Barko, "Dropouts to Nowhere," Reporter, March 29, 1962, p. 34.

[9] A Connecticut study, for instance, found that only 23% of Negro vocational-technical school graduates had completed high school as compared with 34% of whites. (State of Connecticut, Commission on Civil Rights, Training of Negroes in the Skilled Trades, Hartford, Conn., 1954.)

[10] In 1962, for example, the Department of Labor announced that a St. Louis vocational school could no longer be used to train apprentices working on Government projects unless Negroes were admitted to the training programs.

[11] While Negroes have historically had higher rates of unemployment than whites, these unemployment rates reflect in part the Negro's concentration in less skilled occupations, but also his lower seniority because of his more recent commitment to urban occupations. During the 1957-58 recession, for instance, the unemployment rates for white and non-white men rose by about the same amount in the semi-skilled occupations, but the unemployment rate for men increased faster for Negroes than whites in the unskilled category. (U. S., Bureau of Labor Statistics, Notes on the Economic Situation of Negroes in the United States, August, 1959, p. 3.)

[12] Sumner Slichter, Robert Livernash, and James Healy, The Impact of Collective Bargaining on Management (Washington: The Brookings Institute, 1961), p. 69.

[13] National Manpower Council, A Policy for Skilled Manpower (New York: Columbia University Press, 1954), p. 228.

[14] The estimates of the proportions to be supplied to various crafts by 1970 at the present rate of apprenticeship training range from a low of 3 per cent for painters and paperhangers and boilermakers to 36 per cent for electricians. (1961 United States Commission on Civil Rights Report, "Employment," Table 16, p. 232. See also: Louis Ruthenburg, "The Crisis in Apprentice Training," Personnel, July - August, 1959; U. S. Department of Labor, Our Manpower Implications. (Washington: U. S. Government Printing Office, 1957); and National Manpower Council, A Policy for Skilled Manpower.)

[15] New York State Commission Against Discrimination, Apprentices, Skilled Craftsmen and the Negro, 1960, p. 15.

[16] Connecticut Commission on Civil Rights, Training of Negroes in the Skilled Trades (Hartford, 1954), p. 56. See also Appendixes A and B.

[17] See: S. H. Slichter, Union Policies and Industrial Management, (Washington: The Brookings Institute, 1941), pp. 32-34.

[18] See Slichter, Livernash, and Healy, op. cit.; Architectural Record, August, 1946, pp. 95-100; Plastering Industries, November 1951; United Association Journal, May, 1952.

[19] One management official, sympathetic to the upgrading of Negroes described a common management view of the use of Negroes in skilled positions: "Negroes, basically and as a group, with only rare exceptions, are not as well trained for higher skills and jobs as whites. They appear to be excellent for work, usually unskilled, that requires stamina and brawn--and little else. They are unreliable and cannot adjust to the demands of a factory."(J. J. Morrow, "American Negroes, A Wasted Resource," Harvard Business Review, January-February 1957, p. 69) A Connecticut study found that management explained the absence of Negroes from skilled positions in that state on the basis of: Negroes are not "by nature" suited for skilled work and are better suited for heavy, unskilled work; Negroes do not apply for skilled jobs; Negroes do not possess the skills to do the skilled jobs; they lack education. (Connecticut Commission on Civil Rights, op. cit.) A San Francisco study found that employers refused to hire and upgrade Negroes for the following reasons: fear of customer or employee reaction; tradition; Negroes are bad credit risks and get involved in heavy debts: slightly under one-third of the San Francisco employers mentioned physical, mental or social traits that disqualified Negroes for certain jobs, as illustrated by the following statements concerning Negroes: "not orderly--do not have an organized mind," "not intelligent enough to hold higher jobs," "not put in executive training positions because we don't expect them to be and they don't expect to get to be top management," ". . .could not pass the physical examination, especially with regard to venereal disease." "They are not interested in working up because of the grief and responsibility involved. They don't want responsibility." This study concluded that employers who had hired Negroes seemed less prejudiced than those who had not but that there" . . .appeared to be a consensus among some employers that nonwhites lacked motivation for advancement into higher supervisory positions." (Irving Babow and Edward Howden, "A Civil Rights Inventory of San Francisco," Part I, Employment, Council for Civic Unity of San Francisco, p. 109.)

A Birmingham study found that management attributed limited opportunity for Negroes to the following factors: education and training; the inability to use Negroes where they must meet the public; fear of the reaction of white workers (it was found, however, that where whites dominated the work force before unionization, there tended to be more friction after unions came in); Negroes lack a sense of responsibility; separate rest rooms would have to be installed; Negro workers are "well suited" to the type of work they are performing and are more productive than whites in jobs requiring a lot of strength, which are repetitive or require intense heat. (Langston T. Hawley, "Negro Employment in the Birmingham Metropolitan Area," Case Study No. 3, in Selected Studies in Negro Employment in the South, National Planning Association, Committee of the South, Report No. 6, February 1955).

Noland and Bakke found that Negroes were not hired for skilled jobs because of management's belief that Negroes were not acceptable to white workers, had insufficient training, were careless in work habits and were not self-reliant. (E. William Noland and E. Wight Bakke, Workers Wanted: A Study of Employer Hiring Policies, Preferences and Practices, [New York: Harper Brothers, 1949], p. 59) They also found that some managements believed that Negroes were "not as capable as whites for production jobs. Their intelligence is believed to be lower and their training less varied and adequate. . .are slow learners. . .unreliable, irresponsible, lazy, overbearing, sensitive, unambitious, restless, and unpersevering." (Ibid., p. 32) These results were also found in a study by the Urban League in New Orleans and by the New York State Commission Against Discrimination. (Bernard Rosenberg and Penny Chapin, "Management and Minority Groups: A Study of Attitudes and Practices in Hiring and Upgrading," in Discrimination and Low Incomes, New York State Commission Against Discrimination, 1949.)

[20]In the South Carolina textile industry, for example, employment of white males declined from 62% of the total to 51.8% between 1940 and 1945, while the employment of white females increased their proportion from 34% to 43.1% and Negro males from 3.9% to 5.1%. In all manufacturing in South Carolina, which tends to be dominated by low-wage employment, white males declined 9.7% during these years and white females increased 8.2% while Negro males increased .2% and Negro females by 1.3%. (Donald Dewey, "4 Studies of Negro Employment in the Upper South," Case Study No. 2 in Selected Studies of Negro Employment in the South, [Washington: National Planning Association], pp. 190-194.)

[21]Recently, for example, when white operating engineers in Virginia threatened to quit if Negroes were hired, they did not do so after the business agent told them they would be replaced by other Negroes.

[22]In New Orleans, for example, a food-processing plant formerly hired only whites for certain materials handling jobs, but introduced Negroes during the war when whites were not available. This action led to strikes by whites, but they were replaced by Negroes and later other whites were hired for the same jobs.

[23]On the railroads, for example, the unions have prevented Negroes from being hired as firemen and being promoted to engineering positions. Indeed, the Brotherhood of Locomotive Engineers actually negotiated a contract defining Negroes as "non-promotable" and sought to force their elimination from the railroads. (See footnote 25.)

[24]The percentages of Negroes in some of the main building trades for various census years were as follows:

Craft	1950	1940	1930	1920	1910	1900	1890
Bricklayers	10.9%	6.0%	6.9%	8.1%	7.5%	9.0%	6.1%
Carpenters	3.9	3.9	3.5	3.9	4.3	3.7	3.6
Cement finishers	26.2	15.2	15.8	15.4	13.0	10.5	10.3
Electricians	1.0	0.7	0.7	0.6	0.6	n.a.	n.a.
Painters	5.2	3.8	3.6	3.2	2.9	2.1	2.0
Plumbers	3.3	2.2	2.0	1.7	1.7	1.2	1.1

Since most of these Negro craftsmen were in the South, and since the strength of building trades unions increased from 1890 to 1920 and declined from 1920 to 1928 and then rose again until 1950, there would appear to be no statistical proof that unions have excluded Negroes from these occupations.

[25]See: Steele v. L. & N. R. R., 323 (U. S.) 192; Archibald, Cox, "The Duty of Fair Representation," Villanova Law Review, January, 1957, p. 151; Aaron, Benjamin, "Some Aspects of the Union's Duty of Fair Representation," Ohio State Law Journal, Winter, 1961, p. 39; Wallace Corp. v. N.L.R.B., 323 (U. S.) 248; Central of Georgia Ry. v. Jones, 229 F. 2d 648, Cert. denied 352 (U.S.) 848; Richardson v. Texas & New Orleans Ry. Co., 242 F. 2d 230; 77 S. Ct. 230.

[26]See: National War Labor Board, Termination Reports, Vol. I, Ch. 12, and Press Release, Office of War Information, Southeastern Region, June 9, 1945.

[27]Donald Dewey, op. cit., p. 163.

[28]See "International Brotherhood of Electrical Workers Local 35 v. Connecticut Commission on Civil Rights," Connecticut Law Journal, December 29, 1954.

[29]For discussions of these relations, see Ray Marshall, "Some Factors Influencing Union Racial Practices", Proceedings of the Industrial Relations Research Association, 1961, and Idem, "Union Racial Problems in the South," Industrial Relations, May, 1962.

40

Appendix A

Negro Population Changes

The proportion of Negro workers in agriculture was 9.2% in 1960 as compared with 21.1% in 1948, and 15.7% in 1955. The 1960 census reveals that of 20.5 million non-whites in the United States, 14.8 million (or 72%) lived in urban areas; over half of the Negro population lived in rural areas in 1940 and 60% lived in urban areas in 1950. The percentage of employed non-white males engaged in agriculture was 43.3 in 1940, 31.3 in 1944, 22.4 in 1948, 24.6 in 1950, 19.2 in 1952 and 11.5 in 1960.

Census figures also show that Negroes have migrated in large numbers from the South to other areas, especially large northern and western metropolitan areas. The proportions of all non-whites in the United States in the South for various census dates were as follows:

1890	92%
1910	88
1920	85
1930	74
1940	77
1950	64
1960	57

The 1960 figure of 57% really overstates the proportion of non-whites in the South because it is based on the census definition of the South, which includes Maryland, the District of Columbia, and Delaware, with 1,098,146 non-whites. If we make this adjustment, the states of the Southeast and Southwest contained only 10.4 million of the 20.5 million non-whites in the United States. The "South" as defined in this more restricted sense had the following proportions of the non-white population:

1940	73%
1950	60
1960	51

Since this trend has continued and since a large proportion of non-whites in the South are Negroes, it seems safe to conclude that today less than half of the Negroes in the United States are located in the South.

The interregional non-white net migrations between 1950 and 1960 were as follows:

South	-1,457,000
Northeast	541,000
North Central	558,000
West	332,000

The major states to which non-whites migrated were:

California	354,000
New York	282,000
Illinois	189,000
Ohio	133,000
New Jersey	112,000
Florida	101,000

The major states from which Negroes emigrated were:

Mississippi	323,000
Alabama	224,000
South Carolina	218,000
North Carolina	207,000
Georgia	204,000
Arkansas	150,000

The extent to which Negroes have concentrated in large metropolitan areas outside the South is shown in Appendix B, Table 1, which reveals that the proportions of non-white population in the southern cities are declining while the proportions in the non-South are rising. Moreover, the Negro's rising political power as he moves to Northern metropolitan areas will undoubtedly continue to be used to help him solve his economic problems. The extent of this power is indicated by the fact that in 1960 non-whites accounted for 18 per cent of the voting age population in the twenty-five largest cities. The percentages of non-whites in the total voting-age population for some major cities in 1960 were as follows:

New York	13.2%
Chicago	20.2
Los Angeles	15.4
Philadelphia	23.9
Detroit	25.7
Baltimore	30.5
Houston	21.7
Cleveland	26.0
St. Louis	24.7
San Francisco	15.2
New Orleans	32.8
Pittsburgh	15.3
Buffalo	11.7
Cincinnati	19.8
Memphis	33.7
Atlanta	34.9
Minneapolis	19.0
Indianapolis	17.6

42

Appendix B

Negro Employment Patterns

While the basic pattern of Negro employment has not changed significantly since
1940, there have been some important trends within various employment categories.
As seen in Appendix B, Table 2, there has been a decline in the proportion of non-
whites in the relatively less skilled groups from 68. 3% in 1940 to 64. 4% in 1960; however, these are still almost double the white proportions of 38. 6% and 33. 5% for these
years.

Table 3 shows the proportion of total non-white males to total male employment
in each of these categories. The influence of full employment conditions during World
War II is demonstrated by the increase in the non-white proportion of the total to 9. 8%,
a proportion which has not subsequently been equaled; non-whites also increased their
proportions of every job category between 1940 and 1944, and declined subsequently,
though there was an increase in the proportion of professional, technical, and kindred
jobs held by Negroes between 1952 and 1959 and a significant increase in the clerical
and kindred workers categories. There has likewise been a significant decline in the
proportion of private household jobs held by non-white males. Table 4 indicates the
trend in total (male and female) employment between 1948 and 1961.

While the 1960 census figures for detailed occupations are not complete at this
writing, Table 5 gives the number and proportion of Negroes in selected crafts for
1950, and Table 6 reveals that Negroes held even lower proportions of apprenticeship
in various crafts. Table 7 shows the number of non-whites in various occupations for
1950 in the United States. These figures reveal that a large proportion of bakers,
blacksmiths, forgemen, bricklayers, carpenters, linemen, railroad firemen, painters,
paperhangers, and glaziers were in the South and relatively low proportions of the
other categories were in that region. There is also some evidence that these proportions might have declined in the South between 1950 and 1960. For example, the percentage changes in non-white male proportions of total employment between 1950 and
1960 were as follows in certain occupational categories for which statistics are available in fifteen southern and border states (Alabama, Arkansas, Florida, Georgia,
Louisiana, Mississippi, North Carolina, Oklahoma, South Carolina, Tennessee,
Texas, Virginia, Kentucky, West Virginia, and Delaware):

Managers and officials-- salaried	-36%
Retail sales workers	0
Foremen	4
Mechanics and repairmen	0
Metal craftsmen	-7
Construction craftsmen	-3
Other craftsmen	-24
Drivers and deliverymen	10
Operatives--manufacturing	-13

Source: U.S., Censuses of Population, 1950 and 1960.

These figures probably reflect the migration of skilled Negroes out of the South and
the migration of white mechanics into that region.

While the Negro proportions of jobs in the skilled categories are relatively low,
they are even lower in the technical fields. In 1950, for example, only 1.1% of male
chemists and 0.3% of male technical engineers were Negroes; the proportions for females were 0.2% and 2.3%, respectively.

Becker[1] constructs an index of occupational position of Negro and white males in the North and South by placing Negroes in skilled, semi-skilled, and unskilled classifications, estimating the relative income positions in 1939 and multiplying this by the percentages of Negro and whites in each classification for various years. He concludes from this index that the occupational position of Negroes in the South actually deteriorated between 1910 and 1950; the ratio of the Negro to white "indexes of occupational position" in the South was .67 in 1910, .63 in 1940, and .65 in 1950. The corresponding ratios for the North were: .73 in 1910, .74 in 1940, and .77 in 1950. These results cause Becker to conclude "very tentatively" that "almost all the increases in the absolute occupational position of Negroes was caused by forces increasing the position of whites as well" and that there have been "neither striking increases nor striking decreases in discrimination against Negroes...during the last four decades."

However, Becker's conclusions are subject to the limitations of any index number. Rayack, for example, objects to Becker's calculations on the ground that the use of 1939 weights of relative income position fails to take into consideration the narrowing of the income spread between skilled and unskilled workers.[2] Since Negroes are concentrated in semi-skilled and unskilled occupations, he argues, Becker's use of constant weights underestimates the improvement in the Negro's position. Rayack applies relative income position indexes for 1939 and 1951 and concludes that Negroes improved their positions relative to whites in the North from .74 in 1940 to .85 in 1950 and in the South from .68 in 1940 to .80 in 1950. It may not, however, be very meaningful to construct different indexes for the North and South because of the influence of population changes. Rayack's more meaningful figures for the United States as a whole found the ratio of Negroes to whites was .68 in 1940, .82 in 1950, and .80 in 1958. The decline in the Negro's relative position since 1948 reflects his higher rate of unemployment and the disappearance of those labor shortages that made it possible for him to improve his position relative to whites during the war.

It should be noted, however, that Rayack's criticism is not entirely valid because he and Becker have different purposes. Becker used constant weights in order to abstract from the wage leveling process that affected whites and non-whites alike; he was interested in determining changes in discrimination alone, while Rayack is interested in the changing conditions of Negroes for whatever reason.

It may, however, be objected that Becker's definition attributes differences in occupational position to discrimination which might actually be due to other factors such as the relative supplies of skilled and unskilled labor. For instance, Negro positions relative to whites may be improved by migration from the South, which has a relative surplus of unskilled labor to the North, but this would not necessarily mean a decline in discrimination in either region. Similarly, Negroes might improve their position by overcoming some of the cultural handicaps they face and this would have little to do with changes in discrimination. In short, unless we assume that Negroes and whites have the same amounts of seniority, education and training, and are similarly motivated, we cannot attribute differences in position to discrimination.

Other evidence from various industries and areas of the country confirm the general pattern established by census data for the whole country. For example, a compliance review survey of defense contractors in twenty-four standard metropolitan areas in 1957 probably is indicative of the general racial patterns in major manufacturing concerns in the United States. While the statistics must be accepted with reservation because of the size of the sample (522,375 workers), some obvious inconsistencies in the "skilled" categories, and inaccurate reporting by some enumerators, the totals appear to be approximately correct. Of the twenty-four SMA's surveyed, ten were in the South; the southern firms had 92,136 employees, 13,417 (or 14.6 per cent) of whom were Negroes. The non-southern metropolitan areas had 430,717 employees, 31,990 (7.4 per cent) of whom were Negroes.

The compliance review figures for SMA's are broken down by occupational category, but the skilled category for the Birmingham area was so grossly exaggerated (it was reported that 97.3 per cent of Negroes in the survey were skilled and that 36.4 per cent of the skilled jobs in Birmingham were held by Negroes) that the proportions are not very useful. However, the white-collar category is probably accurate and showed that in the South 21.1 per cent of whites and only 0.5 per cent of Negroes held white-collar jobs. The figures for the non-South were 36.7 per cent of whites and only 4.1 per cent of Negroes.

The compliance review survey for sixteen non-southern and seven southern automobile plants in 1957, seven southern and eleven non-southern plants in 1958, and four southern and twelve non-southern plants for 1959 revealed the following:

	1957 South	1957 Non-South	1958 South	1958 Non-South	1959 South	1959 Non-South
Total employees	14,825	83,058	11,306	49,606	7,572	30,426
Total Negro employees	495	15,045	424	7,426	335	5,008
Per cent Negro	3.3	18.1	3.8	15.0	4.4	16.5
Negroes salaried	1	129	1	73	1	71
Per cent total	---	1.0	---	0.7	---	1.8
Negroes hourly	334	14,886	423	7,353	334	4,937
Per cent total	3.1	21.6	4.6	19.2	5.3	18.8

If these figures are representative of the industry, and they probably are, they support the conclusion that there are very few Negroes in white-collar positions and relatively few in production jobs in the South. Indeed, independent observations in the South reveal very few Negroes in the automobile industry in other than menial classifications, though a few have been upgraded in recent years under pressure from Negro organizations and government contract committees. The only Negro found in a white collar position in the South by 1959 was at Louisville, Kentucky. Moreover, there were no white-collar workers in the automobile plants surveyed in Baltimore, Cincinnati, Los Angeles, Philadelphia, San Francisco, and St. Louis in 1959; of the seventy-one Negro white-collar workers in the survey, forty-one were in a Chicago plant and twenty-one in another in the New York area. If these figures are representative, however, they reveal that the Negro has been able to slightly increase his proportion of total employment in the industry in spite of declining employment. This probably reflects the fact that Negroes have been in this industry long enough to acquire seniority protection in lay-offs, and were integrated into a wide variety of jobs during the war.

The oil industry surveys for 1957 and 1958 give different results for the North-South comparison. Unlike the automobile industry, a large proportion of petroleum refining is in the South and the southern refineries are older relatively than southern automobile plants. As would be expected, therefore, Negroes have a larger proportion of oil industry jobs in the South than in the non-South, where there are relatively few Negroes. The results of the 1957 and 1958 surveys were as follows:

	1957 South	1957 Non-South	1958 South	1958 Non-South
Total employees	24,124	35,619	19,121	33,336
Total Negro	3,129	983	2,281	1,024
Per cent Negro	13.0	2.8	11.9	3.1
Negro white collar	8	14	8	15
Per cent Negro	0.1	0.2	0.3	0.2
Negro production	3,121	969	2,273	1,009
Per cent Negro	17.1	3.5	14.7	14.7

These figures show that there are relatively few Negro white-collar workers in the petroleum refining industry and that in the South the Negro has not been able to hold his proportion of the jobs in the industry, mainly because Negroes have been concentrated disproportionately in those jobs most subject to technological displacement. Independent investigations reveal that while there has been considerable agitation by Negroes in the South to break down segregated lines of progression, few Negroes have moved into previously all-white jobs because of declining production worker employment in the industry. Negroes have traditionally been hired in this industry into "labor pools" which did not promote into the "mechanical" and "operating" jobs held by whites. Since 1950, however, previously all-white lines of progression have been altered in most of the major refineries of the area, but the employment pattern has not changed very much. These formal changes could be significant, however, if production expanded or Negroes applied for white-collar jobs. Recently, for example, a number of Southern refineries have expressed a willingness to hire Negro white-collar workers and a number have been hired in Houston. In the meantime, however, the proportion of Negroes in many major refineries has continued to decline because of segregated seniority rosters and the concentration of Negroes in the lowest categories. At Esso (Standard of New Jersey) in Baton Rouge, Louisiana, there were 7,527 employees in June, 1950, 1,200 (16 per cent) of whom were Negroes; in 1960 there were about 6,000 employees, 700 (11 per cent) of whom were Negroes.

These general conclusions concerning Negro employment patterns have been confirmed for various places throughout the United States by numerous investigations. [3] In the South, Negroes have rarely been hired for jobs which would either make it possible for them to be promoted to supervisory positions over white workers or to work in the same jobs with whites. Where the latter occurs, it is usually in unskilled jobs, where whites are working temporarily, or in jobs like materials handling or longshoring. Sometimes young whites will work with Negroes while they are learning the trade. The South's leading industry, textiles, has very few Negro workers and these are confined to a few large plants and are in housekeeping or maintenance jobs. Negroes sometimes occupy skilled jobs in the South, but where they do it is usually either in the Negro community or in occupations traditionally reserved for Negroes. Examples of the latter are foundry work, cement finishing, plastering, and bricklaying. It has also been commonly assumed that mechanized operators - even when they replace "Negro" jobs - will go to whites. These jobs are frequently difficult, hot or dirty. In some areas Negroes have been awarded "quotas" of well-defined, high-paying jobs by what is generally known as "informal" or "traditional" seniority. Such informal agreements usually limit the Negro's opportunities, but in some cases Negroes have more jobs under these arrangements than if jobs were opened on the basis of seniority and qualifications. [4]

As we have seen, there are relatively large numbers of Negroes in some construction jobs in the South, especially in the so-called "trowel trades," where Negroes have a long tradition and have sufficient supplies of labor to perpetuate themselves. It is rare, however, to find Negro electricians or plumbers working on larger construction projects. They usually work for Negro contractors and in Negro neighborhoods or on small projects. These contractors are usually too small to have apprentices, and even if they did they might find it difficult for their colored apprentices to get the necessary academic work in trade schools. Moreover, in some cases, Negro contractors have blocked the expansion of Negroes into skilled trades because those contractors usually have protected markets which they want to perpetuate.

Racial employment patterns in the non-South differ from those of the South in degree, but the general pattern is essentially the same. Nor is this surprising, since employment opportunities tend to be influenced by family and social relationships and past individual experiences. Moreover, any immigrant group tends to concentrate in narrow occupational categories for ethnic reasons. In the Negroes' case, however, racial discrimination is a factor in the non-South as well as in the South. Negro supervisors of white workers are rare in the North, but more common than in the South. While Negroes in the North occupy a disproportionate number of unskilled and menial jobs, it is increasingly more common for Negroes and whites to work side-by-side at the same jobs. In this respect, the South appears to be at about the same position now as many northern plants were before Negroes broke into numerous industries during World War II. But the rate of change in the South will probably be slower than in the North because of declining job opportunities and the fact that, unlike the North, the prevailing moral sentiment in the South has been opposed to racial job equality, though there is less resistance here than to social integration.

It should also be noted that there are some jobs for which Negroes have advantages in the South as compared with other areas. Indeed, Negro bricklayers, cement finishers, iron workers, and others have been barred from practicing their trades when they migrated North. Other jobs for which Negroes have advantage in the South (as compared with other areas) include: roofers, longshoremen, hotel and restaurant workers, bartenders, some trucking classifications, railroad firemen, and others where Negroes have had a long tradition in the South. For example, Negroes predominate in general longshoring in the South while they have been barred from such jobs in New York, Philadelphia, Portland, Oregon, and New Jersey. Similarly, Negroes have traditionally worked as bartenders, cooks, and waiters in the South, but have been excluded from such jobs in San Francisco. Generally, however, the jobs for which Negroes have advantages in the South are menial or have other disagreeable features.

Appendix B Notes

[1]Gary Becker, The Economics of Discrimination (Chicago, The University of Chicago Press, 1959), p. 113.

[2]Elton Rayack, "Discrimination and the Occupational Progress of Negroes," Review of Economics and Statistics, May, 1961.

[3]V. W. Henderson concluded, concerning the employment situation for Negroes in Nashville, that when Negroes were employed above the unskilled level it was mainly in the Negro community and that 80 per cent of the Negroes in Nashville were in menial, unskilled occupations. He found, further, that there were very few Negroes in manufacturing jobs, that 80 per cent of the unskilled jobs in Nashville were held by Negroes, that Negroes were not employed as managers, clerks, or supervisors in bus companies though Negroes furnish the most lucrative markets for the companies. He noted, however, that Negroes were hired as bus drivers in Nashville in 1960. Most Nashville Negroes also attended segregated schools, and vocational training was available to them only in jobs they customarily held. The white school offered courses in electronics, IBM, refrigeration, air conditioning, drafting, radio, television and electronics. The Negro school offered courses in tailoring, bricklaying, cabinetmaking, and diversified occupations to include cook, maid, maintenance and dietetics. ("Employment Opportunity for Nashville Negroes," Community Conference on Employment Opportunity, Fisk University, April 22-23, 1960.)

A New York study of Negro participation in apprenticeship training concluded that "Negroes have not been (and are not) represented in the most minimal way in the New York State apprenticeship programs." This study also found that with only 5.4 per cent of the total work force in New York, Negroes had only 2.9 per cent of the jobs as craftsmen, but the following proportions of unskilled jobs: domestics, 39.8 per cent; service, 14.7 per cent and 12.8 per cent of the jobs as laborers. (New York State Commission Against Discrimination, Apprentices, Skilled Craftsmen and the Negro, April, 1960, pp. 16, 115.)

A Connecticut study found that 92 per cent of the Negroes working for larger employers in that state were in unskilled or semiskilled jobs, and only 4 per cent were in the skilled trades. (H. G. Stetler, "Minority Group Integration by Labor and Management," Connecticut Commission on Civil Rights, 1953, p. 19.) According to another Connecticut survey, ". . .it seems fairly conclusive that Negroes possess decidedly limited opportunities for acquiring training in certain of the highly skilled trades." (Connecticut Commission on Civil Rights, Training of Negroes in the Skilled Trades, 1954, p. 56.)

A San Francisco study found that Negroes were concentrated in semiskilled and unskilled jobs, but that even many of these were closed to Negroes and Orientals. With respect to participation in the skilled trades, this study found that "The data on the building trades unions suggests that while there were Negro and Oriental members in a number of them, there remained a serious problem of upgrading, and in a number of unions a problem of admission. While it appears that these skilled crafts were difficult for anyone to enter, the practices of some unions and of many contractors made the problems of training and entry even greater for Negroes than for white workers." (Irving Babow and Edward Howden, "A Civil Rights Inventory of San Francisco," Part I, Employment, San Francisco, Council for Civic Unity, 1958.)

A St. Louis study found that a high proportion of Negroes were working below their educational levels and that the craftsman was the most underemployed of all. St. Louis Negroes were concentrated in unskilled jobs. (Irwin Sobel, Werner Z. Hirsch, and Harvey C. Harris, The Negro in the St. Louis Economy, 1954, Urban League of St. Louis, 1954.)

Negroes were concentrated disproportionately in the lower job classifications in companies like International Harvester, which has one of the most equalitarian racial positions of any company in the South. In a study of three Harvester plants in the South, John Hope II found that "Except in one maintenance shop at the Louisville works, the plants have not employed Negro journeymen or journeymen's helpers in the apprenticeable trades, nor have they enrolled Negroes in company sponsored apprentice programs." ("3 Southern Plants of International Harvester Company," in Selected Studies of Negro Employment in the South, Washington; National Planning Association, 1955.)

In "4 Studies of Negro Employment in the Upper South," (in ibid.) Donald Dewey found that not only were Negroes restricted largely to unskilled or menial jobs for the most part but that the pattern had remained fairly constant before and after World War II. (See also, Donald Dewey, "Negro Employment in Southern Industry," Journal of Political Economy, August 1952.)

Similarly, in "Negro Employment in the Birmingham Metropolitan Area," (in <u>Selected Studies of Negro Employment in the South</u>), Langston Hawley found that a great majority of Negroes were located in unskilled and semiskilled jobs and in personal and building service occupations and that "For the most part, both the occupations filled by Negroes and the nature of the work required was found to have been remarkably stable since 1939." (P. 247.)

William H. Wesson's study of "Negro Employment Practices in the Chattanooga Area," (in ibid.) found that 18.3 per cent of the 2,276 Negro workers surveyed in that area were in skilled jobs, but that these were mainly in foundry work or in the trowel trades in construction. Wesson found strong regional industry and company biases against using Negroes in higher occupational categories outside these traditional areas.

A Cincinnati study found very small percentages of Negroes in professional or skilled categories. (City of Cincinnati, The Mayor's Friendly Relations Committee, <u>Racial Discrimination in Employment in the Cincinnati Area</u>, April, 1953.) This study found that Negroes were usually confined to the least desirable types of work, that 60 per cent of casuals were Negroes and that in clerical and sales jobs, "whites are 9 times as likely to be placed as Negroes; 5 times as likely to be placed in skilled industrial work; and 6 times as likely in semiskilled work. In service jobs, Negroes are 5 times as likely to be placed as whites, but this reversal of the ratio by no means offsets their low placement rate in longer lasting and generally more desirable jobs." (P. 5.) It was also found that in Cincinnati, many avenues of training were closed to Negroes regardless of ability.

The "1961 Report of the United States Commission on Civil Rights" concluded that "Although their occupational levels have risen considerably during the past 20 years, Negro workers continue to be concentrated in the less skilled jobs. And it is largely because of this concentration in the ranks of the unskilled and semiskilled, the groups most severely affected by both economic layoffs and technological changes, that Negroes are also disproportionately represented among the unemployed." (<u>Employment</u>, p. 153.)

A report by the Ohio Governor's Advisory Commission on Civil Rights (1958) covering employment practices in that state concluded that "While Negroes are able to upgrade from unskilled to skilled occupations in some plants, Negro skilled workers still remain very few in number in a majority of plants." (P. 20.) Furthermore, it was found that "Very few plants promote Negroes from factory jobs to supervision, but there are scattered instances of this in most metropolitan communities." (<u>Ibid.</u>) Finally, "It is in the apprenticeship field that the suggestion of a pattern of discrimination is most strong with no Negroes enrolled through the state in some programs. . . Negro apprentices are permitted in many industrial unions, although the number of Negro apprentices is extremely small." (Pp. 19-20.)

Recent studies by the Southern Regional Council in Chattanooga, and Houston, confirm the Negro's concentration in low-paying jobs and discrimination in opportunities for education and training to perpetuate this pattern.

[4]In 1957, for instance, Negroes at the Firestone plant in Memphis brought legal action to abolish separate lines of progression, but, while Negroes moved into previously all-white jobs, they lost more jobs than they gained because they were "bumped" by senior whites.

Appendix B, Table 1. Population Changes, Selected Southern and Non-Southern States, by Color, 1950 and 1960

Standard Metropolitan Areas:	Population 1960	Per cent increase in population 1950-60	Per cent non-white population	Population 1950	Per cent increase in population 1940-50	Per cent non-white population	Per cent change in non-white 1950-60
I. South							
Atlanta	1,017,008	46.4	22.8	694,699	29.7	24.7	-1.9
Birmingham	634,864	13.6	34.6	558,928	21.5	37.3	-2.7
Dallas	1,083,601	76.2	14.6	614,799	54.3	13.6	+1.0
Houston	1,243,158	54.1	20.1	806,701	52.5	18.7	+1.4
Louisville	725,139	25.7	11.6	576,900	27.8	11.5	+0.1
Memphis	627,019	30.0	36.4	482,393	34.7	37.4	-7.4
Nashville	399,743	24.2	19.2	321,758	24.1	20.0	-0.8
New Orleans	868,480	26.7	31.0	685,405	24.1	29.3	+1.7
Norfolk-Portsmouth	578,507	29.6	26.4	446,200	72.3	27.5	-1.1
Richmond	408,494	24.5	26.4	328,050	24.7	26.6	-0.2
Southern averages	748,601	37.5		551,583			
II. Non-South							
Baltimore	1,727,021	29.1	22.2	1,337,373	23.5	19.9	+3.0
Cleveland	1,769,595	20.7	14.5	1,465,511	15.6	10.5	+4.0
Chicago	6,220,913	13.2	14.8	5,495,364	13.9	11.0	+4.8
Cincinnati	1,071,624	18.5	12.1	904,402	14.9	10.6	+1.5
Detroit	3,764,131	24.8	15.1	3,016,197	26.9	12.0	+5.1
Indianapolis	697,567	26.4	14.4	551,777	19.7	11.8	+2.6
Kansas City	1,040,454	27.8	11.4	814,357	18.6	10.8	+0.6
Los Angeles	6,746,356	54.4	8.8	4,367,911	49.8	6.3	+2.5
Philadelphia	4,343,524	18.3	15.7	3,671,048	14.7	13.2	+5.1
Pittsburgh	2,406,301	8.7	6.8	2,213,236	6.3	6.2	+0.6
St. Louis	2,060,614	22.6	14.5	1,681,281	17.4	12.9	+1.6
San Francisco	2,783,355	24.2	12.5	2,240,767	53.3	9.4	+3.1
Washington	1,989,377	35.9	24.9	1,464,089	51.3	23.4	+1.5
Non-South averages	2,816,987	25.3		2,247,947			

Source: U. S. Censuses, 1950 and 1960.

Appendix B, Table 2. Percentage Distribution of Employed Males, by Color, 1940-1960

| | Percentage distribution | | | | | |
| | Non-white | | | White | | |
Occupational group	1960	1950	1940	1960	1950	1940
Total employed men	100.0	100.0	100.0	100.0	100.0	100.0
Professional, technical and kindred workers	3.9	2.2	1.9	11.0	7.9	6.6
Managers, officials and proprietors, except farm	2.3	2.0	1.6	11.5	11.6	10.6
Clerical and kindred workers	5.0	3.4	1.2	7.1	6.8	6.5
Sales workers	1.5	1.5	1.0	7.4	6.6	6.8
Craftsmen, foremen and kindred workers	10.2	7.6	4.4	20.5	19.3	15.9
Operatives and kindred workers	23.5	20.8	12.4	19.5	20.0	18.7
Private household workers	0.7	0.8	2.3	0.1	0.1	0.1
Service, except private household	13.7	12.5	12.3	6.0	4.9	5.2
Laborers, except farm and mine	19.4	23.1	21.3	5.6	6.6	7.6
Other	8.4	1.3	0.6	4.2	1.2	0.7
Total Non-farm	88.5	75.2	58.9	92.1		
Farmers and farm managers	4.4	13.5	21.1	5.6	10.5	14.2
Farm laborers and foremen	7.1	11.3	20.0	2.3	4.4	7.0
Total farm	11.5	24.8	41.1	7.9	14.9	21.2

Source: U. S., Bureau of the Census.

Appendix B, Table 3. Proportion of Non-white to Total Males in Each Occupational Group, 1940-1959[1]

| | Percentage distribution | | | | | |
Occupational group	1959	1952	1950	1948	1944	1940
Total employed men	9.2	8.9	8.3	8.4	9.8	9.0
Professional, technical and kindred workers	3.0	2.5	2.6	2.6	3.3	3.1
Managers, officials and proprietors, except farm	1.5	1.6	1.9	1.8	2.1	1.5
Clerical and kindred workers	6.5	[2]3.4	[2]2.8	[2]2.3	[2]2.8	1.6
Sales workers	1.8					1.4
Craftsmen, foremen and kindred workers	4.2	4.0	3.9	3.7	3.6	2.7
Operatives and kindred workers	10.7	10.4	8.5	10.1	10.1	6.1
Private household workers	37.7	31.6	51.3	53.7	75.2	61.8
Service, except private household	20.6	21.7	21.4	20.7	21.9	17.4
Laborers, except farm and mine	29.5	26.9	21.4	23.6	27.6	21.2
Farmers and farm managers	8.2	10.7	10.5	9.8	11.0	13.1
Farm laborers and foremen	24.0	16.2	19.8	15.8	21.1	22.5

Source: U. S., Bureau of the Census.

[1]April of selected years.

[2]Includes Sales 1944-1952.

Appendix B, Table 4. Occupational Group of Employed Persons, by Color,
1948-1961

	Percentage distribution									
Occupational group	1948		1955		1957		May 1959		October 1961	
	N-W	W	N-W	W	N-W	W	N-W	W	N-W	W
Total employed men	100	100	100	100	100	100	100	100	100	100
Professional, technical and kindred workers	2.4	7.2	3.5	9.8	3.6	10.7	4.6	11.6	3.9	12.2
Farmers and farm managers	8.5	7.8	5.0	6.0	4.1	5.2	4.0	5.0	2.6	4.1
Managers, officials and proprietors, except farm	2.3	11.6	2.3	11.1	2.1	11.2	2.1	11.5	2.8	11.5
Clerical and kindred workers	3.3	13.6	4.9	14.2	5.9	15.0	5.4	14.7	7.4	15.4
Sales workers	1.1	6.7	1.3	6.9	1.2	7.0	1.2	7.1	1.5	6.9
Craftsmen, foremen and kindred workers	5.3	14.6	5.2	14.1	5.6	14.2	6.2	13.9	6.1	13.8
Operatives and kindred workers	20.1	21.0	20.9	20.2	20.9	19.1	18.6	17.9	19.2	17.7
Private household workers	15.6	1.5	14.8	1.8	14.9	1.9	15.2	2.1	13.1	2.1
Service workers, except private household	14.7	6.4	16.8	7.2	17.1	7.5	16.7	7.8	17.6	8.5
Farm laborers and farm foremen	12.5	4.6	9.5	3.9	9.7	3.6	10.7	3.7	12.4	3.3
Laborers, except farm and mine	14.3	4.9	15.8	4.7	14.9	4.6	15.2	4.7	13.4	4.4

Sources: 1959 and 1961 from U.S. Department of Labor, Employment and
Earnings, June 1959 and October 1961.
1948--1957, U.S. Department of Commerce, Bureau of the Census,
Statistical Abstract, 1958.

Appendix B, Table 5. Male Employment in Selected Crafts, by Race, 1950

Detailed occupations	Total	White	Negro	Other	Percentage distribution		
					White	Negro	Other
Boilermakers	34,950	34,140	810	—	97.7	2.3	—
Brickmasons, stone-masons, tile setters	163,650	145,530	17,910	210	88.9	10.9	0.1
Cabinet makers	71,280	69,480	1,680	120	97.5	2.4	0.1
Carpenters	898,140	861,780	34,860	1,500	96.0	3.9	0.1
Cement finishers	28,200	20,790	7,380	30	73.82	26.17	0.01
Electricians[1]	307,013	303,429	3,236	348	98.2	1.0	0.8
Machinists	496,320	488,130	7,530	660	98.4	1.5	0.1
Mechanics and repairmen	1,670,370	1,596,450	69,990	3,930	96.6	4.2	0.2
Millwrights	58,980	57,720	1,170	90	97.9	2.0	0.1
Painters	381,150	360,600	19,860	690	94.6	5.2	0.2
Paperhangers	17,760	15,780	1,980	—	88.9	11.1	—
Plumbers & pipe fitters	271,530	262,940	8,800	210	96.69	3.24	0.07
Roofers & slaters	43,200	40,170	3,000	30	93.0	6.94	0.06
Structural metal workers	64,650	59,940	4,650	60	92.7	7.2	0.1
Sheet metal workers	117,270	116,040	990	240	99.14	0.84	0.2
Tool and die makers	152,940	152,520	420	—	99.7	0.3	—

Source: U.S., Bureau of the Census.

[1] Male and female.

Appendix B, Table 6. Apprentices in Selected Crafts, by Race, 1950

Detailed occupation	Total	Male White	Negro	Other	Percentage distribution		
					White	Negro	Other
Auto mechanics	3,600	3,510	90	----	97.6	2.4	----
Bricklayers	6,510	6,240	270	----	95.9	4.1	----
Carpenters	9,930	9,870	60	----	99.4	0.6	----
Electricians	9,360	9,240	90	30	98.7	1.0	0.3
Machinists and toolmakers	14,550	14,430	60	60	99.2	0.4	0.4
Mechanics	6,720	6,480	210	30	96.4	3.1	0.5
Plumbers and pipe fitters	11,010	10,890	90	30	98.9	0.8	0.3
Building trades (nec)	2,690	2,510	150	30	95.1	4.1	0.8
Metal working trades	7,170	7,020	150	----	97.9	2.1	----
Trades not specified	13,440	13,320	90	30	99.1	0.7	0.2

Source: U.S., Bureau of the Census.

Appendix B, Table 7. Employed Males in Selected Crafts, by Race and Region, 1950

Occupation	United States			South		
	Total	White	Non-white	Total	White	Non-white
Bakers	101,880	95,460	6,420	18,716	15,214	3,502
Blacksmiths and forgemen	54,240	51,510	2,730	15,494	13,568	1,926
Boilermakers	34,950	34,140	810	9,347	8,956	391
Bricklayers	172,530	154,140	18,390	53,467	39,883	13,584
Carpenters	898,140	861,780	36,360	325,509	297,480	28,029
Compositors and typesetters	160,560	157,620	2,940	29,826	29,026	800
Electricians	302,340	298,920	3,420	79,652	78,300	1,352
Foremen (nec)	773,100·	762,690	10,410	185,697	182,153	3,544
Linemen (telephone, telegraph, power)	205,230	203,130	2,100	56,126	54,806	1,320
Railroad firemen	53,310	51,180	2,130	14,514	12,690	1,824
Machinists and job setters	520,440	511,890	8,550	84,178	81,875	2,303
Mechanics	1,699,020	1,624,980	74,040	439,324	404,659	34,655
Millwrights	58,980	57,720	1,260	9,002	8,629	373
Molders, metal	57,150	46,650	10,500	8,484	6,877	1,607
Painters, paperhangers, and glaziers	409,290	386,460	22,830	124,253	110,761	13,492
Plasterers and cement finishers	88,380	70,950	17,430	27,413	15,628	11,785
Plumbers and pipefitters	271,530	262,440	9,090	50,655	48,039	2,616
Pressmen (except #6)	47,910	47,220	690	13,988	13,457	531
Stationary engineers	212,580	207,600	4,980	55,286	53,225	2,061
Structural metal workers	48,180	46,680	1,500	12,416	11,676	740
Other	71,730	67,530	4,200	109,716	99,784	9,932
Apprentices (operative)	111,750	109,590	2,160	20,308	19,150	1,158

Source: U.S., Bureau of the Census.

Appendix B, Table 8. Unemployment Rates, by Major Occupation Group and Color, 1960

Major occupation group	Unemployed as per cent of civilian labor force in category	
	White	Non-white
Total unemployed	5.0	10.2
Professional, technical, and kindred workers	1.7	2.9
Farmers and farm managers	.3	.5
Managers, officials, and proprietors, except farm	1.3	2.7
Clerical and kindred workers	3.6	7.3
Sales workers	3.6	5.9
Craftsmen, foremen, and kindred workers	5.0	9.6
Operatives and kindred workers	7.5	11.2
Private household workers	3.4	6.6
Service workers, except private household	5.1	9.1
Farm laborers and foremen	4.1	8.2
Laborers, except farm and mine	11.5	15.0

Source: U.S. Department of Labor, Bureau of Labor Statistics, "Special Labor Force Report," Monthly Labor Review, April, 1961, A40, Table F-5.

Female Labor Force Mobility
and Its Simulation

by John Korbel

Inadequacy of conventional labor force participation rates for female labor supply estimate - Need to test Woytinsky's and Long's hypotheses - Limitations implicit in using available data - Using simulation techniques - Factors affecting probabilities of labor force entry and exit - The 1961 Wisconsin Survey Research Laboratory data - Its limitations - Concentration of mobility - Characteristics of the highly mobile group - "Voluntary" compared with "involuntary" determinants - Symbolic statement - Results

LABOR FORCE PARTICIPATION RATES AND MOBILITY

Empirical investigations of labor supply have usually relied heavily on measurements of labor force participation rates. It will be my contention that this measure, in the case of female labor force participation, fails to reveal many of the things which we need to know about the subject.

A labor force participation rate is a proportion -- specifically, that proportion of individuals with certain specified characteristics who can be classified as being in the labor force at some moment of time, usually the calendar week preceding an enumeration by census interviewers. [1]

Obviously such a proportion in itself tells us nothing about the extent to which it is composed of the same individuals at different points in time. In other words, it reveals nothing about labor force turnover or mobility, the amount and characteristics of movement into and out of the labor force during a given time interval. But for many practical purposes this flow type of information is more important than the participation rate, which yields simply the stock of individuals in the labor force. For example, whenever we are concerned with the matching of job seekers with job openings during any time interval we are concerned with this flow type of consideration. In fact, it is this flow aspect of labor supply which is most closely related to the usual concept of supply as applied to commodities where we consider the amount of a given commodity that would become available over a specified time interval under various prevailing prices, costs, and other circumstances.

So long as males comprised the great bulk of the labor force the

Author's note. A very substantial contribution was made by George Schink and Kevin Winch in data preparation and computer programing. Funds were provided by the Social Systems Research Institute and the Research Committee of the Graduate School of the University of Wisconsin. The author is also appreciative of helpful advice by Professor Guy H. Orcutt.

purely stock aspect of labor force participation rates was not such a
serious deficiency. Males tend to stay in the labor force once they get
in and out of it once they leave. Thus, given the participation rates for
the various age groups and the age structure of the male population, one
has a fairly reliable basis for estimating the major part of the annual
additions to and departures from the male labor force. But for females,
labor force participation is far more intermittent. "Whereas for males
25-64 gross changes are practically negligible, for females of this age
entrances and exits constitute about one-fifth of the total in the labor
force, a proportion 6 to 9 times as great as that for males."[2] Much of
this movement is of course associated with the family responsibilities of
marriage and child rearing. Thus, as the female labor force grows, and
particularly its married component, which today comprises almost two-
thirds of the female labor force, the total volume of such movements increases. Table 1 shows that even ten years ago almost two-thirds of all
monthly labor force entrants were women. Now the proportion is
probably even greater because the steady increase in the participation of
married women that began with World War II has continued.

In spite of the large total amount of female labor force mobility and
its importance with respect to labor supply in the flow sense, we have
very little positive knowledge regarding its determinants and the
strengths of their effects. Detailed data on the subject, data that would
permit multivariate statistical analysis, are practically non-existent.
Even the monthly volume of female entrances and exits for the nation as

Table 1. Average Monthly Gross Changes in the Labor Force

Year	Additions		Reductions	
	Total (thousands)	Per cent female	Total (thousands)	Per cent female
1949	2,823	63.1	2,767	62.9
1950	2,978	63.2	2,938	62.0
1951	3,320	64.1	3,307	62.9
1952	3,193	62.8	3,173	62.9

Source: Derived from U. S., Bureau of the Census, Current Population Reports, Series
P-50, Nos. 19, 31, 40 and 45.

a whole, by age groups, has not been published since the appearance of
the sources cited in Table 1. As a result, the various hypotheses that
have been proposed to explain variations in the phenomenon remain
largely untested. Woytinsky's "additional worker" theory of the 1930's
-- that unemployment of husbands induces wives and other family members to enter the labor force in greater numbers -- still holds the field
along with Long's countersuggestion that "even though additional work-
seekers may appear in the market in periods of slack demand, their
influence upon the unemployment totals will be offset, either completely,
or in part, by those who become too discouraged to continue looking for
work."[3] On the basis of a study of gross flows into and out of unemployment -- over the period 1948 through 1959 -- Hansen concludes that "the
separate 'additional worker' and 'discouraged worker' flows do occur as
suggested by Woytinsky and Long, respectively. However, Long's
version, which includes both kinds of flows and assumes that they tend to
be offsetting, is far more consistent with the experience of the last decade
or so than the cruder version of Woytinsky. Still this is not to suggest
that under conditions of severe recession or deep depression the inflows
and outflows can be expected to balance out as in the past."[4]

Clearly, what is needed for a fuller understanding of the phenomenon is more detailed information on the characteristics of the individuals who stay out of the labor force, who move into it, who stay in it, and who move out of it under various circumstances of family income, family composition, aspiration levels, employment opportunities, wages offered, and so forth.

Aggregative time series on gross flows into and out of the labor force for the nation as a whole cannot provide us with such information. Nor can the usual single-shot, cross-section type of survey in which individuals are observed at only one point in time; such surveys can and do yield labor force participation rates, but have severe limitations for the study of mobility. What is needed are survey data in which the same individuals are observed at successive points in time, so-called panel data. The Current Population Survey (C. P. S.) of the Bureau of the Census — the source of the gross change data that has been published to date — does have some of the features of a panel. Three-quarters of the households in the sample one month are also in the sample the next month and each household appears in the sample for four successive corresponding months in two successive years. Unfortunately the Bureau is not satisfied that it has solved the problems of bias control in these data as it relates to labor force entry and exit. In particular there is the unsolved problem of how to ascertain the labor force status of those respondents who move their place of residence between surveys.

In spite of these unsolved problems, the Current Population Survey is still potentially the most promising source of data on labor force mobility. Not only is it a scientifically designed and administered panel with a sample size that would be difficult to match with private funds, but it uses an area sample of such design that the boundaries of the areas in practically all cases correspond with the labor market area boundaries specified by the Bureau of Employment Security. Since the Bureau of Employment Security collects data on labor market conditions in these areas on a continuing basis, the combined use of the C. P. S. and the Employment Security data should permit the empirical analysis of labor force mobility under varying labor market conditions. The C. P. S. interviews could reveal not only the movements of individuals into and out of the labor force, but also the characteristics of the mobile and of the immobile individuals, while the Employment Security data on labor market areas could reveal the labor market conditions impinging on the individuals. Thus one could ascertain not only the numbers of individuals moving in both directions, but also the characteristics of those who do the moving under varying labor-demand conditions.

THE DETERMINANTS OF LABOR FORCE MOBILITY

However, these sources of data are not available for use in this way at present. Therefore, in order to investigate female labor force mobility, we have resorted to the use of a small survey conducted in November 1961 by the University of Wisconsin Survey Research Laboratory. Although the data provided by this survey did not permit any testing of the hypotheses of Woytinsky or Long, they nevertheless permitted some preliminary investigation of the effects of certain demographic variables upon female labor force mobility. Such preliminary investigation serves two useful purposes. In the first place "any investigation at the individual or household level must take explicit account of

. . . demographic variables (which have sometimes been appropriately characterized as 'nuisance' variables) since the 'noise' from such variables if they are ignored altogether may very easily lead to bias in the estimated relationships among other factors."[5] Thus, any light thrown on the influences of demographic variables upon labor force mobility may contribute to the elimination of "noise" in later testing of the more interesting hypotheses about the subject. In the second place, the investigation of the effects of the demographic variables is a necessary step in the determination of so-called "operating characteristics"[6] for use in computer simulations of the economic activity of households.

These operating characteristics are analogous to actuarial tables of death probabilities. The characteristics such as sex, age, and race which appear in the stubs of a mortality table are examples of "inputs,"[7] while death, the probability of which is given by the table entries, is an "output".[8] Thus one might conceive of similar tables in which the entries give the probabilities of entering the labor force, with characteristics such as sex, age, and marital status as the inputs or determining factors. Corresponding tables would give as entries the probabilities of leaving the labor force.

Simulation of Labor Force Entry and Exit

The use of such tables or operating characteristics in simulation experiments may be briefly sketched as follows: Imagine that we have a magnetic tape on which are recorded the characteristics of a sample of individuals. Each tape record contains the sex, age, household status, education, and so forth, of an individual. This tape is read into the computer, one record at a time. A single reading and processing of all the records corresponds to a short time period, such as a month, and simulates the aging of the population by one month. Processing of each record consists essentially of the following. While the characteristics of an individual, as read from this tape record, are in the computer's core storage, a series of random numbers are generated. Each random number is used to determine an event, or output, for this individual. The determination of the event by means of the random number is by the Monte Carlo method. Thus, suppose a four-digit random number which has just been generated is to be used to determine whether or not the event, death, is to occur for the individual at hand. An operating characteristic table for the output, death, is entered on the basis of the individual's characteristics and the appropriate probability is selected, which may be, say, 0004. If the random number is greater than 0003, the individual is considered to have survived, while if random numbers 0000, 0001, 0002, or 0003, have been generated, the individual is considered to have died and is erased from the tape.

Simulation of labor force participation will be carried out in essentially the same manner. We will therefore need to know what the probabilities are that a woman who is out of the labor force will enter it in any given month and the probabilities that a woman who is in the labor force will leave it in any given month. These probabilities constitute the operating characteristics that we need.

What would the inputs be for such operating characteristics? Ideally, they would include all of the principal determinants of female labor force entry and exit. It would perhaps be conventional to classify them under the headings of supply factors and demand factors. The supply factors would include the obvious personal characteristics that are believed to affect a woman's readiness and ability to enter or leave the labor force -- such characteristics as age, health, marital status, presence or absence of small children in the household, assets, husband's income, education, color, various attitudinal dimensions, and prior labor force participation. There is a considerable body of evidence, both in the form of census tabulations and multivariate statistical analyses, that these variables influence female labor force participation.[9] It is fairly safe, therefore, to presume that they also influence labor force entry and exit rates.

The demand factors would include some sort of characterization of the employment opportunities in the woman's labor market area[10] and the level of wages available to her.[11] These variables are known to affect labor force participation and hence probably also labor force entry and and exit rates.

Some variables on which it is operationally feasible to get data relating to labor force entry and exit may not fit conveniently into such a

Table 2. Monthly Probabilities of Labor Force Entry for Females, by Age and Month of the Year

Month	Age				
	14-19	20-24	25-44	45-64	65+
January	.0495	.0411	.0431	.0340	.0094
February	.0661	.0610	.0472	.0447	.0123
March	.0586	.0603	.0503	.0439	.0133
April	.0621	.0529	.0478	.0428	.0156
May	.0907	.0836	.0627	.0546	.0192
June	.1756	.0735	.0560	.0481	.0168
July	.1454	.0673	.0477	.0457	.0133
August	.1173	.0710	.0584	.0438	.0193
September	.0973	.0868	.0658	.0559	.0165
October	.0995	.0695	.0629	.0515	.0161
November	.0779	.0697	.0585	.0533	.0202
December	.0727	.0578	.0430	.0430	.0140

Source: See text.

Table 3. Monthly Probabilities of Labor Force Departure for Females, by Age and Month of the Year

Month	Age				
	14-19	20-24	25-44	45-64	65+
January	.1997	.0893	.0939	.1045	.1695
February	.1209	.0772	.0824	.0865	.1830
March	.1474	.0649	.0821	.0863	.1973
April	.1487	.0819	.0870	.0879	.1566
May	.1353	.0703	.0812	.1026	.1661
June	.1312	.0802	.0912	.1038	.1829
July	.1732	.0934	.1076	.1052	.1842
August	.1839	.0897	.0991	.0920	.1467
September	.2796	.0824	.0912	.0861	.1499
October	.1903	.0864	.0874	.0881	.1677
November	.1701	.0729	.0872	.0864	.1655
December	.1624	.0807	.0840	.1102	.1965

Source: See text.

60

traditional supply versus demand classification. For example, the
seasonal factor, which is known to have a great effect on entry and exit
rates, is partly a supply-induced phenomenon and partly a demand-
induced phenomenon: school vacations operate on the supply side and
harvesting and the seasonal pattern of retail activity operate on the demand side.

We are in no position at present to incorporate such an extended list
of inputs in our labor force entry and exit operating characteristics. In
fact, in view of the extreme scarcity of data on the subject, we must be
content with very modest beginnings, both in the number of inputs that
can be incorporated and in the accuracy of the probabilities estimated.

As a first crude approximation to the operating characteristics
required, consider Tables 2 and 3. Table 2 may be interpreted as
providing the monthly probabilities that women who are out of the labor
force will enter it, with age and month of the year as arguments or
inputs. Table 3 may be similarly interpreted as providing probabilities
of labor force departures. The probabilities of staying out of the labor
force or in the labor force are the complements of the entries in Tables
2 and 3, respectively. The entries in these tables were derived from the
gross flows by sex, age, and month that were published by the Bureau of
the Census for the years 1949, 1950, 1951, and 1952 in the Current
Population Reports cited in Table 1. Of course, it would be desirable to
have more years of data on which to base our estimates of the probabilities
of labor force entry and exit but, as mentioned above, these are the only
years for which gross flows were published in such detail. Some conso-
lation may be derived from Hansen's findings cited above and from the
following comment of the Bureau of the Census.

> Turnover was about as extensive in 1950 and in 1949 and in other
> years for which data are available, in spite of changing levels of
> business activity. This fact suggests that labor force turnover is
> more a reflection of seasonal movements than of changing economic
> conditions. Such factors as the fluctuating demand for labor in
> agriculture, trade, and construction, and the pattern of school
> vacations are alone responsible for a continuing large volume of
> labor force entries and withdrawals each month. Although turnover
> was substantial throughout 1950, the number of labor force entries
> ranged over a wide scale, from a seasonal low of 2.2 million
> at the beginning of the year, when both farm and nonfarm activity
> were at their midwinter low, to a peak of 4.4 million between May
> and June, when large numbers of students entered the labor
> market to take a look for summer jobs. Similarly, monthly with-
> drawals from the labor force ranged fron an estimated 2.2 million
> to about 4.2 million over the course of the year. [12]

To compute the entries for Table 2, the reported labor force "additions"[13] from one monthly enumeration to the next, for each age group of
females, were divided by the numbers of females in the corresponding
age groups who were out of the labor force in the first enumeration. This
was done for all of the forty-eight months from 1949 through 1952 for
which additions were published. The interpretation of such ratios as
proportions of those in a given age group who entered the labor force is
slightly inaccurate due to the fact that the additions for a certain age
group may include those who aged into it from a younger age bracket

between enumerations. But in view of the fact that we are using these figures as first approximations, this inaccuracy is negligible. Also, of course, it may be noted that the definition of additions implies that movements into and out of the labor force between enumerations fail to get counted. But a good case can be made for neglecting such tenuous labor force attachments as these.

The actual entries in Table 2 are simply the means for each age group of the four January-to-February proportions, the four February-to-March proportions, and so forth. The January-to-February mean is shown under January, and the February-to-March mean under February because the C. P. S. was taken every month during the week that included the eighth day. Therefore most of the interval between two enumeration weeks was in the month which included the first week.

Table 3 was derived in a similar manner, by averaging the ratios of monthly labor force "reductions"[14] to numbers in the labor force.

Limitations of Available Data

Even if the mean proportions shown in Tables 2 and 3 were accurate estimates of the probabilities that they are supposed to represent, they would be unsatisfactory for our purposes for two reasons. In the first place, they fail to take account of many of the inputs that we shall be concerned with -- inputs such as marital status and the others mentioned above. Failing to take account of these inputs in simulation experiments would result in non-representative distributions of work experience -- or numbers of months worked during the year. For example, we have evidence that the labor force mobility or turnover of married women is higher than that of single women within the same age bracket; that is, the monthly entry and exit probabilities are higher for married women. But if we were to proceed on the assumption that these probabilities are the same for single and married women, as we would be doing if we used Tables 2 and 3 for our operating characteristics, we would have married women staying in the labor force in our simulation for excessively long durations and single women for excessively short durations. This would distort the distribution of months worked or work experience for both single and married women. And since annual earnings are a function of months worked during the year, this would distort the distributions of earned income for both single and married women.

Of course, the distortion of work experience and income distributions would not be the only effects of neglecting the additional variables referred to. With the same probabilities of entry and exit for single and married women within each age bracket, our simulation would generate highly unrealistic proportions of married versus single women in the labor force. This is because it is most unlikely that entry and exit proportions bear the same relationship to each other for single and for married women, as would be implied by the use of common probabilities for both marital statuses. Suppose, as is probably the case, that many single women who are in the labor force keep their jobs after they are married, until perhaps their first pregnancy. This means that the labor force entry rate for single women in the 20-24 age bracket will be higher than the exit rate, while for married women in this bracket the exit rate is likely to be higher than the entry rate. Therefore, if in our simulation the same over-all rates are applied to both marital statuses, we will be over-estimating the entry of married women and underestimating the entry of single women. Also, we will be underestimating the exit of married women and overestimating the exit of. single women. Thus, for married women entry would be overestimated and exit underestimated while for single women entry would be underestimated and exit overestimated. This would obviously result in too many married women in the labor force and too few single women in the labor force, even if we started out with the correct proportions or participation rates.

The second deficiency of Tables 2 and 3 for our purposes is that they fail to take account of the woman's labor force participation prior to the outset of any given month. Thus, a woman who had been in the labor force for one month would be assigned the same probability of leaving as another in the same age bracket who had been in the labor force for two years. But we have evidence that this is unrealistic. Firmness of attachment to the labor force varies considerably even among women of the same age. This is revealed by Lorenz curves which show considerable concentration of months worked among small percentages of women within the same classifications by age. The effect upon our simulation of neglecting this concentration of work experience or variability of attachment within the finer breakdowns would be to further distort the distribution of work experience and therefore of income.

SOME SURVEY FINDINGS ON LABOR FORCE MOBILITY

The survey conducted by the University of Wisconsin Survey Research Laboratory in November 1961 has permitted some investigations bearing on those deficiencies of Tables 2 and 3. The schedule used in this survey contained a set of questions on work histories dating back to January 1950. Respondents were asked what they were doing in January 1950 and their responses were coded into the following categories:
1. Working
2. Unemployed
3. Not in the labor force, i. e. keeping house, attending school, retired, disabled, ill, or etc.

Respondents were also asked for the dates of each subsequent change of status up to the time of the survey, November 1961. Thus we were able to ascertain (subject, of course, to recall and response error) the dates of all of the types of status changes shown in Figure 1. The schedules also contained questions on certain demographic sociological and economic characteristics of the respondents.

Obviously the most straightforward way to investigate labor force entry and exit probabilities would have been to study the characteristics of those entering and leaving the labor force each month. However, the sample was so small that the entrances and exits in any given month were too few to make this feasible. The total sample size was only one thousand adult respondents, 574 of whom were women. [15] Two hundred and forty of these women worked at some time during the twelve months from December 1, 1960, to November 30, 1961. The number who worked during all of the twelve months was 107 and the number who worked at some time during the year but did not work in one or more of the twelve months was 128 (5 of the 240 could not be classified in either of these categories because of faulty responses). For the 128 women who did not work all months, the numbers moving in and out of the working status each month are shown in Figure 2. This chart summarizes the responses to a question on the schedule which asked specifically for the numbers of

63

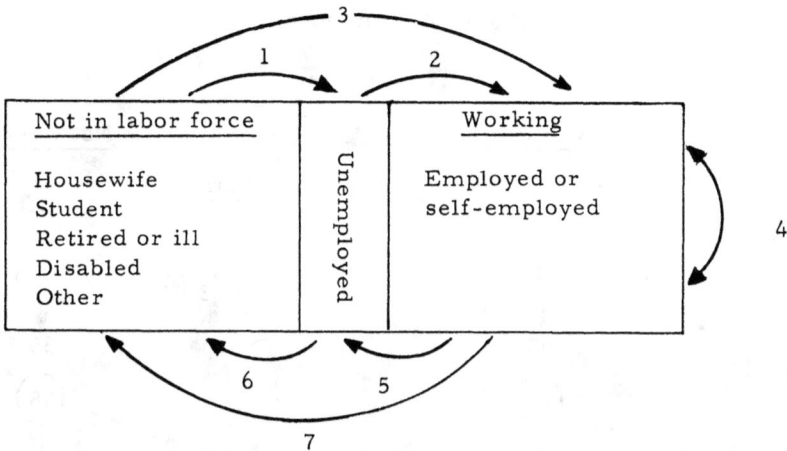

Figure 1. Types of labor force status change. Type 1 and Type 3 changes comprise entrances into the labor force. Type 6 and Type 7 changes comprise departures from the labor force.

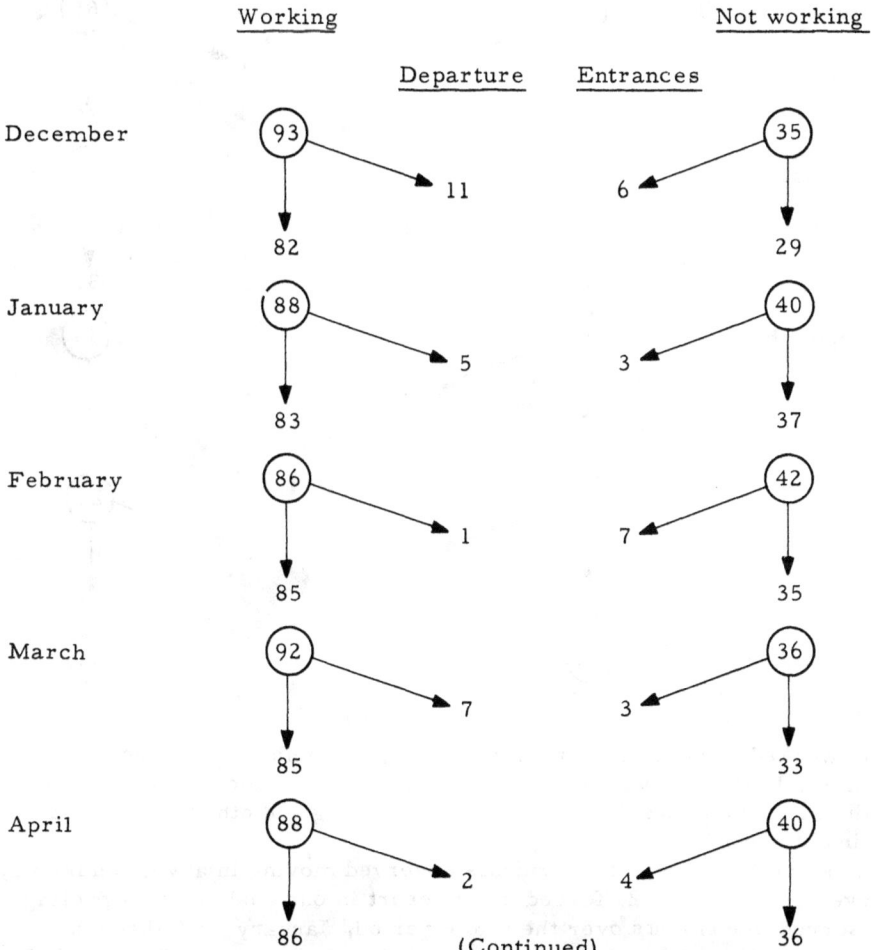

Figure 2. Numbers of women moving in and out of the working status between months from December 1960 to November 1961.

Figure 2 (Continued)

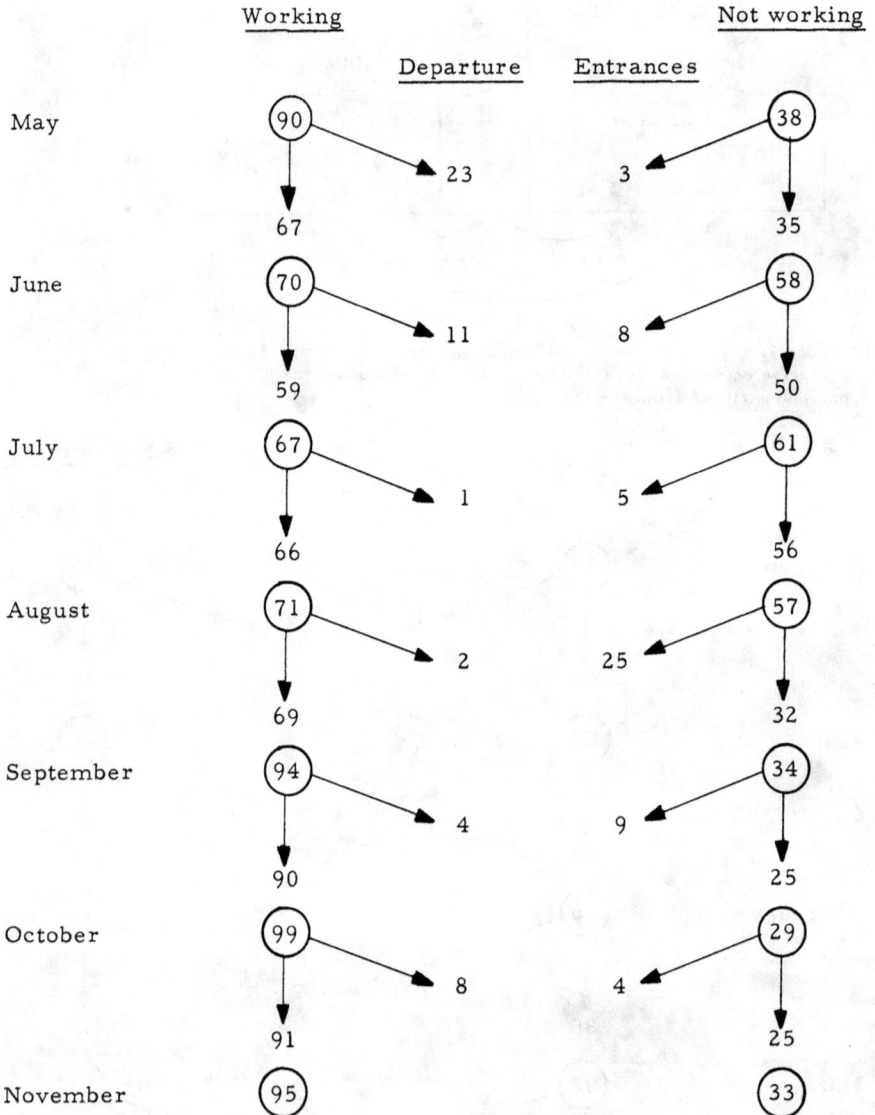

weeks worked in each of the twelve months, December 1960 through November 1961. If a woman said she worked one or more weeks in any month she was counted for that month as working and otherwise as not working.

The small number of individuals observed moving in any given month, as revealed by Figure 2, forced us to resort in our analysis to a pooling of observed movements over the whole period, January 1950 through November 1961. Labor force status over the whole period, January 1950 through November 1961, was recorded in usable form for 558 of the 574 women in the sample. Two hundred and seventy-eight of the 558 usable responses showed one or more labor force status changes over this period and the remaining 280 responses showed no change. The distribution of the no-change responses by status and age in November 1961 is shown in Table 4.

Table 4. Distribution of Respondents Who had no Change in Labor Force Status from January 1950 to November 1961, by Status and Age in November 1961

Status	Age					
	21-30	31-40	41-50	51-60	61 +	Total
Out of labor force	8	40	50	28	81	207
Working	5	13	18	15	22	73
Total	13	53	68	43	103	280

The 278 women who had one or more status changes over the period had a total of 824 changes. The distribution of these changes by status before and after change is shown in Table 5. The relatively small number of recorded movements into and out of unemployment indicates a serious under-reporting of this status by the respondents.

Table 5. Number of Recorded Status Changes, January 1950 - November 1961

Status After Change·	Status Before Change					
	Not in Labor Force		Unemployed		Working	
Not in labor force			(Type 6)	0	(Type 7)	178
Unemployed	(Type 1)	1			(Type 5)	24
Working	(Type 3)	203	(Type 2)	27	(Type 4)	376

Note: Status before and/or after change was not ascertained in the case of 15 changes of status.

This is probably largely due to the fact that the distinction between being out of the labor force and being unemployed is harder to define than the distinction between working and not working. It took the Bureau of the Census a considerable length of time to evolve satisfactory criteria for distinguishing the unemployed from those not in the labor force. Further-more, the Current Population Survey interviewers have considerably more training and experience in applying these criteria than the inter-viewers that were used in our survey. Also the distinction is probably more nebulous for women than for men since many women are not under the same imperatives as men to find a job outside the home. Thus when a woman tries to recall her statuses over an eleven-year period, as she was called upon to do in this survey, she may more easily report periods of not working as periods of being engaged in housework, whereas a man would be more likely to remember such periods as periods of unemploy-ment.

In view of this obvious under-reporting of unemployment, we shall confine our analysis of mobility to movements between working and not working; Type 3 and Type 7 changes.

One of the first things revealed by these tabulations is that there is a considerable degree of concentration of mobility. Over the twelve-year period for which work histories were recorded, less than half of the women in the sample accounted for all of the recorded changes of all

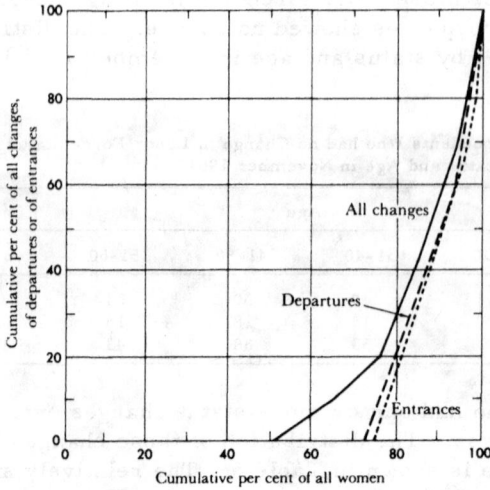

Figure 3. Cumulative percentages of all changes, of all departures and of all entrances plotted against cumulative percentages of all women.

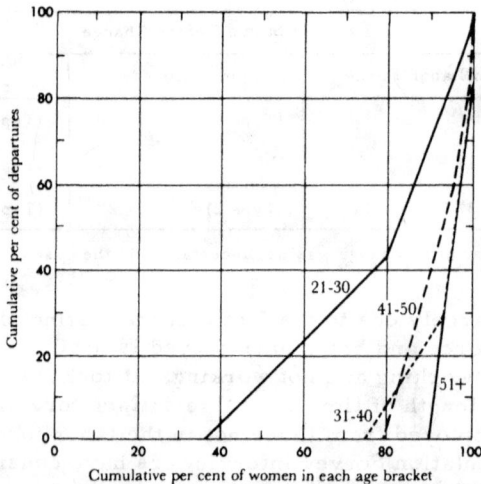

Figure 4. Cumulative percentages of labor force departures plotted against cumulative percentages of women, by age groupings according to age, in November 1961.

types. Departures from and entrances into the working status are even more concentrated. The Lorenz curves in Figure 3 serve to illustrate these points.

We can go on to note that entrances into and departures from the work status become more concentrated with age. This is shown for departures in Table 7 and by the Lorenz curves in Figure 4.

Related to this concentration of mobility is the fact that work experience is highly correlated over time. This aspect of the phenomenon is revealed by the correlation coefficients shown in Table 8. These coefficients represent the sample correlations between the hours worked by women in 1961 and in 1960; 1961 and 1959; 1961 and 1958; and so forth.

A consequence of these rather high correlations is of course that the

Table 6. Cumulative Numbers of Women, of Changes, of Departures, and of Entries

Cumulative No. of Women	Cumulative No. of Changes	Cumulative Women	Cumulative Number of Labor Force Departures	Cumulative Women	Cumulative Number of Labor Force Entrances
280	0	401	0	414	0
366	86	518	117	529	115
426	206	549	179	550	157
474	350	556	200	558	178
508	486	558	204		
524	566				
534	626				
544	696				
551	752				
554	779				
556	798				
557	810				
558	824				

Table 7. Cumulative Numbers of Women and Cumulative Numbers of Departures, by Age Groupings According to Age in November 1961

Age 21-30		Age 31-40	
Cumulative No. of Women	Cumulative No. of Departures	Cumulative No. of Women	Cumulative No. of Departures
43	0	80	0
96	53	103	20
117	95	107	28
121	107	109	34
122	111		

Age 41-50		Age $\geq$51	
Cumulative No. of Women	Cumulative No. of Departures	Cumulative No. of Women	Cumulative No. of Departures
93	0	185	0
113	20	206	21
117	28	208	25
119	34		

amount of a woman's work experience in any given year is largely explainable given her prior experience. Furthermore, the explanatory power of prior experience is almost entirely contained in the experience of the immediately prior year. This is shown by the value of R^2 as we successively regress hours worked in 1961 upon hours worked in 1960; 1960 and 1959; 1960, 1959 and 1958; and so forth. These values of R^2 are shown in Table 9. A value of R^2 of .43, or the ability to explain 43 per cent of the variance, may not at first appear to be particularly high. But when this is compared with the explanatory power of cross-sectional type variables it becomes more impressive. Table 10 shows the values of R^2 which were obtained when we regressed, whether or not the woman worked in 1961, upon a succession of characteristics of the woman and her environment in 1961. If the woman worked at any time in 1961 the de-

Table 8. Sample Correlation Coefficients Between Hours Worked in 1961 and Hours Worked in Prior Years

Prior Year	Age					
	All Women	21-30	31-40	41-50	51-60	61 +
1960	.6584	.4678	.6537	.6669	.8021	.5794
1959	.6207	.4238	.6242	.7178	.7670	.5353
1958	.5923	.3968	.5987	.6658	.7450	.5268
1957	.5769	.3162	.6208	.5918	.7449	.5546
1956	.5976	.2208*	.5820	.6434	.7479	.5596
1955	.5534	.0982*	.4558	.6793	.7323	.6675
1954	.5093	-.0183*	.3988	.6560	.7095	.6727
1953	.4859	-.0054*	.3245	.5722	.7383	.6507
1952	.4552	-.0107*	.2754*	.5034	.7238	.6595
1951	.4168	-.1058*	.1677*	.5060	.7161	.6520
1950	.4028	-.1536*	.1528*	.5261	.7090	.6342
Number of observations	457	79	91	102	65	120

*Not significantly different from 0 at the .95 confidence level.

Table 9. Values of R^2 for Hours Worked in 1961 Regressed upon Hours Worked in Prior Years

Years	(Corrected $\underline{R}^2$ for Degrees of Freedom)
1960	.4322
1960, '59	.4311
1960, '59, '58	.4302
1960, '59, '58, '57	.4302
1960, '59, '58, '57, '56	.4330
1960, '59, '58, '57, '56, '55	.4413
1960, '59, '58, '57, '56, '55, '54	.4435
1960, '59, '58, '57, '56, '55, '54, '53	.4427
1960, '59, '58, '57, '56, '55, '54, '53, '52	.4422
1960, '59, '58, '57, '56, '55, '54, '53, '52, '51	.4411
1960, '59, '58, '57, '56, '55, '54, '53, '52, '51, '50	.4404

pendent variable took the value one, and if she did not work at any time in 1961 the dependent variable took the value zero.

These findings are consistent with those of Mahoney with respect to the superior explanatory power of prior work experience. He found "that past employment experience is the single variable most predictive of current participation; the married women employed in any time period tend to be those who were employed in the previous time period."[16]

Some indication of the relative importance of personal versus environmental factors in the determination of movements out of employment is provided by the results tabulated in Table 11.

The "other voluntary" reasons enumerated in the fourth row of Table 11 include: inadequate wages; poor working conditions; too long or inconvenient hours; inconvenient work location; dull, uninteresting work; work too hard; work too dangerous; and "family reasons." The "involuntary reasons" enumerated in row 6 of Table 11 include: laid off, lack of work, business shut down, and fired or discharged. In general, the so-called involuntary reasons tend to reflect the influence of circumstances outside the individual, except perhaps for the last one, "fired or discharged," whereas the so-called voluntary reasons reflect the

Table 10. Values of R^2 for Whether Woman Worked in 1961 Regressed upon Characteristics of the Woman in November 1961[1]

Variables	(Corrected R^2 for Degrees of Freedom)
Age	.0539
Age and marital status	.1241
Age, marital status and child status	.1430
Above variables plus education	.1439
Above variables plus husband's income	.1533
Above variables plus urbanization	.1570
Above variables plus unemployment rate in respondent's county	.1568

Note: There were only eleven non-white women in the sample, so these were excluded from the computations.

[1] The explanatory variables were introduced into the computations in 0 - 1 dummy variable form using the technique explained by Daniel B. Suits, "Use of Dummy Variables in Regression Equations," Journal of the American Statistical Association, 52 (December, 1957), 548-55. In our computations, women were classified by: (1) Age as being in one of the following intervals; 21-30, 31-40, 41-50, 51-60, 61 +.

The classifications for the other variables were as follows: (2) Three classifications for marital status; single, married, and separated, widowed or divorced. (3) Three classifications for child status according to the age of the youngest child; 0-5, 6-12, 13-18. (4) Three classifications for education; 4 years of college or more, 1-3 years of college, high school graduate, non-high school graduate. (5) Five classifications of husband's income; < $2000, $2000 - $3999, $4000 - $5999, $6000 - $7999, $8000 and over. (6) Four classifications of urbanization, rural, town population; 2,500 - 24,999, 25,000 - 99,999, 100,000 and over. (7) Four classifications of county unemployment rate; 3.8 - 4.2 per cent, 4.5 - 6.1 per cent, 6.3 - 7.3 per cent, 8.9 - 9.9 per cent.

Table 11. Frequency Count of Reasons Given by Respondents for Leaving Jobs

Reason	Type 4 changes - changing jobs	Type 5 changes - becoming unemployed	Type 7 changes - leaving the labor force
Pregnancy	13	0	48
Getting married	18	0	22
Moving	41	4	9
Other "voluntary" reasons	172	5	47
Total "voluntary"	244	9	126
"Involuntary" reasons	122	15	49
Reason not ascertained	10	0	3
Total	376	24	178

influence of characteristics or actions of the individual herself. A comparison of the frequencies in rows 5 and 6 of the last two columns of the table is interesting as a crude indication of the relative importance of personal as compared to environmental factors in the determination of departures from the labor force. Apparently personal factors are more influential in the determination of departures than are environmental factors, except, of course, in the case of becoming unemployed.

AN ATTEMPT TO ADJUST THE PROBABILITIES OF LABOR FORCE MOBILITY INPUTS

What can these findings contribute to the estimation of the probabilities of entrance into and departure from the labor force? Obviously, in view of the smallness of the sample, and the likelihood of bias due to the heavy reliance on the respondent's powers of recall, we cannot expect very much. Clearly, then, it would be advantageous to put primary reliance on the Current Population Survey data presented in Tables 2 and 3 and consider whether our survey provides any basis for adjusting those estimates in any way so as to incorporate the effects of other variables besides age and month of the year.

In view of the considerable concentration of mobility revealed in Figures 3 and 4, it would obviously be desirable in any simulation of female labor force participation to avoid assigning the same probabilities of labor force entrance and exit to all of the women in any particular age bracket. It would be desirable somehow to take account of the fact that the probabilities of entrance and exit must be higher for the more mobile women, that is, for the women that tend to spend shorter lengths of time both in and out of the labor force. Our survey does provide some basis for doing this.

The figures in parentheses in the cells of Table 12 represent the usable observed departures from the working status, cross classified by age of woman at time of departure and months in labor force prior to departure. The cell entries that are not in parentheses represent the corresponding relative frequencies, which for any given age interval shall be designated q_1, q_2, and q_3. Thus for the 25-44 age interval $q_1 = .37$.

The figures in parentheses in the cells of Table 13 were obtained by cross classifying the women in the labor force in November 1961 by age and by length of time that they had been employed. The cell entries that are not in parentheses represent the corresponding relative frequencies, which for any given age interval shall be designated x_1, x_2, and x_3. Thus x_1 for the 25-44 age interval is .11.

Now let N = number of women 25-44 years old in the labor force

and p_d = probability that a woman who is 25-44 years old will leave the labor force in a given month, say October, which is .0874, according to Table 3.

Then Np_dq_1 = number of women in the 25-44 year-old age bracket leaving the labor force in October who have been in 1-24 months

and Nx_1 = number of women in the 25-44 year-old age bracket who have been in the labor force 1-24 months.

So that the probability of departure for a woman in the 25-44 year-old age bracket who has been in the labor force 1-24 months would be

$$\frac{Np_dq_1}{Nx_1} = p_d\left(\frac{q_1}{x_1}\right)$$

Similarly, the probabilities of entrance into the labor force for 25-44 year-old women who had been out for 1-24 months would be $p_e\left(\frac{r_1}{y_1}\right)$

where p_e = probability that a woman who is 25-44 years old will enter the labor force in a given month, say October, which is .0629 according to Table 2.

r_1 = fraction of all 25-44 year old women entering the labor force who have been out 1-24 months.

y_1 = the fraction of all 25-44 year old women out of the labor force who have been out 1-24 months.

Thus the ratios $\frac{q_i}{x_i}$ and $\frac{r_i}{y_i}$ are age specific adjustment factors to be applied to the entries of Tables 3 and 2 respectively to obtain labor force departure and entry probabilities corrected for interval spent in or out of the labor force prior to movement. Table 14 shows the values of the age specific $\frac{q_i}{x_i}$ computed from Tables 12 and 13. Table 17 shows the values of the age specific $\frac{r_i}{y_i}$ computed from Tables 15 and 16.

Table 12. Distributions of Lengths of Time in Labor Force Prior to Departure (Type 7 change) for Each of Three Age Groups: Yielding Estimates of the q_i

Months in labor force prior to departure	Age at time of departure						Totals	
	≤ 24		25 - 44		45			
1 - 24	.40	(20)	.37	(16)	.12	(3)	(39)	
25 - 48	.14	(7)	.10	(4)	.04	(1)	(12)	
$\geq$ 49	.46	(23)	.53	(23)	.84	(21)	(67)	
Totals	1.00	(50)	1.00	(43)	1.00	(25)	(118)	

Note: Figures in parentheses are absolute frequencies; others are relative frequencies, the q_i. Combining the first two rows, $\chi^2 = 10.10$.

$\chi^2_{.05}$ for 2 d.f. = 5.991.

Table 13. Distribution of Women in the Labor Force in November 1961, by Length of Time in Labor Force and Age: Yielding Estimates of the x_i

Months in labor force	Age in November 1961						Totals	
	21 - 24		25 - 44		≥ 45			
1 - 24	.26	(6)	.11	(11)	.06	(6)	(23)	
25 - 48	.04	(1)	.07	(7)	.04	(4)	(12)	
$\geq$ 49	.70	(16)	.82	(81)	.90	(83)	(180)	
Totals	1.00	(23)	1.00	(99)	1.00	(93)	(215)	

Note: Combining first two rows, $\chi^2 = 6.23$.

Table 14. Table 3 Modifiers, Values of q_i/x_i for Each of Three Age Groups (Based on the Relative Frequencies in Tables 12 and 13)

Months in labor force prior to departure	Age at time of departure		
	$\leq 24*$	25 - 44	≥ 45
1 - 24	1.54	3.36	2.00
25 - 48	3.50	1.43	1.00
$\geq$ 49	.66	.65	.93

*The q_i used to derive these ratios q_i/x_i were derived from observations on women who might have been as young as 9 years old at time of departure, whereas the x_i were derived from observations on women aged 21-24. This probably biases q_1/x_1 upward (x_1 too small) and q_3/x_3 downward (x_3 too large).

Table 15. Distributions of Lengths of Time out of Labor Force Prior to Entrance (Type 3 Change), Relative Frequencies not Age Specific: Yielding Estimates of the r_i

Months out of labor force prior to entrance	Age at time of entrance						Totals	
	≤ 24		25 - 44		≥ 45			
1 - 24	. 36	(26)	. 36	(12)	. 36	(6)	(44)	
25 - 48	. 14	(3)	. 14	(11)	. 14	(3)	(17)	
$\geq$ 49	. 50	(26)	. 50	(28)	. 50	(7)	(61)	
Totals	1. 00	(55)	1. 00	(51)	1. 00	(16)	(122)	

Note: Combining first two rows, $x^2 = 0.9696$, therefore the three distributions for the age groups were pooled in computing the relative frequencies of lengths of time out of the labor force.

Table 16. Distribution of Women out of the Labor Force in November 1961 by Lengths of Time out of the Labor Force and Age: Yielding Estimates of the y_i

Months out of labor force	Age in November 1961						Totals	
	21 - 24		25 - 44		≥ 45			
1 - 24	. 30	(3)	. 11	(12)	. 04	(6)	(21)	
25 - 48	. 20	(2)	. 04	(5)	. 03	(5)	(12)	
$\geq$ 49	. 50	(5)	. 85	(94)	. 93	(153)	(252)	
Totals	1. 00	(10)	1. 00	(111)	1. 00	(164)	(285)	

Note: Combining first two rows, $x^2 = 30.26$.

Table 17. Table 2 Modifiers, Values of r_i/y_i for Each of Three Age Groups (Based on the Relative Frequencies in Tables 15 and 16)

Months out of labor force prior to entrance	Age at time of entrance		
	$\leq 24^*$	25 - 44	≥ 45
1 - 24	1. 20	3. 27	9. 00
25 - 48	. 70	3. 50	4. 67
$\geq$ 49	1. 00	. 59	. 54

*The r_i used to derived these ratios r_i/y_i were derived from observations on women who might have been as young as 9 years old at time of entrance whereas the y_i were derived from observations on women aged 21-24. This probably biases r_1/y_1 downward (y_1 too large) and r_3/y_3 upward (y_3 too small).

Further Needs for Testing Assumptions

It is clear that the method of estimating the probabilities of labor force entry and departure developed above represents a serious compromise with data deficiencies and leaves a great deal to be desired. Even if the estimates of the q_i, the r_i, the x_i, and the y_i had substantive value, their use in adjusting the Table 2 and 3 entries implies a number of assumptions whose validity needs checking. One of these assumptions is that the composition of the flows into and out of the labor force are constant over time with respect to the durations that individuals are out and in prior to movement. Similarly, it is assumed that the distributions of women in and out of the labor force are constant over time with respect to the lengths of time that they have been in and out of the labor force respectively.

In order to investigate the validity of these assumptions and develop more satisfactory estimates of the probabilities of labor force entry and departure, we are going to need considerably larger samples drawn from a wider section of the country. Also it would be well to place less reliance on the respondent's recall to determine when movement occurred. And in order to ascertain movement into and out of the labor force we should adopt the same questionnaire phrasing as that used by the Bureau of the Census in its determination of labor force status.

In stating these data requirements it becomes obvious that the most satisfactory way of meeting them would be to use the Current Population Survey interviews. It is hoped that the problem of preserving the confidentiality of the C. P. S. respondents can be solved so that this can be accomplished.

74

NOTES

[1] For definitions of the criteria used by the Bureau of the Census in classifying persons as in or out of the labor force, see the Bureau's Current Population Report, Series P-23, No. 5. The history of the development of these criteria is presented by L. J. Ducoff and M. J. Hagood in Labor Force Definition and Measurement, Social Science Research Council (New York, 1957). See also, Clarence D. Long, The Labor Force Under Changing Income and Employment (Princeton: Princeton University Press, 1958), Appendix E; and Gertrude Bancroft, The American Labor Force: Its Growth and Changing Composition (New York: Wiley, 1958), Appendix C.

[2] Philip M. Hauser, "Mobility in Labor Force Participation," in E. Wight Bakke, et al., Labor Mobility and Economic Opportunity (New York: Wiley, 1954), p. 37.

[3] W. Lee Hansen, "The Cyclical Sensitivity of the Labor Supply," The American Economic Review (June, 1961), LI-3, p. 300.

[4] Ibid., p. 308.

[5] T. P. Hill, "An Analysis of the Distribution of Wages and Salaries in Great Britain," Econometrica, July, 1959, 27-3, p. 368.

[6] For a full definition of the term, see Guy H. Orcutt, et al., Microanalysis of Socioeconomic Systems: A Simulation Study (New York: Harper & Bros., 1961), p. 15.

[7] Ibid., p. 15.

[8] Ibid., p. 14.

[9] See, for example, Gertrude Bancroft, op. cit.; Clarence D. Long, op. cit.; Thomas A. Mahoney, "Factors Determining the Labor-Force Participation of Married Women," Industrial and Labor Relations Review (July, 1961), 14-4, pp. 563-577; Richard Rosett, Working Wives: An Econometric Study, Cowles Foundation Discussion Paper, N. 35.

[10] See, for example, Nedra Bartlett Belloc, "Labor Force Participation and Employment Opportunities for Women," Journal of American Statistical Association, 45 (September, 1950), 400-410.

[11] Jacob Mincer, "Labor Force Participation of Married Women," Conference on Labor Economics, National Bureau of Economic Research, Preliminary mimeographed report dated April 22 and 23, 1960.

[12] U. S., Bureau of the Census, Current Population Reports, Series P-50, No. 31, p. 4.

[13] "Additions to the number in a given employment status between one month and the next consists of those with that status in the survey week of the preceding month." (Current Population Reports, Series P-50, No. 19, p. 10.)

[14] "Reductions in the number in a given employment status between one month and the next consist of those with that status in the survey week one month who were in another status in the survey week of the following month." (Ibid.)

[15] The sample design was an area probability sample of households in Wisconsin with one adult respondent selected randomly from each household chosen.

[16] Thomas A. Mahoney, "Factors Determining the Labor-Force Participation of Married Women," Industrial and Labor Relations Review, July, 1961, Vol. 14, No. 4, p. 574.

The White / Non-White Unemployment Differential

Introduction - Last-hired, first-fired proposition - How it can be interpreted and tested - Empirical examination of white/non-white unemployment differential - Timing of unemployment series turning points - Variability of the two rates - Comparisons of cycle changes in the inter-race differential - Regional differences - Conclusions.

An aspect of unemployment that has received little attention from economists until recently is that of the differential incidence of unemployment for various subgroups (classified by age, sex, color, occupation, etc.) within the labor force.[1] In the past, these unemployment differentials have been principally the concern of government agencies and scholars in other disciplines.[2]

This study is concerned with one aspect of one of the differentials; it seeks to examine systematically the behavior of the white/non-white unemployment differential over the course of the business cycle and to relate such behavior to other aspects of market discrimination against non-whites that have been revealed in other works.[3]

STATEMENT OF THE PROBLEM

The proposition that members of minority groups are "the last to be hired and first to be fired" (hereafter referred to as the "lhff" proposition) has gained wide acceptance.[4] However, no one, to my knowledge, has specified precisely what kind of market behavior the proposition im-

Author's note: This paper is part of a larger study in preparation for my doctoral dissertation at the University of Chicago on Discrimination and the White-Non-white Unemployment Differential. In the early period the study was supported financially by the Labor Workshop of the University of Chicago.

I want to thank the officers and the staff of the National Bureau of Economic Research for the facilities and computational aid extended to me during the last year and a half. I would also like to thank Albert Rees, A. C. Harberger, Gary S. Becker, Geoffrey H. Moore, and Robert Lipsey for their help and suggestions. My very special thanks go to H. Gregg Lewis and Jacob Mincer for their invaluable contributions in both analysis and methodology for the entire work.

plies. Generally, this proposition is presented without supporting evidence. Occasionally it is documented with a comparison of white and non-white aggregate unemployment rates. [5]

Given the ambiguity of the "lhff" proposition, it is not surprising to find that it is subjected to a variety of tests and to the drawing of contradictory conclusions from these tests. For some purposes, such as the interpretation of white/non-white differentials in the incidence of certain social problems, it may be unnecessary to decide whether the proposition implies current racial discrimination in hiring and firing. For these purposes the researcher may not be too interested in finding out whether the proposition would be true if he standardized the data for, say, differences in skill. He could test the validity of the proposition with a comparison of the white and non-white aggregate unemployment rates and observe whether their differential was wider at troughs than at peaks of business cycles. His findings would depend upon the particular measure he selected for this differential. With reference to U. S. data, if he chose to observe the behavior of the ratio of non-white to white unemployment rates, he would conclude that the proposition is false. [6] If he emphasized the behavior of the absolute difference in their unemployment rates, [7] he would conclude that the proposition is true. [8]

For the purpose of analyzing market behavior it is important to distinguish the effects of discrimination on, for example, the level of educational attainment and the occupational distribution from its effects on unemployment rates given the level of skill. Policies directed toward the reduction of the economic effects of discrimination, if they are to succeed, must be based upon the knowledge of the particular areas wherein discrimination occurs. Also, an employer who may not wish to practice discrimination may regret the fact that prior limitations prevented non-whites from acquiring higher skills, but this regret would be irrelevant for his hiring and firing decision; at the same money wages for equal skill he would be indifferent between white and non-white labor. Furthermore, I believe that the proponents of the "lhff" proposition intend to describe discriminatory market behavior in current hiring and firing. The proposition implies, for instance, that because of discrimination employers will first discharge members of minority groups when reduction in the input of labor is necessitated by a decrease in the demand for labor. In this context, if the phrase "because of discrimination" is to have meaning, it must apply to hiring and firing of labor of equal skill. A researcher desiring to test the validity of the proposition under this formulation can no longer do so by comparing aggregate unemployment figures. He must first adjust the data for differences in such things as skill.

The purpose of this study is to determine whether the greater absolute increase in the non-white aggregate unemployment rate during troughs of business cycles can be accounted for by some characteristics of the non-white labor force which also produce higher increases in the unemployment rates of subgroups of the white labor force. The non-white labor force has, for instance, a substantially different occupational distribution. Non-white workers also have lower educational attainment (see Table 1) and less on-the-job training [9] than do white workers.

On the basis of differential cyclical variability in demand for the product, economists have long predicted and observed greater cyclical fluctuation in unemployment of workers in durable good industries. If non-whites were disproportionately concentrated in durable good industries, one would expect their aggregate unemployment rate to show greater cyclical variability. This would occur in the absence of differential hiring and firing practices for white and non-white workers within given industries.

Recently, economists have also taken note of the differences in the cyclical pattern of unemployment rates between occupations, or levels of skill. They have formulated hypotheses concerning the inverse relationship between levels of skill and variability in unemployment rates. [10] Since non-whites are predominantly concentrated in lower skill occupations, their aggregate unemployment rate will show greater cyclical variability (see Tables 1 and 4, and Figure 2). Again, this will occur even if within a given occupation employers are indifferent between white and non-white workers.

The unfavorable industrial, or occupational, distributions or the lower level of educational and skill attainment of non-whites are probably themselves results of discrimination. [11] However, the analysis here will differentiate between the effects of discrimination on unemployment through an unfavorable occupational distribution and the effects of discrimination on unemployment through differential lay-off rates within given occupations.

ANALYSIS OF THE PROBLEM

The existence of racial discrimination need not, in the absence of skill or other economic differences, produce a higher unemployment rate for non-whites. Moreover, the fact that non-whites may have a higher unemployment rate during the peak of the business cycle does not imply that they will also have a greater increase in their unemployment rate during the downswing in business activity. Presumably, discrimination can be reflected in ways other than increased unemployment for minority groups. If whites could be compensated — bribed in the form of differential wages — for working with non-whites, reduced demand for labor need not result in a higher lay-off rate for non-whites. As long as preferences for discrimination are not infinite there could exist an equilibrating set of wage differentials at which white and non-white labor in a given occupation and locality will have the same unemployment rate. Thus we cannot speak of hiring and firing decisions outside the context of the wage rates for white and non-white labor.

The analysis here begins with the utility maximization postulate utilized by Gary Becker to explain market behavior with regard to the employment of minority groups. [12] Under this postulate, given the preferences for discrimination, members of minority groups will be employed only under terms that will compensate discriminating employers, fellow employees, and consumers for the non-pecuniary disadvantage of associating with them. Because of differences in the preferences for discrimination, in Becker's terms, differences in the coefficients of discrimination (CD), among members of the majority group, the equilibrium wage ratio, $\frac{\text{White wages}}{\text{Non-white wages}}$, at which employers will be indifferent between hiring white and non-white labor will depend, among other things, on the ratio of whites to non-whites within the labor force. If there were only a few non-whites seeking work, they could find employment with those employers who have small coefficients of discrimination. They could then be employed under terms approximating those of whites. As the number of non-whites seeking employment increases, they must work for employers who have increasingly higher discriminating coefficients. For the larger number to find employment, the equilibrium ratio of white to non-white wages must rise. There could obviously exist a wage differential at which all non-whites desiring work could find employment.

Suppose that the coefficients of discrimination could not be compensated for completely in the form of lower wages being paid to members of minority groups. [13] We might then observe a higher unemployment rate for non-whites. This will occur because some employers are now no longer receiving sufficient compensation for their coefficients of discrimination. Such employers may now refuse to employ non-whites. Limiting the number of employers willing to employ non-whites could result in higher rates as well as in longer average duration of unemployment of non-whites.

However, it must be kept in mind that the existence of wage differentials cannot be verified by looking at the wages of white and non-white labor performing the same tasks. Wage compensation for discriminatory preferences could occur by paying a non-white worker having a higher marginal physical product the same wage as a white worker with a lower one. Thus the unfavorable occupational distribution of non-whites may be due in part to the inability of discriminators to receive compensation for their preferences in the form of differential wages, i.e. occupational downgrading may be a means of getting greater wage flexibility.

Wage rigidities may also produce differences in the industrial distribution of whites and non-whites of a given occupation. Suppose that employment of a given occupation in highly seasonal and cyclically sensitive industries were less desirable. Then, if non-white labor could not be paid lower wages, one would expect them to end up in these less desirable industries. Differential industrial distribution may be another way of making wages more flexible. Differential industrial and occupational distributions would of course tend to produce greater cyclical variability in non-white aggregate unemployment rates. But within a given occupation and industry employers could be indifferent between whites and non-whites.

Thus far the hypothesis may provide for a differential incidence in unemployment between whites and non-whites in a given occupation, at a given time. It also provides for greater cyclical variability in the non-white aggregate unemployment rate. It does not provide for differences in the differential within occupations and industries between peaks and troughs of business cycles, for those employers who did hire non-whites were compensated for their discriminating coefficients. They should have no particular incentive to discharge non-whites first.

Differential cyclical behavior within occupations could take place under one or more of the following conditions. It could take place if the desire for discrimination became more intense during recessions. This view is in accord with the popular conception of discrimination. It holds that whites do not particularly object to working with non-whites during periods of full employment. They would object to working with them during periods of recession.

The parallel economic argument would be that the pattern of wage differentials established during periods of full employment does not provide sufficient compensation during periods of recession — periods of greater desire for discrimination. To this, one would have to add the assumption that wages are rigid on the downward side. Only then is the

wage ratio prevented from adjusting itself to satisfy the new pattern of
preferences. If wages are prevented from performing the task of allocating the available jobs, other factors, namely preferences for discrimination, will perform this task. Non-whites would then experience a more
than proportional increase in their unemployment rate.

Even with this formulation it is necessary to assume that employer
preferences for discrimination increase during recessions along with
those of white workers. For suppose that only employee preferences
changed. Then white workers might not like to see any non-whites being
retained as long as some whites were being laid-off. However, they may
have difficulty in acting on their preferences. Recessions are characterized by paucity in alternative employment opportunities.[14] The disutility
in working with non-whites may be viewed in the same manner as other
disutilities associated with particular jobs. Each of them represents a
cost to the worker. They can thus be compensated for these disutilities
in the form of higher wages. If the wages in a particular firm do not
compensate employees for these non-pecuniary disadvantages, one would
expect them to seek alternative employment. One may view normal turnover as attempts to compensate, in part, for all the non-pecuniary disadvantages associated with a particular job. We know that the quit rate
declines during periods of recession. And, there is no reason, on a
priori grounds, to expect differential behavior with respect to the particular non-pecuniary disadvantage associated with discrimination. It is one
thing to dislike some particular condition, it is quite another thing to
assume costs to correct it.[15]

There is yet another way in which one could accept the validity of the
"lhff" proposition. With the same preferences for discrimination over
the course of the business cycle, non-whites would tend to experience
greater cyclical variability in their unemployment if their wages were
more rigid than those of whites. It is not wage rigidity but differential
cyclical wage rigidity which would bring about this result. If non-white
wages were more rigid on the downward side, the wage differential
established during the peak would tend to narrow in the trough of the
business cycle. The earlier wage differential was necessary to assure
the higher rate of employment of non-whites. The lower differential in
the trough would no longer satisfy many employers or white employees.
Under these conditions the existing discriminating coefficients may tend
to bring about a more than proportionate increase in non-white unemployment.

It may be argued that we do indeed have differential wage rigidity.
Non-whites, because of their lower wages, are more likely than whites
to hit the floor of statutory and union minimum wage laws during recessions. It becomes a question of the effect of such minimum wages on the
downward rigidity of non-white wages. It would seem that at most it
would present a problem for the unskilled laborers. Even for this group
one would have to note that non-whites are predominant in the construction, agricultural, and service industries — industries not subject to
legal minima. If we took into account other factors that may be responsible for the downward wage rigidity (degree of unionization and possession
of liquid assets income) we would expect white wages to be more rigid on
the downward side.

For both of the earlier formulations there is one additional factor that
seems relevant for our understanding of the problem. It is this:
Discrimination is costly. If the employer lays off a non-white worker he

is laying off someone whom he most likely pays a lower wage. The consumption of the commodity "discrimination" depends upon its income elasticity of demand and upon the income of the individuals consuming it. If we also assume that it is not an inferior commodity, its consumption should be positively correlated with income. [16] Since the income of consumers of discrimination tend to decline during recessions, its consumption should, ceteris paribus, decline during these periods. Even if the preferences for discrimination on the part of employers were to increase during recessions, its consumption may not increase. Thus, the validity of the "lhff" proposition must rest on the assumption that employer preference for the consumption of discrimination not only increases during recessions but increases sufficiently to offset the possible negative income effects.

Finally, there are a number of factors other than racial discrimination that would tend to cause non-whites to have greater cyclical fluctuations in unemployment. The non-whites have had greater shifts from agriculture to other industries. To the extent that they enter unionized industries, they are likely to be adversely affected by seniority policies. Also, suppose we had some upward wage rigidity. Again due to differential mobility and other factors, non-whites are likely to have somewhat lower levels of skill in non-agricultural industries. Under these conditions they could have greater cyclical fluctuations in unemployment in the absence of discrimination.

It is now obvious that we can neither definitely reject nor accept the "lhff" proposition on the basis of a priori economic reasoning. Lack of definite knowledge about differential wage rigidities, the degree of monopsony power during recessions, the pattern of preferences for discrimination over the course of the business cycle, and the income elasticity of demand for discrimination all prevent us from making definite assertions on the subject. If we assume the same employer preferences throughout the cycle and that discrimination is a normal good, the proposition must be rejected. On the contrary, the reverse should be true. An examination of the data should reveal the answers to some of these questions.

VARIABILITY IN THE NON-WHITE/WHITE
UNEMPLOYMENT DIFFERENTIAL

This part of the study is concerned with the empirical analysis of the behavior of the white/non-white unemployment differential. The analysis is conducted mainly with the U. S. unemployment data generated by the monthly sample for the Current Population Survey of the Bureau of the Census. [17] Whenever necessary and useful, I utilize data from the 1940 or 1950 decennial census as well. The period covered is that for which data with the relevant breakdowns are available. Unemployment rates by sex, color, and major occupations have been tabulated by the Bureau of the Census for January, April, July, and October of each year beginning with October, 1953. Thus, the comparison of white and non-white male unemployment rates by major occupations extends over the period October, 1953, to October, 1961. [18] Beginning with January, 1957, unemployment rates have become available by color, sex, and intermediate occupation groups. For the intermediate occupational breakdown, the analysis covers the years 1957-1961. [19]

Primarily, I am concerned with the fluctuations in the non-white and white unemployment differential over the various phases of the business cycle. But I shall also compare total variability in this differential. The analysis is presented in the following three sections which are organized around three separate tests, each suggested by a different formulation of the "lhff" proposition. The first section deals with timing of the cyclical turning points in the respective series; the second analyzes total variability in the non-white and white series; and the last compares peak-to-trough and trough-to-peak changes in the unemployment rates of white and non-white males. [20]

Timing of Cyclical Turning Points

If the "lhff" proposition reflects discriminatory market behavior, the effects of such behavior should be strongest during the upswings and downswings of business cycles. This is because expansions are characterized by greater than usual hiring and declining unemployment rates and recessions by greater than usual lay-off rates and rising unemployment rates. One should therefore be able to observe the discriminatory hiring and firing in the changing pattern of unemployment rates during the various phases of the cycle. [21]

The first test of the validity of the "lhff" proposition is suggested by formulating it so that the words, "last" and "first" are emphasized. If non-whites are hired last, they should be the last to experience a decline in their unemployment rate. If they are fired first, they should be the first to experience an increase in their unemployment rate. We should therefore observe that the non-white unemployment rate series lags the white series during the upturn and leads the white unemployment rate series during the downturn of the business cycle.

Figure 1 shows the seasonally adjusted unemployment rates for white and non-white male, experienced, wage and salary workers only. The

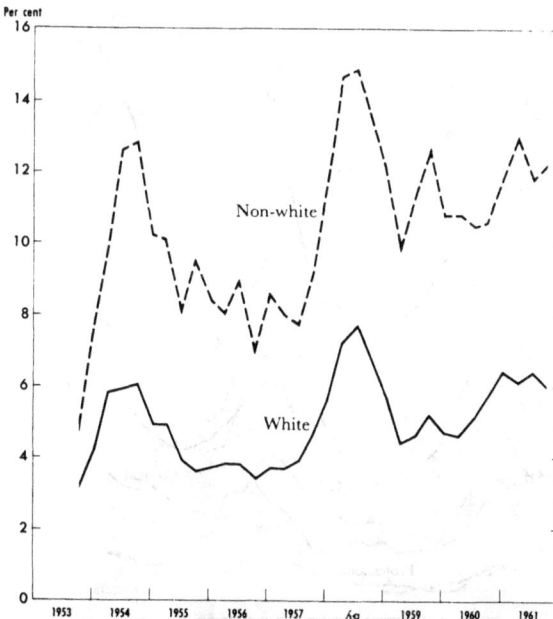

Figure 1. Unemployment rates of male, experienced wage and salary workers, seasonally adjusted data, October 1953-October 1961.

unemployment rates shown are the January, April, July, and October observations for each year from October, 1953, through October, 1961. On the basis of these data the "first fired" aspect of the proposition would have to be rejected. If anything, in both the 1957 and 1960 business downturns the white rather than the non-white series displays a slight lead. There also appears to have been little or no lag in the non-white series in both the 1954 and 1958 business upturns.

A similar conclusion was reached in the New York Commission Against Discrimination study. [22] The Commission notes that "changes in white and non-white unemployment rates tend usually to happen simultaneously. When a decline begins, the non-white unemployment rate generally does not begin to rise before the white unemployment rate. . . . Non-whites do not begin to lose their jobs earlier, but lose them in larger numbers once lay-offs begin." These conclusions are drawn from comparisons of the white and non-white unemployment rates for both sexes combined and for males and females separately for the period 1947-1958.

Unfortunately, that study as well as mine is based on only four observations per year. And these series may be too short to reflect an average lead or lag of only one month.

Unemployment rates by color and sex are actually available on a monthly basis. For these groups this test could have been conducted with monthly data. [23] My reason for limiting this analysis to the particular four observations per year is that the principal purpose of this study is to observe the behavior of the unemployment rates of given skill or occupation groups. And prior to 1959, unemployment rates for these groups, by color, were tabulated by the Bureau of Labor Statistics only for these particular months.

The coincidence in the aggregate unemployment rate series in the downturn of the cycle is difficult to explain. Market discrimination aside, there is reason to expect the non-white aggregate unemployment

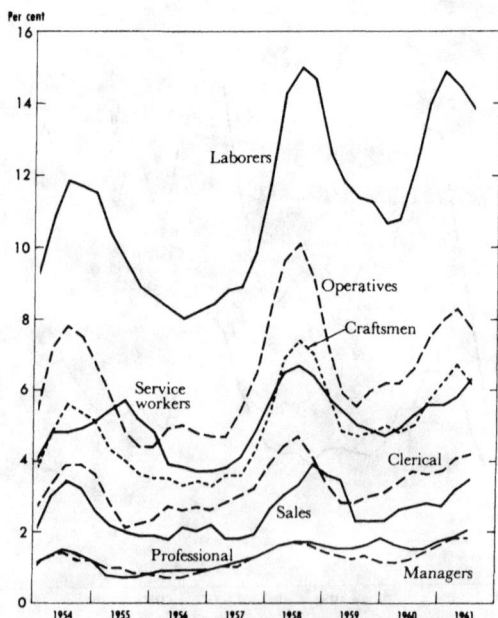

Figure 2. White male unemployment rates, by major occupation group, seasonally adjusted and smoothed (3-point moving average) data, 1954-1961.

Table 1. Percentage Distribution and Median Number of School Years Completed by the Experienced, Male, Labor Force, by Major Occupation Group, and Color

Major occupation group	Percentage Distribution[1]		Median number of school years completed, 1950[2]	
	White	Non-white	White	Non-white
Total	100	100		
Professional and kindred workers	10.5	2.8	16.+	15.3
Farmers and farm managers	6.7	5.2	12.2	8.4
Managers and proprietors, exc. farm	13.7	2.4	12.3	12.0
Clerical and kindred workers	6.8	4.6	12.3	9.7
Sales workers	6.3	1.2	9.6	8.0
Craftsmen, foremen and kindred workers	20.3	9.5	9.0	7.1
Operatives and kindred workers	19.5	24.2	8.7	7.0
Private household workers	0.1	0.6		
Service workers, exc. private household	5.5	14.9	8.9	8.2
Farm laborers and foremen	3.3	8.7	8.3	4.9
Laborers, exc. farm and mine	7.3	25.9	8.5	5.9

[1] Percentage distribution was computed from the unpublished BLS tabulation of the Current Population Survey. Figures are based on the average distribution for January, April, July, and October 1958.

[2] 1950 U.S. Census of Population, Occupational Characteristics, Special Report P-E No. 1B, Tables 10 and 11.

Table 2. Specific Cycle Turning Points of Unemployment Rate Series of the White Male Experienced Labor Force, by Major Occupation Groups, 1954-1961

Major occupation group	Trough	Peak	Trough	Peak	Trough
Professional and kindred workers	July '54	Oct. '55	Oct. '58	Oct. '60	July '61
Managers and proprietors, exc. farm	July '54	July '56	July '58	April '60	July '61
Clerical and kindred workers	Oct. '54	July '55	Oct. '58	July '59	July '61
Sales workers	July '54	July '57	Oct. '58	Jan. '60	July '61
Craftsmen, foreman and kindred workers	July '54	Jan. '57	July '58	July '59	April '61
Operatives, foreman and kindred workers	July '54	Jan. '56	July '58	July '59	April '61
Service workers, exc. private household	July '55	Jan. '57	July '58	Jan. '60	July '61
Laborers, exc. farm and mine	July '54	July '56	July '58	Jan. '60	Jan. '61

Source: Based on Figure 2, above.

84

series to lead the white in the business downturns. This result is expected simply on the basis of the differential skill distribution of the two
groups. Lower skilled occupations are expected both to have larger lay-
off rates and to lead the higher skilled occupations in the downturn of the
cycle. 24 Table 1 shows the distribution of the white and non-white male,
experienced, labor forces between major occupations. This table also
shows the median number of school years completed by these respective
groups. It is apparent that non-whites are concentrated in the unskilled
or semi-skilled occupations. Also, they have lower educational attainment within all of the major occupations. Thus, based on this differential skill distribution alone, the non-white aggregate unemployment rate
series should lead the white in the downturn of the cycle. This should
occur in the absence of differential firing within occupations. The fact
that the two series are coincident in the downturn suggests that within
occupations non-whites may be fired last rather than first.

On the basis of differential skill distribution we would expect the
non-white aggregate unemployment rate series to lead the white in the
upturn of the cycle. The unskilled are expected not only to be fired first
but also to be rehired first. The fact that the two series are coincident in
the upturn does suggest that within occupations non-whites may be hired
last. On the basis of the aggregate unemployment rate series one may
be tempted to conclude that within occupations non-whites are both hired
and fired last.

Figure 2 shows the series of the unemployment rates for white male,
experienced workers by major occupations. The differences in the cyclical behavior of the various series generally conform to expectations. The
lower the average level of skill of a major occupation group, the larger
its cyclical variation in unemployment rates. The dates of the specific
cycle turning points of these series are recorded in Table 2. This table
does not show the expected lead of the unemployment rates for the lower
skilled occupations in the turning points of 1954, 1957, and 1958. Only
in the turning points of 1960 and 1961 do we observe a clear lead of lower
skilled over higher skilled unemployment rates.

Figures 3 through 7 show white and non-white male unemployment
rate series for several of the major occupation groups. The classification by occupation produces some standardization for differences in skill.
However, since the unemployment rates here are based on smaller
samples, these series contain greater random variability. Keeping this
factor in mind, I would conclude that these series too are generally coincident in the timing of their turning points. If there is a slight difference
in their timing, it occurs in the upturn of the cycle. Since we still do not
have complete standardization for skill, the implications developed earlier
are relevant for the major occupation groups as well; the feeling persists
that non-whites may have a slower rate of recovery (last hired) but they
are not the first to be laid off.

There are, of course, some differences in the timing pattern both
between occupations and between turning points of a given major occupation group. For laborers (Fig. 3), the non-white series lags the white
series in the 1954 and 1960 upturns but leads the white series in the upturn of 1958. The two series are coincident during the downturns. For
operatives (Fig. 5) the non-white series lags the white series in the 1954
upturn. The other two upturns are coincident. The non-white possibly
leads the white in the first downturn but lags in the second one. And so on.

The real problem with the timing test is that one is not in a position
to adjust for the differences in the variability of the white and non-white

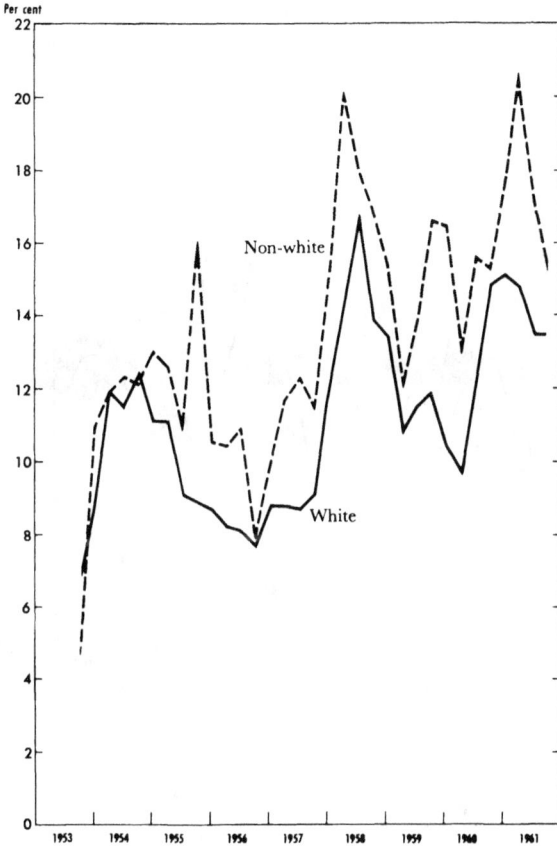

Figure 3. Unemployment rates of male laborers, except farm and mine, seasonally adjusted data, October 1953-October 1961.

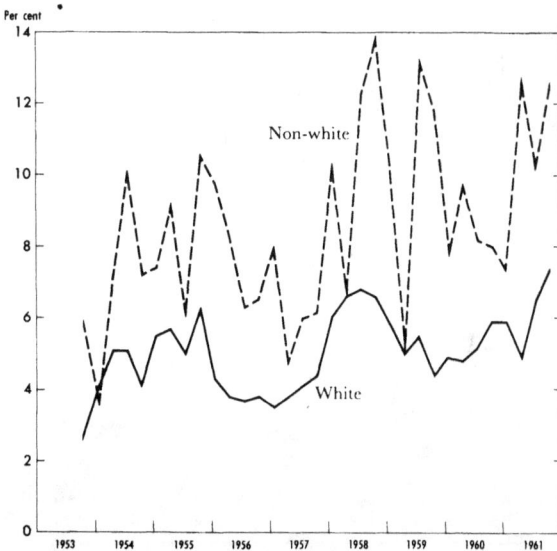

Figure 4. Unemployment rates of male service workers, except private household, seasonally adjusted data, October 1953-October 1961.

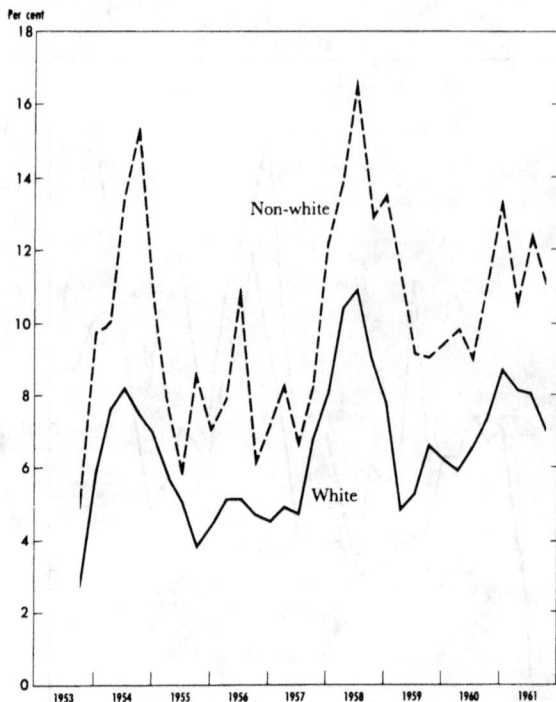

Figure 5. Unemployment rates of male operatives, seasonally adjusted data, October 1953-October 1961.

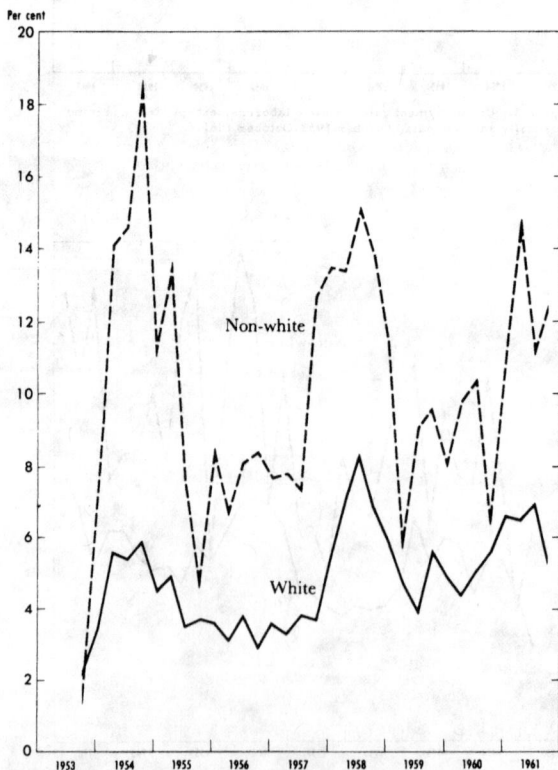

Figure 6. Unemployment rates of male craftsmen and kindred workers, seasonally adjusted data, October 1953-October 1961.

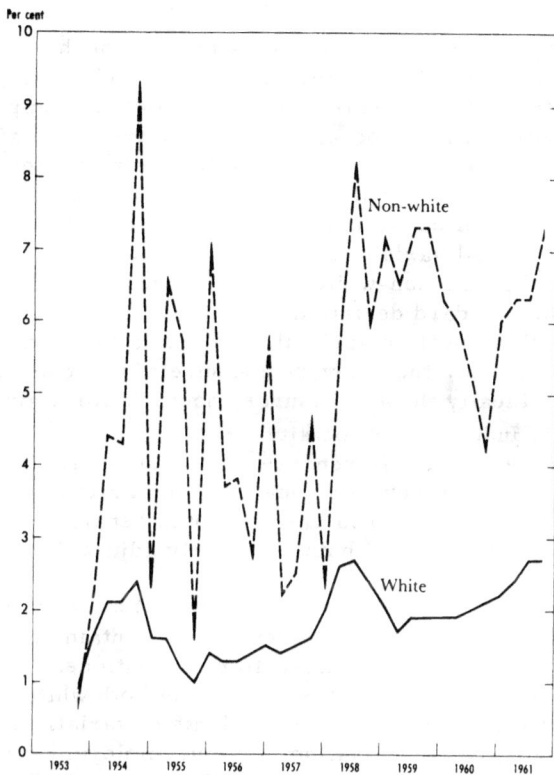

Figure 7. Unemployment rates of male white-collar workers, seasonally adjusted data, October 1953-October 1961.

series which may be due to the differences in their sample sizes. This differential random variability would increase with increased standardization for skill. This problem is easier to deal with in the next two sections where my tests are for differences in variation of amplitude in the two series rather than for differences in their timing.

Total Variability of White and Non-white Unemployment Rates

The first test was suggested by the particular formulation of the "lhff" proposition. I emphasized the words "last" and "first." A literal interpretation of this proposition would, of course, require little testing. It could be rejected at the outset, for such an interpretation would require that within a given occupation and locality we have no involuntary white unemployment before all non-whites have been laid off. This formulation is obviously too extreme.

It may be that the formulation for the first test -- that of timing -- also is too extreme. We may reformulate the proposition so that it has no implications for leads or lags. We could suppose that the proposition intends to say only that whenever there is a need to reduce the input of labor, non-whites will experience a disproportionate increase in their unemployment rate. They would also experience a greater decrease in their unemployment rate with expansions in demand for labor.

In this formulation I do not limit myself to measuring changes in the unemployment rate during the upswing and downswing phases of the cycle. Hiring and firing goes on at all times and the non-whites may be adversely affected in all of these situations. Thus we need measures which are capable of detecting differential hiring and firing over time. The standard deviations and coefficients of variation of particular unemployment series are two such measures. The arguments in favor of these measures are based on the assumption that differential hiring and firing practices will tend to be reflected in differential variability in the unemployment rates. If non-whites within a given occupation are differentially affected by fluctuations in demand, this should be reflected in greater variability of their unemployment rates over time. The standard deviation of a given time series, while reflecting mainly specific cycle amplitudes, [25] will also include additional variability. Thus for any given occupation, we want to see whether the non-white unemployment rate varies more, as measured by the standard deviation, than the white unemployment rate. If one assumes that absolute variability is also a function of the average level of unemployment, the relevant measure of a series is its coefficient of variation. [26] Ideally these measures should be for a given level of skill within a specific industry and locality.

Table 3 presents the differences[27] in the standard deviations and coefficients of variation between non-white and white males, by major occupation. [28] These measures are presented separately for the raw data, seasonally adjusted data, and both seasonally adjusted and smoothed data. [29]

The differences in the coefficients of variation tend to be small or negative for blue-collar workers (occupations containing most of the non-whites), but are positive for white-collar occupations. However, these differences tend to diminish with smoothing of both white and non-white series. Thus on the basis of the coefficients of variation, the non-whites do not experience greater variation in the unemployment rate than the whites in the lower skilled occupations. Even for the higher skilled occupations the differences diminish or disappear with smoothing.

For reasons that I shall develop in the next section, the standard deviation is the more relevant of the two measures for testing the validity of the "lhff" proposition. The standard deviations of the non-white series are consistently higher than those for the white series. Here, too, the differences between the non-white and white standard deviations diminish with smoothing. Again the effect of smoothing is greater on the higher skilled occupations.

Up to this point we would have to conclude that the non-whites have greater absolute variability in their unemployment rates and that they are probably adversely affected by variations in the demand for labor. However, by comparing standard deviations of major occupation groups I have not standardized sufficiently for skill. It is true that within a major occupation the non-whites are concentrated in the lower skilled, detailed occupations. Even for whites the standard deviations are greater the lower the level of skill (see Appendix Table 1).

Some evidence of the unequal distribution of white and non-white males within major occupations is presented in Table 4. This table shows the fraction of white and non-white males of a given major occupation who are employed in the detailed occupations of the lowest level of skill. The lowest quartile of detailed occupations on the basis of skill are the 25 per cent of detailed occupations earning the lowest incomes. The earnings are those of white males only. I assume that for white males there is a positive correlation between income and the level of skill. After finding the detailed occupations that make up the lowest quartile of skill, I calculated the fraction of white and non-white males that are employed in these occupations.

It can be seen from Table 4 that a larger fraction of non-white males

Table 3. Differences (NW-W) in the Standard Deviations and Coefficients of Variation of Male Unemployment Series, by Major Occupation, October 1953-October 1961.

Major occupation group	Raw data		Deseasonalized data		Deseasonalized and smoothed data	
	$S_{NW}-S_W$	$\left(\frac{S}{\bar X}\right)_{NW} - \left(\frac{S}{\bar X}\right)_W$	$S_{NW}-S_W$	$\left(\frac{S}{\bar X}\right)_{NW} - \left(\frac{S}{\bar X}\right)_W$	$S_{NW}-S_W$	$\left(\frac{S}{\bar X}\right)_{NW} - \left(\frac{S}{\bar X}\right)_W$
Professional and kindred workers	1.65	0.31	1.51	0.24	0.42	-0.05
Managers and proprietors, except farm	1.79	0.51	1.86	0.52	0.92	0.46
Clerical workers	2.58	0.22	2.74	0.25	1.67	0.11
Sales workers	5.03	0.57	6.14	0.69	4.41	0.46
Craftsmen and foremen	1.96	0.00	2.22	0.07	1.64	0.03
Operatives	0.89	-0.02	0.99	0.00	0.68	-0.02
Service workers, exc. private households	1.57	0.09	1.58	0.10	0.86	0.04
Farm labor	1.73	0.10	2.11	0.20	1.24	0.11
Laborers, exc. farm and mine	0.47	-0.03	0.92	0.03	0.52	0.00

Source: Appendix Table 1.

Table 4. Distribution of the White and Non-white Male, Experienced, Labor Force, by Major Occupation and Level of Skill, 1950

Major occupation	Per cent of major occupation in lowest quartile of skill		Non-white/ White
	White	Non-white	
Professional and kindred workers	14.3	39.3	2.7
Managers, officials, and proprietors	14.4	40.3	2.8
Sales workers	56.7	74.0	1.3
Clerical workers	13.8	24.8	1.8
Craftsmen, foremen and kindred workers	33.5	48.9	1.5
Operatives and kindred workers	25.7	34.4	1.3
Service workers, exc. household	18.5	23.2	1.3
Laborers, exc. farm and mine	57.3	59.2	1.03

Source: Ranking of skill is based on income figures for white males in 1949, by detailed occupation. 1950 U.S. Census of Population, Occupational Characteristics, Special Report P-E No. 1B, Table 19. Distribution of White and Non-white Experienced Civilian Labor Force, By Detailed Occupation was computed from Table No. 3 in the same volume.

is in the lowest skilled occupations of each major occupation. Also, the inequality in the distribution becomes more pronounced the higher the average level of skill of the major occupation.

Another factor which is responsible for the greater variability of non-white unemployment series is that their unemployment rates are based on smaller sample sizes (see Table 5). Even if their true unemployment rates were no higher than those for the whites, over time they will show greater variability. This is already apparent in the fact that smoothing of both white and non-white series reduces the difference in their standard deviations. The non-white series have greater randomness and the random component is larger the smaller the sample size. Before we can conclude from a comparison of standard deviations that within a given skill non-whites are more adversely affected by fluctuations in the demand for labor, we must have greater standardization for skill. We must also remove that part of the variability of a series which can be attributed to the smaller sample size. The results presented in Table 5 are adjusted, to some extent, for both of these factors.

This table presents the differences in the standard deviations and coefficients of variation of white and non-white male unemployment rate series, by intermediate occupation groups.[30] Columns 1 and 2 show these differences for the unadjusted data. Columns 3 and 4 present the differences in standard deviations and coefficients of variation after removing that part of the variability which is due to differences in sample size. The random component has been estimated in the following manner:

$$S^2 \underset{\text{observed}}{} = S^2 \underset{\text{deflated}}{(t)} + S^2 \underset{\text{sampling}}{(u)} \qquad (1)$$

The second component of the variance is that component which would exist even if the true unemployment rate did not change. It is due simply to taking different samples from the same population. Its magnitude depends on the sample size and can be computed with the aid of the standard formula for sampling variance of sample means (proportions) of independent simple random samples drawn from a binomial population whose mean is P.[31] In formula (2), P is the average unemployment rate

$$S^2 (u) = \frac{P(1-P)}{n} \qquad (2)$$

for the entire period and n is the estimated sample size. Thus:

$$S^2 (t) = S^2 (\text{observed}) - S^2 (\text{sampling}) \qquad (3)$$
$$S^2 (\text{observed}) - \frac{P(1-P)}{n} \qquad (4)$$

Columns 1 and 2 of Table 5 present a pattern that is not unlike that of the major occupations. The differences in standard deviations tend to be higher for the higher skilled occupations. This is equally true for the differences in coefficients of variation except that a larger fraction is either close to zero or is actually negative. When I adjust the variability for sample size (columns 3 and 4), the differences in the coefficients of variation (column 4) are negative or zero in eighteen out of the twenty-five occupations. In the case of standard deviations (column 3), the differences in nine out of the twenty-five occupations are negative. For

Table 5. Differences (NW-W) in Standard Deviations and Coefficients of Variation for Male Unemployment Rates, by Intermediate Occupation Group, 1957-1961

Intermediate occupation group	Raw Data		Adjusted for sample size		Estimated sample sizes[1]	
	$S_{NW}-S_W$ (1)	$\left(\frac{S}{X}\right)_{NW}-\left(\frac{S}{X}\right)_W$ (2)	$S_{NW}-S_W$ (3)	$\left(\frac{S}{X}\right)_{NW}-\left(\frac{S}{X}\right)_W$ (4)	White	Non-white
Technical engineers	4.03	2.39	*	*	648	11
Medical - salaried only	4.56	0.57	-	-	121	9
Other prof. salaried only	2.08	0.34	-	-	1,556	47
Managers salaried only	3.07	0.46	-	-	1,933	21
Stenos, sec. and typists	10.19	1.02	-	-	45	3
Other clerical	2.35	0.18	1.52	0.10	1,875	143
Salesmen - retail trade	6.15	0.51	3.16	0.23	727	23
Salesmen - other trade	6.59	0.88	3.08	0.36	1,081	16
Machinist and other metal craftsmen	4.70	0.30	1.14	-0.06	724	24
Mechanics and repairmen	1.27	0.02	-	-	1,315	88
Construction craftsmen	2.44	0.00	1.33	-0.07	1,781	96
Other craftsmen	5.13	0.25	-	0.04	1,082	53
Foremen	3.29	1.18	1.62	-	688	11
Drivers and deliverymen	1.03	-0.03	0.44	-0.09	1,425	232
Mine and durable goods operatives	1.88	-0.01	1.26	-0.05	1,867	207
Non-durable goods operatives	2.36	0.11	1.13	0.00	985	104
Non-manufacturing operatives	1.00	0.05	0.10	-0.05	1,153	182
Private household workers	2.20	-0.06	-	-	20	15
Protective service workers	6.25	0.71	1.78	0.10	485	18
Waiters, cooks and bartenders	2.64	0.26	1.92	0.19	272	71
Other service workers	1.47	0.06	1.22	0.05	797	339
Paid farm labor	0.84	0.02	0.58	-0.01	672	222
Construction labor	-0.47	-0.08	-0.84	-0.10	469	217
Manufacturing labor	0.94	0.02	0.59	-0.01	597	223
Other labor	0.66	-0.04	0.11	-0.07	900	271

Source: Calculated from unpublished data of the BLS. See Appendix Table 2.
*The negative signs if not followed by figures indicate that the non-whites variances after removing the component due to sample size are negative.

[1]Samples are estimated from census estimates of the ratio of population to sample size.

these occupation or industry groups the non-whites have lower absolute variability in their unemployment rates than do the whites. In the majority of cases, however, (64 per cent) the non-whites do have greater absolute variability. However, I am not in a position to make additional standardization for skill, and undoubtedly the non-whites are concentrated at the lower end within these intermediate occupation groups. I can only conclude that standardization for skill reduces substantially the differences in variability.

Peak-to-Trough and Trough-to-Peak Changes

Findings based on a comparison of the standard deviations of two series are somewhat difficult to interpret. Even if the standard deviations for non-whites were consistently higher than those for the white series, one could not draw strong conclusions from these findings. This is so because the discriminatory behavior implied in the "lhff" proposition refers to differential behavior in demand. The standard deviation will also include variation in unemployment rates, some of which may be due to other factors. For one, the non-white series contain larger trend components. If we look at changes in employment, we observe that non-whites have had greater increases in employment in all of the major occupation groups. Racial discrimination aside, we expect unemployment rates to be positively related to increases in the labor force. Indeed, the ratios of non-white to white unemployment rates have risen over time. Since cyclical fluctuations in unemployment rates are more likely to be purely demand phenomena, we want a measure limited to variation over the cycle. We could suppose that the proponents of the proposition intend only to say that during recessions non-whites will be laid off in disproportionate numbers. Under this formulation we want to observe whether, in going from peak to trough of a business cycle, non-whites experience a greater increase in their unemployment rates. They should also experience a greater decrease in their unemployment rates from trough to peak.

There is a question whether the comparison should be between the ratio or the difference in unemployment rates in the peak and trough. I believe that a careful reading of the proposition will provide the answer to this question. Suppose that during the peak of the cycle the respective unemployment rates of whites and non-whites were 3 per cent and 6 per

	W	NW	NW/W	NW-W
Peak	3%	6%	2	3%
Trough	6%	12%	2	6%

cent. If both rates doubled (6 per cent and 12 per cent) during the trough their ratio would be unaffected; it would remain at 2. If one used the ratio as the measure, the proposition would be rejected. Yet, in going from peak to trough the non-whites had twice the lay-off rate of the whites. Their increase in unemployment was greater than the increase for the whites and the proposition should therefore be accepted. The crucial point here is that in this test I am not particularly interested in the fact that non-whites always have a higher unemployment rate. What I want to determine is whether an increase in unemployment affects them more adversely. To determine that, the difference is the relevant measure, i. e. , the proposition will be rejected if the difference in the two rates remains constant throughout the cycle. [32] These considerations are also relevant for the choice between the standard deviation and coefficient of variation discussed earlier.

Table 6 presents the peak-to-trough and trough-to-peak changes in absolute unemployment rates of white and non-white groups. The relevant row for our purposes is the one labeled "NW-W." If non-whites have a larger increase in their unemployment rate during the downturn, the difference in the two rates should widen and will be positive from peak to trough of the cycle. [33] The non-white rate should experience a larger decrease, the difference between the two should narrow, and the change in the difference will be negative in going from trough to peak.

For each of the series, I present the results based on a comparison of the single highest and lowest point as well as on a comparison of averages of three points for both the peak and trough of the cycle. [34] It is clear that in all of the series presented in this table the non-whites are more adversely affected by declines in demand for labor. [35] They are helped by upturns. The adverse effect is stronger if we consider males alone.

I am including a comparison of male experienced workers to see whether the differential behavior over the cycle is significantly different for this group than for all males. When I standardize for occupation my analysis can only be conducted for experienced workers. And it could be argued that the results would be different had I included new entrants into the labor force. [36] However, in the aggregate at least, while both whites and non-whites tend to have lower unemployment rates in both peaks and troughs when only the experienced workers are considered, the pattern over the cycle shows little difference between experienced workers and all males. [37]

In Table 7, I present the results of the same tests described in conjunction with Table 6. This time they are presented separately for several major occupation groups. The peaks and trough here are those of the specific cycles of the white and non-white series, respectively. Since specific cycle turning points of major occupation groups that have small sample sizes are difficult to identify, I have combined all white-collar workers into one group.

For the non-whites this and subsequent standardizations will result in our data being based on smaller and smaller sample sizes. The sample sizes become very small for higher skilled occupations where non-whites have small participation rates. A series based on small samples will of course tend to contain a large random component. As a result, we have the problem of determining how much smoothing we have to do before we can identify peaks and troughs of a cycle. I have handled this problem in two separate ways and the results are given in each column under -1- and -2- separately. Under -1-, I present the results with equal smoothing of the white and non-white series. [38] This method has obvious shortcomings. If there is any justification for the smoothing of a series it is the need to remove its random variability. The random variability will depend on the sample size. Since non-whites have smaller samples, I would have to do more smoothing of their series to make the white and non-white series equally free of random variability. Under 2, I present the results with differential smoothing.

For the differential smoothing I utilized the method of removing random variability described in Section B. I smoothed the non-white series until the difference in the standard deviations between the non-white and white series was equal to the difference in these standard devia-

94

Table 6. Peak-to-Trough and Trough-to-Peak (Specific Cycles) Changes in Unemployment Rates, by Color, 1948-1961

Unemployment series, seasonally adjusted	P-to-T 1948-1949 1	P-to-T 1948-1949 2	T-to-P 1949-1953 1	T-to-P 1949-1953 2	P-to-T 1953-1954 1	P-to-T 1953-1954 2	T-to-P 1954-1957 1	T-to-P 1954-1957 2	P-to-T 1957-1958 1	P-to-T 1957-1958 2
Both sexes										
NW	9.1	5.8	-9.3	-6.2	6.5	4.4	-3.4	-1.8	6.6	4.7
W	4.0	3.4	-4.8	-4.2	3.1	2.4	-2.0	-1.3	3.2	2.8
NW-W	5.1	2.4	-4.5	-2.0	3.4	2.0	-1.4	-0.5	3.4	1.9
Male										
NW	11.1	7.3	-12.6	-8.6	7.8	5.9	-4.3	-2.7	8.0	5.7
W	4.5	3.4	-5.5	-4.2	3.1	2.3	-2.1	-1.5	3.7	3.3
NW-W	6.6	3.9	-7.1	-4.4	4.7	3.6	-2.2	-1.2	4.3	2.4
Male, experienced workers only										
NW	10.9	7.0	-12.3	-8.1	7.9	5.9	-4.9	-3.1	7.7	6.1
W	4.4	3.4	-5.3	-4.1	3.0	2.2	-2.0	-1.5	3.3	2.9
NW-W	6.5	3.6	-7.0	-4.0	4.9	3.7	-2.9	-1.6	4.4	3.2
Male, experienced wage and salary workers only										
NW					8.1	5.8	-5.9	-3.7	8.0	6.2
W					2.9	1.9	-2.6	-2.0	4.3	3.6
NW-W					5.2	3.9	-3.3	-1.7	3.7	2.6

Unemployment series, seasonally adjusted	T-to-P 1958-1960 1	T-to-P 1958-1960 2	P-to-T 1960-1961 1	P-to-T 1960-1961 2	Avg. P-to-T 1	Avg. P-to-T 2	Avg. T-to-P 1	Avg. T-to-P 2
Both sexes								
NW	-4.0	-2.5	3.2	2.7	6.4	4.4	-5.6	-3.5
W	-2.2	-1.8	1.7	1.5	3.0	2.5	-3.0	-2.4
NW-W	-1.8	-0.7	1.5	1.2	3.4	1.9	-2.6	-1.1
Male								
NW	-4.2	-2.2	3.1	1.4	7.5	5.1	-7.0	-4.5
W	-2.6	-2.2	1.8	1.5	3.3	2.6	-3.4	-2.6
NW-W	-1.8	-0.0	1.3	-0.1	4.2	2.5	-3.6	-1.9
Male, experienced workers only								
NW	-4.6	-3.1	3.3	1.2	7.5	5.1	-7.3	-4.8
W	-2.5	-2.0	1.6	1.4	3.1	2.5	-3.3	-2.5
NW-W	-2.1	-1.1	1.7	-0.2	4.4	2.6	-4.0	-2.3
Male, experienced, wage and salary workers only								
NW	-5.1	-3.3	3.2	1.1	6.4	4.4	-5.5	-3.5
W	-3.3	-2.3	2.0	1.3	3.1	2.3	-3.0	-2.2
NW-W	-1.8	-1.0	1.2	-0.2	3.3	2.1	-2.5	-1.3

Source: Based on Appendix Table 3.

[1] Comparisons are based on single highest and lowest observation of each series.

[2] Comparisons are based on averages of three observations for each peak and trough of the respective series.

Table 7. Peak-to-Trough and Trough-to-Peak (Specific Cycles) Changes in Male Unemployment Rates, by Color, and Major Occupation, October 1953-October 1961

Major occupation		Avg. P-to-T		Avg. T-to-P		No. of Correct Signs P-to-T		No. of Correct Signs T-to-P	
		1	2	1	2	1	2	1	2
White-collar workers	W	1.0	1.4	-0.9	-1.2	3	3	2	2
	NW	3.1	3.3	-2.5	-2.3	3	3	2	2
	NW-W	2.1	1.9	-1.6	-1.1	3	3	2	2
Craftsmen and foremen	W	2.6	4.0	-2.5	-3.7	3	3	2	2
	NW	6.8	5.9	-7.6	-6.7	3	3	2	2
	NW-W	4.2	1.9	-5.1	-3.0	3	3	2	2
Operatives	W	3.6	5.4	-4.0	-5.3	3	3	2	2
	NW	5.0	5.5	-5.6	-6.0	3	3	2	2
	NW-W	1.4	0.1	-1.6	-0.7	2	2	2	1
Service workers, exc. private household	W	2.1	3.3	-2.0	-2.6	3	3	2	2
	NW	4.9	4.6	-4.2	-3.8	3	3	2	2
	NW-W	2.8	1.3	-2.2	-1.2	3	2	2	2
Laborers, exc. farm and mine	W	4.6	6.7	-4.1	-5.9	3	3	2	2
	NW	5.7	5.3	-4.0	-3.4	3	3	2	2
	NW-W	1.1	-1.4	0.1	2.5	3	0	1	0

Source: Appendix Table 4.

[1] Based on a comparison of single points from smoothed deseasonalized series (3-point moving average).

[2] Based on differential smoothing for the white and non-white series.

tions after adjusting for differences in the size of the samples.[39] With this technique the difference in the smoothing of a white and non-white series for given occupations depended on their differences in sample sizes. For some occupation groups, the requirement of equality of the random component necessitated smoothing the non-white series three times as much as the white series.[40] In such instances, my peak and trough comparisons were made on the basis of the single highest and lowest observations for the white series as against averages of three observations for the non-white series.

Analysis of the difference rows (NW-W) reveals a pattern similar to that of the differences in standard deviations. The differential cyclical behavior revealed here by the positive sign in peak-to-trough and negative sign in trough-to-peak movements is most noticeable for the higher skilled groups. The difference is highest here for the craftsmen. With differential smoothing the differences are diminished, they become very small for operatives and are actually reversed for the laborers.

Columns 3 and 4 show the number of correct signs in each series. In the period under analysis[41] there are three peak-to-trough and two trough-to-peak movements. I count it as a correct sign each time the non-whites experience a greater change in the unemployment rate than the whites. This test should tell us whether the average results recorded in columns 1 and 2 are due to single extreme observations. Thus if we observe three correct signs for peak-to-trough movements we know that in all three cycles the non-whites had a larger increase in their unemployment rates during the business downturn. The reverse holds for the trough-to-peak movement. This test strengthens our findings of columns 1 and 2. For the non-skilled or semi-skilled workers, the results of the sign test are somewhat weaker even without differential smoothing. They become weaker or go the other way with differential smoothing.

Table 8 presents the results of the same tests but this time they were conducted with reference to the turning points of the aggregate unemployment rate series rather than those of specific cycles.[42] For those occupations that are listed in both Tables 7 and 8, the results are pretty much the same. This is not too surprising since the white and non-white series of those occupation groups are fairly coincident.[43] For the other two occupation groups -- Private Household and Farm Workers -- the results on the average go in the other direction.

Table 9 attempts to present a picture of the distribution of white and non-white males of given major occupation groups between various industry groups.[44] We want to determine from this table whether non-whites of a given major occupation are concentrated in cyclically sensitive industries. The results are not based on detailed enough information and are therefore difficult to interpret. However, for both craftsmen and laborers non-whites do have a somewhat larger fraction in the construction industry which is subject to greater cyclical swings. In the operative group, however, they have a somewhat smaller fraction in the durable goods industries.

Table 10 presents the results of the tests described in Tables 7 and 8, but for intermediate occupations. These results are more significant for our purposes because they are for more homogeneous groups, both with respect to skill and industry. Also, the differential smoothing is more important here because of the small samples for non-whites. The analysis here is on the basis of 20 observations; January, April, July and October for the years 1957-1961.

Table 8. Peak-to-Trough and Trough-to-Peak Changes in Male Unemployment Rates, by Color, and Major Occupation (Total Unemployment Reference Points), October 1953-October 1961

Major occupation	Series	Avg. P-to-T		Avg. T-to-P		No. of correct signs P-to-T		No. of correct signs T-to-P	
		1	2	1	2	1	2	1	2
White-collar workers	W	0.7	1.0	-0.6	-0.7	3	3	2	2
	NW	1.9	2.4	-1.0	-1.2	2	2	2	2
	NW-W	1.2	1.4	-0.4	-0.5	2	2	1	1
Craftsmen and foremen	W	2.3	3.4	-2.1	-3.0	3	3	2	2
	NW	5.5	5.5	-6.1	-6.1	3	3	2	2
	NW-W	3.2	2.1	-4.0	-3.1	3	3	2	2
Operatives	W	3.2	4.2	-3.4	-3.7	3	3	2	2
	NW	4.9	5.3	-5.4	-5.9	3	3	2	2
	NW-W	1.7	1.1	-2.0	-2.2	3	3	2	2
Private household workers	W	3.9		-4.2		3		2	
	NW	-1.1		3.7		2		1	
	NW-W	-5.0		7.9		1		1	
Service workers, exc. private household	W	1.6	1.5	-1.6	-1.1	3	2	2	2
	NW	2.6	3.4	-1.6	-2.1	3	3	2	2
	NW-W	1.0	1.9	0.0	-1.0	2	3	1	2
Farm laborers and foremen	W	1.9	2.8	-1.8	-1.8	2	3	1	1
	NW	0.9	1.5	-0.2	0.0	2	2	1	1
	NW-W	-1.0	-1.3	1.6	1.8	2	2	1	1
Laborers, exc. farm and mine	W	4.2	6.0	-3.6	-5.0	3	3	2	2
	NW	4.4	4.4	-2.1	-2.1	3	3	2	2
	NW-W	0.2	-1.6	1.5	2.9	2	0	0	0

Source: See Appendix Table 5.

[1] See Appendix Table 5.
[2] See Appendix Table 5.

Table 9. Percentage Distribution of the White and Non-White Male Labor Force, by Major Occupation and Industry Group

Major occupation and industry group	CPS average 1960		1950 U.S. census		Notes
	White	Non-white	White	Non-white	
Craftsmen	19.9	9.4	19.8	7.8	Male Labor Force = 100
1) Construction	31.7	37.1			Male Craftsmen = 100
2) Machinists, jobsetters and metal	13.1	10.0			Male Craftsmen = 100
3) Mechanics and repairmen	23.2	29.2			Male Craftsmen = 100
4) Other	19.5	18.6			Male Craftsmen = 100
Operatives	19.5	23.8	20.1	20.5	Male Labor Force = 100
1) Drivers and deliverymen	26.4	30.1	20.6	24.1	Male operatives = 100
2) Manufacturing and mining	52.5	44.5	47.0	39.5	Male operatives = 100
Durable goods and mining	53.7	30.5	29.2	25.1	Male operatives = 100
Non-durable goods	18.9	14.0	17.6	14.1	Male operatives = 100
3) Non-mfg., exc. D & D and mining	21.1	25.5	32.7	36.7	Male operatives = 100
Construction			0.8	0.9	Male operatives = 100
Transportation and utilities			2.0	2.5	Male operatives = 100
Trade			2.1	3.3	Male operatives = 100
Service workers, exc. private household	5.7	14.6	5.1	13.2	Male Labor Force = 100
1) Protective	29.8	5.4			Male Service workers = 100
2) Waiters, cooks and bartenders	17.3	16.0	23.6	21.0	Male Service workers = 100
3) Other	53.1	78.5			Male Service workers = 100
Laborers, exc. farm and mine	7.1	24.3	7.0	23.3	Male Labor Force = 100
1) Construction	23.1	27.9	20.7	20.7	Male laborers = 100
2) Mfg.	30.6	30.2	30.5	29.4	Male laborers = 100
Durable goods			19.5	19.4	Male laborers = 100
Non-dur. goods			11.0	10.0	Male laborers = 100
3) Other	46.3	41.9	48.4	49.6	Male laborers = 100
Transportation and utilities			14.5	14.5	Male laborers = 100
Trade			9.2	7.2	Male laborers = 100

Under -1-, the data are given on the basis of the average unemployment rate for January, April, and July of 1957, 1958, 1960, and 1961. The averages for 1957 and 1960 are identified as peaks. The 1958 and 1961 averages are identified as troughs. The results recorded under -2- are based on differential smoothing. [45] In these instances the white peaks and troughs are identified by the single lowest and highest observation. The white series are here deseasonalized. [46] The maximum number of correct signs is three.

The results recorded under -1- in columns 2, 4, and 6 still indicate greater absolute cyclical variability for non-whites. However, here, even without differential smoothing, only in sixteen out of twenty-five of the occupations listed do the non-whites have a larger decrease in their unemployment rates from trough-to-peak. The average of the two peak-to-trough movements shows that in eighteen of the twenty-five occupations the non-whites had a larger increase in unemployment during the downturn of the cycle. On the basis of the sign test, fifty-five out of a maximum of seventy-five signs were correct.

The picture changes substantially with differential smoothing. Here only seven out of twenty-four trough-to-peak changes (entry -2-, column 2) are in the right direction and in only nine of the twenty-four occupations do the non-whites experience a greater lay-off rate during the trough of the cycle (entry -2-, column 4). The sign test confirms this finding. Only twenty-eight out of seventy-five signs are in the direction that would be predicted on the basis of the "lhff" proposition.

Table 11 presents a comparison of the male unemployment differentials by major occupations in 1940 and 1950. This is a comparison of the differential in a period of high unemployment with one of relatively low unemployment. As such it can shed additional light on the question of whether non-whites are more adversely affected during periods of recession. Columns 3, 6, and 9 are the relevant columns in the test. If the non-whites are more adversely affected in periods of high unemployment, the difference between the non-white and white unemployment rates should be greater in 1940 than in 1950. The 1940 difference minus the 1950 difference should then be positive. In columns 3 and 6, the results are presented with the 1940 data not counting those on government emergency projects as belonging to the labor force. In the results of column 9, emergency workers are counted as unemployed.

When I exclude from the labor force those on emergency work, the differences in 1940 are smaller than those in 1950 in seven of the ten major occupations. When I include under the unemployed those on emergency work, the differences between the non-white and white unemployment rates in 1940 widen. They widen sufficiently to reverse the signs in column 9 for several major occupations. Still in three of the ten major occupations, the non-whites had a greater differential in 1950. And in two more, service workers and operatives, the difference in the differential between the two periods is small. It may be said that the behavior of the unemployment rates for the non-skilled or semi-skilled workers does not support the "lhff" proposition. These also happen to be the groups with relatively large sample sizes for the non-whites and with a narrow spectrum of skill within them.

Regional Differences. Thus far all of the data presented in this paper have been for the United States as a whole. It is desirable to see whether the pattern would change if we were able to standardize for locality. It is

Table 10. Differential changes in Male (NW-W) Unemployment Rates, by Intermediate Occupation Group and Cycle Phase, 1957-1961

Intermediate occupation group	Peak-to-trough 1957-1958		Trough-to-peak 1957-1958		Peak-to-trough 1960-1961		Avg. peak-to-trough		Avg. difference 1957-1961	No. of correct signs	
	1	2	1	2	1	2	1	2		1	2
Technical engineers	1.0	-6.6	0.0	1.5	1.6	-0.9	1.3	-3.8	0.4	2	0
Medical - salaried	8.2	1.2	-8.2	1.3	7.6	1.4	7.9	1.3	2.6	3	2
Other professional	-2.4	1.9	2.8	1.9	1.2	1.4	-0.6	1.7	2.5	1	2
Managers, salaried	-6.3	-6.7	5.9	4.7	-1.8	-2.2	-4.1	-4.5	2.9	0	0
Steno., sec. and typists	4.0	-5.4	12.7	15.9	-9.9	-11.8	-3.0	-8.6	3.8	1	0
Other clerical	1.0	0.4	3.4	4.7	-2.3	-3.6	-0.7	-1.6	4.3	1	1
Salesmen - retail	11.0	9.5	-9.8	-7.7	2.5	-1.2	6.8	4.2	5.2	3	2
Salesmen - other trade	1.6	1.1	-11.0	-10.3	8.8	7.8	5.2	4.5	4.3	3	3
Craftsmen - const.	3.3	1.2	-3.9	0.0	-1.6	-4.1	0.9	-1.5	6.0	2	0
Craftsmen - machinists and metal	0.1	-1.7	-0.4	2.0	0.7	-1.1	0.4	-1.4	4.6	3	0
Craftsmen - mechanics and repairmen	1.3	-1.4	-3.1	0.0	2.6	-0.5	2.0	-1.0	3.7	3	0
Craftsmen - others	7.9	5.7	-1.1	-1.5	1.5	1.3	4.7	3.5	4.1	3	3
Foremen	1.7	3.0	0.9	-1.1	4.1	5.8	2.9	4.4	1.8	2	3
Drivers and deliverymen	2.3	0.0	-2.6	2.2	1.7	-1.4	2.0	-0.7	3.4	3	0
Mine and durable goods operatives	3.8	3.3	-2.8	-3.2	1.1	-2.2	2.5	0.6	5.3	3	2
Non-durable goods operatives	2.8	1.8	0.3	2.7	-2.8	-5.8	0.0	-2.0	4.7	1	1
Non-mfg. goods operatives	1.2	-0.2	-1.3	0.2	0.5	-1.1	0.8	-0.7	2.3	3	0
Private household workers	-11.7		-2.7		2.9		-4.4		1.8	2	
Protective service workers	11.8	7.6	-10.4	-3.5	4.4	-0.4	8.1	3.6	3.9	3	2
Waiters, cooks and bartenders	3.6	0.5	-6.2	-2.7	4.1	6.0	3.9	3.3	0.6	3	3
Other service workers	3.2	2.1	-1.7	0.7	0.2	-4.8	1.7	-1.4	4.4	3	1
Paid farm workers	2.5	-1.0	3.4	6.2	-5.8	-6.2	-1.7	-3.6	1.0	1	0
Const. laborers	2.0	-1.1	2.9	6.8	-4.1	-9.7	-1.1	-5.4	3.4	1	0
Manufacturing laborers	1.3	-3.5	-1.1	4.7	-0.2	-3.5	0.6	-3.5	2.2	2	0
Other laborers	2.4	0.5	-0.6	0.7	1.8	-1.0	1.7	-0.3	3.3	3	1

Source: Unpublished BLS tabulations. See Appendix Table 6.

[1] Comparison of average of three points for peaks and troughs of white and non-white series.

[2] Comparison of results based on differential smoothing.

101

Table 11. Differences in the NW-W Male Unemployment Differential Between 1940 and 1950, by Major Occupation Group

Major occupation	Experienced labor force			Wage and salary workers only			Experienced labor force with emergency workers counted as unemployed		
	NW-W (1940)[1]	NW-W (1950)[2]	Col. (1)- Col. (2) (3)	NW-W (1940)[3]	NW-W (1950)[4]	Col. (4)- Col. (5) (6)	NW-W (1940)[5]	NW-W (1950)[6]	Col. (7)- Col. (8) (9)
Professional	3.5	1.4	2.1	2.5	1.6	0.9	7.5	1.4	6.1
Managers and proprietors	1.1	1.4	-0.3	1.0	1.6	-0.6	1.3	1.4	-0.1
Clerical workers	2.1	3.4	-1.3	2.6	3.5	-0.9	6.1	3.4	2.7
Sales workers	2.7	3.2	-0.5	3.7	4.1	-0.4	4.3	3.2	1.1
Craftsmen and foremen	4.9	4.4	0.5	5.5	4.9	0.6	8.6	4.4	4.2
Operatives	1.3	2.1	-0.8	1.6	2.0	-0.4	2.7	2.1	0.6
Private household workers	-0.6	1.0	-1.6	-15.3	0.8	-16.1	1.2	1.0	0.2
Service workers, exc. private households	3.0	1.5	1.5	2.4	1.3	1.1	4.0	1.5	2.5
Farm laborers and foremen	-4.7	-0.8	-3.9	-7.3	-0.9	-6.4	-6.9	-0.8	-6.1
Laborers, exc. farm	-2.8	1.0	-3.8	-3.5	0.9	-4.4	-2.7	1.0	-3.7

Source: [1]1940 U. S. Census of Population, The Labor Force, Volume III, Part 1, U. S. Summary, Table No. 62.

[2]1950 U. S. Census of Population, Occupational Characteristics, Special Report P-E No. 1 B, Table No. 3.

[3]1940 U. S. Census of Population, The Labor Force, Occupational Characteristics, Table No. 6.

[4]1950 U. S. Census of Population, Occupational Characteristics, Special Report P-E No. 1 B, Tables 3, 12 and 13.

[5]1940 U. S. Census of Population, The Labor Force, Usual Occupation, Table No. 4.

[6]1950 U. S. Census of Population, Occupational Characteristics, Special Report P-E No. 1B, Tables 3, 12 and 13.

102

particularly important to have a North-South breakdown of the data. Unfortunately there is little that is available by color, sex, occupation and region. The available data are presented in Table 12.

Table 12. Differences in the NW-W Male Unemployment Rates between Peak-to-Trough of the 1960-1961 Business Cycle, by Major Occupation Group, by Region

| Major occupation | North and West | | South | |
	Average Difference 1960-1961	P-to-T	Average Difference 1960-1961	P-to-T
Professional and kindred workers	2.7	-0.9	0.8	1.9
Managers and proprietors	3.0	-2.7	0.9	4.5
Clerical workers	5.4	-1.8	3.4	-3.3
Sales workers	2.6	1.4	5.3	10.9
Craftsmen and foremen	5.0	-0.7	4.1	4.3
Operatives	4.7	0.7	2.8	1.0
Service workers, exc. private household	4.5	3.4	2.6	-0.9
Farm labor	17.2	-11.2	-0.6	-3.9
Labor, exc. farm and mine	8.3	2.4	0.3	-4.2

Source: Unpublished BLS tabulations.

The results in Table 12 are for the last cycle. Beginning with January 1959, unemployment data are available by sex, color, major occupation groups, and region. I have chosen the average of the January, April, and July 1960 rates as the peak and the average of the comparable months of 1961 as the trough of the business cycle. For the United States as a whole the same cycle indicated adverse effects on non-whites when the test was conducted by major occupations. When the data are broken down by region the pattern changes substantially, particularly for the North and West where in five out of the nine major occupation groups the non-white/white differential is greater in the peak of the business cycle. This has occurred in the absence of adjustment for differences in sample sizes.

CONCLUSIONS

This paper was organized around the analysis and testing of the "lhff" proposition. In the process, it was necessary to examine the behavior of non-white/white unemployment differential both over time and over the various phases of the cycle. Thus the actual work and findings need not be linked to this proposition. The work can be viewed as an attempt to answer two questions. One, are the non-whites more adversely affected by declines in demand for labor? Two, if they are, does it occur because of their unfavorable occupational and industrial distribution, or because of differential hiring and firing within given occupations or industries? The answer to the first question is, yes. Non-whites tend to have a disproportionate increase in their aggregate unemployment rate during troughs of business cycles. The answer to the second question is, I believe, that this greater increase in unemployment is due mainly to their unfavorable occupational distribution. This finding is based on the results of the peak-to-trough comparisons for intermediate occupation groups recorded in Table 10. I assume that the true values lie between those derived from equal and those derived from differential smoothing. The finding in Table 10 is strengthened by the results recorded in Tables 11 and 12. The results of the peak-to-trough comparisons for major occupations groups (Tables 7 and 8) do show greater absolute increases in non-white unemployment rates during the downswing in business activity. However, even for these groups the difference is small for the lower skilled major occupation groups -- occupations containing most of the non-whites. Also, major occupation groups, particularly the higher skilled groups are not homogeneous with respect to skill (Table 4). This suggests that comparisons for intermediate groups are more relevant than those for major occupation groups. The results of the tests of total variability also show greater absolute cyclical variability for non-whites (column 3 of Table 5) in sixteen of the twenty-five occupations but these data are not adjusted for differential trend. [47]

In this paper I have not emphasized the cross-sectional aspect of this differential. It is clear, however, that the factors which produce differential cyclical results are also likely to cause cross-sectional differences in unemployment rates. However, in the case of the white/non-white unemployment differential there seem to be some differences in two aspects of the problem. The amount of standardization which was sufficient to remove the differences in cyclical behavior of the white and non-white unemployment rates was not sufficient to eliminate cross-sectional differences in their rates. [48]

In my development section, I suggested that the two aspects of the problem are different. Discrimination need not, first of all, result in a higher unemployment rate. The unemployment differential is greater in the North than in the South (Table 12). This difference could not be explained either on the basis of skill or on the basis of greater discrimination in the North. The differences in educational attainment of the two races are greater in the South than in the North. I further suggested that cross-sectional differences will result in differential cyclical behavior only under specific conditions. Since the data do not support the claim that we have differential cyclical behavior, I would conclude that these specific conditions do not in fact exist.

NOTES

[1]Current interest in differential unemployment rates is illustrated by the growing literature in this field. See, for example, U. S., Congress, Senate, Studies in Unemployment, 86th Cong., 2nd Sess., 1960: Margaret L. Plunket, "Youth -- Its Employment and Occupational Outlook," pp. 75-95; Arthur M. and June N. Ross, "Employment Problems of Older Workers," pp. 97-120; Richard C. Wilcox, "Women in the American Labor Force: Employment and Unemployment," pp. 121-22; John Hope II, "The Problem of Unemployment as It Relates to Negroes," pp. 173-223.

[2]See, for example, Ralph Turner, "Foci of Discrimination in the Employment of Nonwhites," American Journal of Sociology, LVIII (November, 1952), pp. 247-256; Philip M. Hauser, "Differential Unemployment and Characteristics of the Unemployed in the United States, 1940-1954," The Measurement and Behavior of Unemployment (Princeton: Princeton University Press, 1957); Donald J. Bogue, The Population of the United States (Glencoe, Ill.: The Free Press, 1959).

   The list of government publications and reports concerned with unemployment differentials and particularly with the white/non-white differential is too extensive for enumeration. Several of these are listed below: U. S., Department of Labor, Negroes in the United States; Their Employment and Economic Status, Bull. No. 1119 (Washington: U. S. Government Printing Office, 1952); U. S., Department of Labor, Notes on the Economic Situation of Negroes in the United States (May, 1957, rev. May, 1958, rev. August, 1959); U. S., Department of Labor, The Economic Situation of Negroes in the United States, Bull. S-3 (October, 1960); U. S., Bureau of the Census, "Employment of White and Nonwhite Persons: 1955." Current Population Reports, Labor Force, Series P-50, No. 66, March, 1956; U. S., Congress, Unemployment: Terminology, Measurement, and Analysis, Subcommittee on Economic Statistics, Joint Economic Committee, 87th Cong., 1st Sess., 1961; New York, Commission Against Discrimination, Nonwhite Unemployment in the United States, 1947-1958: An Analysis of Trends, April, 1958.

[3]See, for example, Gary S. Becker, The Economics of Discrimination (Chicago: The University of Chicago Press, 1957); Morton Zeman, "A Quantitative Analysis of White-Non-White Income Differentials in the United States in 1939" (unpublished Ph. D. dissertation, Department of Economics, University of Chicago, 1955); Donald Dewey, "Negro Employment in Southern Industry," Journal of Political Economy, LX (August, 1952); NPA Committee of the South, Report No. 6, Selected Studies of Negro Employment in the South (Washington: National Planning Association, 1955).

[4]See, for example, Hope, op. cit., p. 173; Bogue, op. cit., p. 636.

[5]Ibid.

[6]U. S., Congress, Subcommittee on Economic Statistics, op. cit., p. 60.

[7]Non-white unemployment rate minus white unemployment rate.

[8]See note 4, above.

[9]Jacob Mincer, "On-The-Job Training: Costs, Returns; And Some Implications," Capital Investment in Human Beings, NBER Exploratory Conference, 1961; pp. 44-45.

[10]Gary S. Becker, Investment in People, unpublished manuscript; NBER, 1962; Walter Y. Oi, "Labor as a Quasi-Fixed Factor of Production" (unpublished Ph. D. dissertation, Department of Economics, University of Chicago, 1961); J. Mincer, op. cit.

[11]This point is emphasized by Dewey (D. Dewey, op. cit.). See also the studies of Negro Employment in the South (NPA Committee of the South, op. cit.).

[12]Becker, Economics of Discrimination, op. cit.

[13]The authors of the NPA studies (NPA Studies of the South, op. cit., pp. 126, 207, 315, 382, and 482) maintain that even the southern Negro workers are paid the same wages when they perform the same tasks as whites.

[14]For a more detailed discussion about the extent of monopsony power during recessions, see: Albert Rees, "Wage Determination and Involuntary Unemployment," Journal of Political Economy, LIX (April, 1951).

[15]This analysis suggests that during recessions, more than at other times, white workers may attempt to legalize discrimination. Since they are not willing to pay the costs for the satisfaction of their tastes, they may attempt to get government to perform this task for them.

[16]Some may argue that discrimination is an inferior good. If it were, we would have an additional explanation for the belief that non-whites will be fired first. I cannot reject this assumption on a priori grounds, but there are a number of social and economic phenomena which are more easily explained by viewing discrimination as a normal good. One, Zeman (M. Zeman, op. cit., Table 15) shows that the ratio of non-white to white income declines with increasing levels of skill. Two, there are few non-white salesmen in industries such as advertising, stocks and bonds, manufacturing, and wholesale trade. Three, it has long been accepted that the banking and finance industries are most discriminatory in their employment policies (against Jews as well as against non-whites). Since these phenomena suggest greater discrimination by upper-income groups, they are consistent with the assumption that discrimination is a normal good. This assumption is strengthened if we accept the proposition that there is a negative correlation between a person's desire to discriminate and his level of education; therefore, greater discrimination by higher income groups would imply that discrimination is a normal good.

[17]For a detailed discussion on the concepts and methods of sampling, see: U. S., Bureau of the Census, Current Population Reports, Series P-23, No. 5, May 9, 1958.

[18]This period covers three business cycles.

[19]This period includes two business cycles.

[20]All of the series that are presented and analyzed here have been adjusted for the change in the definition of unemployment that occurred in January, 1957. Data for January, 1957, are available under both the old and new definitions by color, sex, and major occupations. The adjustment was made by linking the two periods on the basis of the ratio of the two January, 1957, figures.

[21]This statement is strictly true only under the following conditions: (1) The white and non-white labor force participation rates are identical. (2) These rates do not change or have the identical changes over the cycle. Only then is differential hiring and firing immediately reflected in the unemployment rate. However, since I am working only with male data, differential labor force participation rates over the cycle are not an important phenomenon.

[22]New York, State Commission Against Discrimination (op. cit., p. 3, and Figures 1, 2, and 3).

[23]I examined male unemployment rate series with eight observations per year. For these series, too, the white and non-white turning points were generally coincident.

[24]Becker, Investment in People, op. cit.; Mincer, op. cit.

[25]Extreme points of a series.

[26]Standard deviation ÷ average unemployment rate of the respective series.

[27]Non-white/white.

[28]Each standard deviation or coefficient of variation is a measure of variability between thirty-three monthly unemployment rate observations (Jan., Apr., July, and Oct. of each year) for the period of October, 1953, to October, 1961.

[29]Three-point moving average of seasonally adjusted data.

[30]Unemployment rates in this breakdown by color and sex have been tabulated since January, 1957. My own standard deviations have been computed from twenty monthly observations for the period 1957-1961 (Jan., Apr., July, Oct. observations of each year). This period includes two complete business cycles.

[31]The use of formula (2) to compute the random component of CPS data introduces several biases. However, the two main biases go in opposite directions. I have assumed that they are of approximately equal size and will tend to cancel each other. Thus, to the extent that the sample is stratified, formula (2) overestimates the random component. On the other hand, formula (2) will underestimate the random component because (a) the CPS sample is not a simple random sample of workers but a "cluster" sample of families and (b) the quarterly samples are not completely independent.

[32]If the whites and non-whites had the same unemployment rate during the peak of the cycle, either of these measures would give satisfactory results. Strictly speaking, constancy in the difference of the two rates over the cycle also leaves room for a slight differential increase in the non-white rate. The relevant measure is really the difference in the per cent disemployed. We want the same rate of additional lay-offs, for the two groups. However, the same absolute increase in the white and non-white unemployment rates approximates very closely the same additional lay-off rate.

[33]See the table below, particularly column three, rows four and five, for an illustration of this point.

	W	NW	NW-W
P	3	6	3
T	5	9	4
P	3	6	3
P-to-T	2	3	1
T-to-P	-2	-3	-1

[34]The argument for averaging is based on the usual assumption that a series contains some random elements and that the random component has to be removed prior to conducting analysis of the cyclical behavior. Smoothing is not too important for aggregate unemployment series since they are based on relatively large samples. It will take on greater significance with occupational standardization. The issues here are the same as those discussed in the preceding section of this paper.

[35]There is also a trend in the difference over time. What seems to be happening is that for non-whites the difference between peaks and troughs has tended to narrow over the successive cycles. To a lesser extent the same has occurred for the white males (see Appendix Table 3). A complete analysis of this trend would have to take account of the differential occupational and industrial shifts as well as differential growth of the labor force of the two groups over time. This is beyond the scope of this study.

[36]For analysis of what occurs during the downswing of the cycle we want the disemployment rate (see note 32). The experienced unemployment rate is a better approximation of the disemployment rate than is the total unemployment rate.

[37]This pattern is also observed in the comparison of 1940 and 1950 data by major occupation group presented in Table 11.

[38]Three-point moving average of deseasonalized series.

[39]It may be argued that differential smoothing will introduce a downward bias in the results. Smoothing may do more than remove the random component of a series; it may flatten the peaks and troughs of the cycles. However, the smoothing here is performed under a restraint that the standard deviation of the series not be reduced by more than the random variability. If the smoothing reduced the cyclical variability, it would also reduce the standard deviation. Smoothing could reduce cyclical swings and leave the standard deviation unaffected only if the series retained sufficient random variability to offset the reduction in the standard deviation from the reduction in cyclical variability. The question here is whether the smoothed series (3- or 4-point moving average) retains sufficient random variability to introduce a sizable bias in the results. I am currently reworking the data using tests not subject to the bias of differential smoothing. The results of these tests will appear in my dissertation. Preliminary findings indicate that the differential smoothing technique does not introduce a sizable bias. For the interpretation of the results recorded in Tables 7, 8, and 10, those derived from differential smoothing should be viewed as representing lower limit values.

[40]See Appendix Table 2 for smoothing ratios.

[41]October, 1953 - October, 1961.

[42]The question raised in this test is whether the difference between the non-white and white unemployment rates is greater in the trough as identified by the aggregate unemployment rate than in the peak which is also determined on the basis of the aggregate unemployment rate.

[43]See Figures 3 to 7.

[44]The distribution of occupations cross-classified by industry is not available by color. However, both the decennial census (1940 and 1950) and the current population survey results are tabulated by color and detailed or intermediate occupation groups. Some of these occupation groups are synonymous with industry groups.

[45]See Appendix Table 2 for the various smoothing ratios.

[46]Seasonal adjustments are not required for non-white series since their peaks and troughs are averages of several observations in each year.

[47]The completed study will include the results of several multiple regressions. These regressions will take account of such other factors as differences in training within occupations, differences in age composition of occupation groups, and differences in the industrial distribution between occupation groups.

[48]The cross-sectional aspect of the problem will form part of the dissertation.

Appendix Table 1. Standard Deviations and Coefficients of Variation for Male Unemployment Rate Series, by Color and Major Occupation Group, October, 1953-December, 1961

Major occupation group	Raw Data				Deseasonalized Data				Deseasonalized and smoothed[1] data			
	White		Non-white		White		Non-white		White		Non-white	
	s	S/X̄	s	S/X̄	s	S/X̄	s	S/X̄	s	S/X̄	s	S/X̄
Professional and kindred workers	0.46	0.35	2.11	0.66	0.47	0.36	1.98	0.60	0.39	0.30	0.81	0.25
Managers and proprietors	0.39	0.33	2.18	0.84	0.36	0.30	2.22	0.82	0.31	0.26	1.23	0.72
Clerical workers	0.88	0.27	3.46	0.49	0.84	0.25	3.58	0.50	0.68	0.21	2.35	0.32
Sales workers	0.81	0.31	5.84	0.88	0.87	0.33	7.01	1.02	0.62	0.24	5.03	0.70
Craftsmen and foremen	1.79	0.37	3.75	0.37	1.38	0.28	3.60	0.35	1.19	0.24	2.83	0.27
Operatives	2.10	0.32	2.99	0.30	1.81	0.28	2.80	0.28	1.59	0.24	2.27	0.22
Service workers, exc. private households	1.06	0.21	2.63	0.30	1.07	0.21	2.65	0.31	0.84	0.16	1.70	0.20
Farm laborers	3.02	0.53	4.75	0.63	1.34	0.24	3.45	0.44	1.05	0.18	2.29	0.29
Laborers, exc. farm and mine	3.84	0.35	4.31	0.32	2.46	0.22	3.38	0.25	2.22	0.20	2.74	0.20

Source: Computed from unpublished BLS tabulations.

[1] Three-point moving average of deseasonalized series.

Appendix Table 2. Standard Deviations and Coefficients of Variation for White and Non-White Male Unemployment Series, by Intermediate Occupations, 1957-1961

Intermediate occupation group	Raw data				Adjusted for sample size				Non-white to white smoothing ratio
	White		Non-white		White		Non-white		
	s	$\frac{s}{\bar{x}}$	s	$\frac{s}{\bar{x}}$	s	$\frac{s}{\bar{x}}$	s	$\frac{s}{\bar{x}}$	
Engineers	0.65	0.54	4.68	2.93	0.46	0.38	-	-	4
Medical, salaried	1.32	0.83	5.88	1.40	0.60	0.38	-	-	4
Other professional, salaried	0.44	0.22	2.52	0.56	-	-	-	-	4
Managers, salaried	0.52	0.37	3.59	0.83	0.45	0.32	-	-	4
Secretaries, stenos and typists	2.50	0.69	12.69	1.71	-	-	-	-	4
Other clerical	0.73	0.21	3.08	0.39	0.59	0.17	2.11	2.27	3
Salesmen - retail trade	1.11	0.27	7.26	0.78	0.84	0.20	4.00	0.43	3
Salesmen - other trade	0.52	0.31	7.11	1.19	0.35	0.21	3.43	0.57	3
Machinists and metal craftsmen	1.96	0.40	6.66	0.70	1.79	0.37	2.93	0.31	3
Mechanics and repairmen	1.14	0.31	2.41	0.33	0.97	0.26	-	-	4
Construction craftsmen	3.91	0.42	6.35	0.42	3.85	0.41	5.18	0.34	3
Other craftsmen	1.16	0.36	4.45	0.61	1.03	0.32	2.65	0.36	4
Foremen	0.61	0.29	5.74	1.47	0.27	0.13	-	-	4
Drivers and deliverymen	1.87	0.37	2.90	0.34	1.80	0.35	2.24	0.26	4
Mine and durable goods operatives	3.31	0.37	5.19	0.36	3.24	0.36	4.50	0.31	2
Non-durable goods operatives	1.53	0.26	3.89	0.37	1.33	0.23	2.46	0.23	2
Non-manufacturing operatives	1.87	0.26	2.87	0.31	1.71	0.24	1.81	0.19	3
Private household workers	4.13	1.35	6.33	1.29	1.42	0.46	-	-	3
Protective service workers	1.10	0.39	7.35	1.10	0.80	0.29	2.58	0.39	4
Waiters, cooks, and bartenders	2.27	0.26	4.91	0.52	1.47	0.17	3.39	0.36	2
Other service workers	1.21	0.21	2.68	0.27	0.89	0.16	2.11	0.21	3
Paid farm laborers	4.09	0.53	4.93	0.55	3.96	0.51	4.54	0.50	4
Construction laborers	7.45	0.40	6.98	0.32	7.23	0.39	6.39	0.29	4
Manufacturing laborers	4.26	0.34	5.20	0.36	4.04	0.33	4.63	0.32	2
Other laborers	2.49	0.30	3.15	0.26	2.31	0.27	2.42	0.20	3

Source: Computed from unpublished BLS tabulations.

Appendix Table 3. Seasonally Adjusted Unemployment Rates (First Month of Quarter Observations), by Color and Cycle Phase (Specific Cycles), 1948-1961

Unemployment series, seasonally adjusted	Series	Peak-1948 Single point	Peak-1948 Avg. of 3 points	Trough-1949 Single point	Trough-1949 Avg. of 3 points	Peak-1953 Single point	Peak-1953 Avg. of 3 points	Trough-1954 Single point	Trough-1954 Avg. of 3 points
Both sexes, total[1]	NW	4.3	5.6	13.4	11.4	4.1	5.2	10.6	9.6
	W	3.2	3.4	7.2	6.8	2.4	2.6	5.5	5.0
	NW-W	1.1	2.2	6.2	4.6	1.7	2.6	5.1	4.6
Male, total[1]	NW	4.7	5.9	15.8	13.2	3.2	4.6	11.0	10.5
	W	3.1	3.3	7.6	6.7	2.1	2.5	5.2	4.8
	NW-W	1.6	2.6	8.2	6.5	1.1	2.1	5.8	5.7
Male, experienced[2] workers	NW	4.4	5.5	15.3	12.5	3.0	4.4	10.9	10.3
	W	2.9	3.1	7.3	6.5	2.0	2.4	5.0	4.6
	NW-W	1.5	2.4	8.0	6.0	1.0	2.0	5.9	5.7
Male, experienced[2] wage and salary workers	NW					4.7	36.1	12.8	11.9
	W					3.1	33.7	6.0	5.6
	NW-W					1.6	2.4	6.8	6.3

	Series	Peak-1957 Single point	Peak-1957 Avg. of 3 points	Trough-1958 Single point	Trough-1958 Avg. of 3 points	Peak-1960 Single point	Peak-1960 Avg. of 3 points	Trough-1961 Single point	Trough-1961 Avg. of 3 points
Both sexes, total[1]	NW	7.2	7.8	13.8	12.5	9.8	10.0	13.0	12.7
	W	3.5	3.7	6.7	6.5	4.5	4.7	6.2	6.2
	NW-W	3.7	4.1	7.1	6.0	5.3	5.3	6.8	6.5
Male, total[1]	NW	6.7	7.8	14.7	13.5	10.5	11.3	13.6	12.7
	W	3.1	3.3	6.8	6.6	4.2	4.4	6.0	5.9
	NW-W	3.6	4.5	7.9	6.9	6.3	6.9	7.6	6.8
Male, experienced[2] workers	NW	6.0	7.2	13.7	13.3	9.1	10.2	12.4	11.4
	W	3.0	3.1	6.3	6.0	3.8	4.0	5.4	5.4
	NW-W	3.0	4.1	7.4	7.3	5.3	6.2	7.0	6.0
Male, experienced[2] wage and salary workers	NW	6.9	8.2	14.9	14.4	9.8	11.1	13.0	12.2
	W	3.4	3.6	7.7	7.2	4.4	4.9	6.4	6.2
	NW-W	3.5	4.6	7.2	7.2	5.4	6.2	6.6	6.0

[1]Calculated from: Current Population Reports, Labor Force, U.S. Census Series P-50 and P-57.

[2]Calculated from Unpublished BLS Tabulations of the Current Population Survey of the Bureau of the Census.

[3]These averages are for the July and October 1953 observations only.

Appendix Table 4. Seasonally Adjusted Unemployment Rates, by Color, Major Occupation, and Cycle Phase (Specific Cycle Reference Points), October 1953-October 1961

Major occupation	Series	Peak 1	Peak 2	Trough 1	Trough 2	Peak 1	Peak 2	Trough 1	Trough 2	Peak 1	Peak 2	Trough 1	Trough 2
White-collar workers	W	1.5	0.9	2.2	2.4	1.2	1.0	2.6	2.7	1.8	1.7	2.6	2.7
	NW	2.4	1.4	6.1	6.3	3.1	4.0	7.1	7.1	5.1	4.9	6.6	6.8
	NW-W	0.9	0.5	3.9	3.9	1.9	3.0	4.5	4.4	3.3	3.2	4.0	4.1
Craftsmen and foremen	W	3.8	2.3	5.6	5.9	3.3	2.9	7.4	8.3	4.7	3.9	6.7	6.9
	NW	7.7	7.7	15.8	14.8	6.6	6.9	14.1	14.1	8.1	8.7	12.8	12.1
	NW-W	3.9	5.4	10.2	8.9	3.3	4.0	6.7	5.8	3.4	4.8	6.1	5.2
Operatives	W	5.4	2.9	7.8	8.2	4.4	3.8	10.1	10.9	5.5	4.8	8.3	8.7
	NW	8.2	7.4	13.1	13.7	7.1	7.0	14.4	15.0	9.2	9.7	12.0	12.0
	NW-W	2.8	4.5	5.3	5.5	2.7	3.2	4.3	4.1	3.7	4.9	3.7	3.3
Service, exc. private household	W	4.0	2.7	5.7	6.3	3.7	3.5	6.7	6.8	4.7	4.4	6.3	7.4
	NW	5.4	4.7	9.6	9.3	5.6	5.9	12.1	12.6	7.8	8.4	11.8	10.8
	NW-W	1.4	2.0	3.9	3.0	1.9	2.4	5.4	5.8	3.1	4.0	5.5	3.4
Laborers, exc. farm and mine	W	9.2	6.7	11.9	12.4	8.0	7.7	15.0	16.7	10.7	9.7	14.9	15.1
	NW	9.2	9.2	13.1	12.4	9.6	9.6	18.3	17.8	13.8	13.8	18.4	18.4
	NW-W	0.0	2.5	1.2	0.0	1.6	1.9	3.3	1.1	3.1	4.1	3.5	3.3

Source: Unpublished BLS tabulations.

[1] Based on averages of three observations.

[2] Based on differential smoothing of the white and non-white series.

Appendix Table 5. Seasonally Adjusted Male Unemployment Rates, by Color, Major Occupation, and Cycle Phase (Total Unemployment Reference Points), October, 1953-October, 1961

Major occupation	Series	Peak 1953		Trough 1954		Peak 1957		Trough 1958		Peak 1960		Trough 1961		Avg. Rate Oct. 1953-Oct. 1961
		1	2	1	2	1	2	1	2	1	2	1	2	
White-collar workers	W	1.5	0.9	2.0	2.0	1.5	1.4	2.6	2.7	1.9	1.9	2.4	2.4	1.8
	NW	2.4	1.4	5.3	5.3	3.5	3.2	6.7	6.7	6.5	6.5	6.2	6.2	5.1
	NW-W	0.9	0.5	3.3	3.3	2.0	1.8	4.1	4.0	4.6	4.6	3.8	3.8	3.3
Craftsmen and kindred workers	W	3.8	2.3	5.3	5.9	3.6	3.3	7.4	8.3	5.0	4.9	6.7	6.5	4.8
	NW	7.7	7.7	14.8	14.8	7.6	7.6	14.1	14.1	9.1	9.1	12.1	12.1	10.1
	NW-W	3.9	5.4	9.5	8.9	4.0	4.3	6.7	5.8	4.1	4.2	5.4	5.6	5.3
Operatives	W	5.4	2.9	7.5	7.5	4.7	4.9	10.1	10.9	6.2	6.2	8.3	8.1	6.5
	NW	8.2	7.4	13.1	13.7	7.3	7.6	14.4	15.0	9.4	9.4	12.0	11.6	10.1
	NW-W	2.8	4.5	5.6	6.2	2.6	2.7	4.3	4.1	3.2	3.2	3.7	3.5	3.6
Private household workers	W	0.0	0.0	10.0		2.5		3.2		2.3		3.2		4.2
	NW	0.0	0.0	3.8		12.2		1.0		0.0		4.1		3.7
	NW-W	0.0	0.0	-6.2		9.7		-2.2		-2.3		0.9		-0.5
Service workers, exc. private household	W	4.0	2.7	4.9	4.1	3.8	3.8	6.7	6.8	4.7	4.9	5.8	4.9	5.1
	NW	5.4	4.7	8.2	8.0	6.2	5.9	10.9	11.3	9.8	9.3	10.1	10.8	8.8
	NW-W	1.4	2.0	3.3	3.9	2.4	2.1	4.2	4.5	5.1	4.4	4.3	5.9	3.7
Farm laborers and foremen	W	4.8	2.3	4.5	4.5	4.9	4.9	8.7	8.7	4.7	4.7	7.0	7.0	5.7
	NW	5.1	3.5	7.5	6.5	5.0	4.4	10.7	9.5	12.7	11.6	7.4	8.1	7.6
	NW-W	0.3	1.2	3.0	2.0	0.1	-0.5	2.0	0.8	8.0	6.9	0.4	1.1	1.9
Laborers, exc. farm and mine	W	9.2	6.7	11.7	12.4	8.8	8.8	15.0	16.7	10.7	10.4	14.5	14.8	11.1
	NW	9.2	9.2	12.5	12.5	11.3	11.3	18.3	18.3	15.4	15.4	18.4	18.4	13.5
	NW-W	0.0	2.5	0.8	0.1	2.5	2.5	3.3	1.6	4.7	5.0	3.9	3.6	2.4

1 Based on three-point averages.

2 Based on differential smoothing of white and non-white series.

Appendix Table 6. Male Unemployment Rates, by Color, Intermediate Occupation Group and Cycle Phase, 1957-1961

Intermediate occupation group	Series	Peak 1957		Trough 1958		Peak 1960		Trough 1961	
		1	2	1	2	1	2	1	2
Technical engineers	W	0.5	.3	1.5	1.9	1.5	0.4	1.4	2.4
	NW	0.0	5.0	0.0	0.0	0.0	0.0	1.5	1.1
	NW-W	-0.5	4.7	-1.5	-1.9	-1.5	-0.4	0.1	-1.3
Medical, salaried	W	2.6	0.6	0.5	4.4	2.0	0.0	2.2	4.7
	NW	0.0	0.0	6.7	5.0	0.0	1.9	7.8	8.0
	NW-W	-2.6	-0.6	6.2	0.6	-2.0	1.9	5.6	3.3
Other professional, salaried	W	1.5	1.4	2.3	2.6	2.1	1.8	2.4	2.7
	NW	3.6	3.7	2.0	3.0	4.6	4.1	6.1	6.4
	NW-W	2.1	2.3	-0.3	0.4	2.5	2.3	3.7	3.7
Managers, salaried	W	0.8	0.6	2.1	2.3	1.1	1.0	1.8	2.2
	NW	6.1	6.6	1.1	1.6	6.0	5.0	4.9	4.0
	NW-W	5.3	6.0	-1.0	-0.7	4.9	4.0	3.1	1.8
Stenos, secretaries and typists	W	1.6	0.0	7.1	9.5	2.7	0.0	3.6	5.0
	NW	0.0	3.1	9.5	7.2	18.1	13.6	9.1	6.8
	NW-W	-1.6	3.1	2.4	-2.3	15.4	13.6	5.5	1.8
Other clerical workers	W	2.9	2.5	4.5	4.7	3.5	2.4	4.3	4.5
	NW	5.0	5.0	7.6	7.6	10.0	10.0	8.5	8.5
	NW-W	2.1	2.5	3.1	2.9	6.5	7.6	4.2	4.0
Salesmen - retail trade	W	3.2	2.5	4.8	5.6	4.6	3.3	4.5	6.9
	NW	4.2	4.2	16.8	16.8	6.8	6.8	9.2	9.2
	NW-W	1.0	1.7	12.0	11.2	2.2	3.5	4.7	2.3
Salesmen - others	W	1.1	0.9	2.0	2.3	1.6	1.2	2.3	2.9
	NW	8.9	8.9	11.4	11.4	0.0	0.0	9.5	9.5
	NW-W	7.8	8.0	9.4	9.1	-1.6	-1.2	7.2	6.6
Craftsmen - construction	W	7.5	6.3	11.4	12.3	9.8	6.8	13.7	13.2
	NW	13.6	13.6	20.8	20.8	15.3	15.3	17.6	17.6
	NW-W	6.1	7.3	9.4	8.5	5.5	8.5	3.9	4.4
Craftsmen - machinists and metal	W	2.5	2.0	7.1	8.4	4.0	2.9	6.6	7.3
	NW	9.2	9.2	13.9	13.9	10.4	10.4	13.7	13.7
	NW-W	6.7	7.2	6.8	5.5	6.4	7.5	7.1	6.4
Craftsmen - mechanics and repairmen	W	2.6	2.5	5.2	6.4	3.2	2.3	4.5	5.6
	NW	6.1	7.3	10.0	9.8	4.9	5.7	8.8	8.5
	NW-W	3.5	4.8	4.8	3.4	1.7	3.4	4.3	2.9
Craftsmen - others	W	2.3	1.8	5.0	5.6	2.9	2.6	3.7	3.9
	NW	1.5	2.4	12.1	11.9	8.9	7.4	11.2	10.0
	NW-W	-0.8	0.6	7.1	6.3	6.0	4.8	7.5	6.1
Foremen	W	1.5	1.2	2.8	3.0	1.8	1.5	2.4	2.5
	NW	0.0	0.0	3.0	4.8	2.9	2.2	7.6	9.0
	NW-W	-1.5	-1.2	0.2	1.8	1.1	0.7	5.2	6.5
Operatives - drivers and deliverymen	W	3.9	3.3	6.6	7.6	5.2	2.9	6.9	6.7
	NW	6.1	6.0	11.1	10.3	7.1	7.8	10.5	10.2
	NW-W	2.2	2.7	4.5	2.7	1.9	4.9	3.6	3.5
Operatives - mining and durable goods	W	5.5	4.2	14.8	16.7	8.2	6.7	11.5	11.4
	NW	9.1	9.5	22.2	25.3	12.8	12.1	17.2	14.6
	NW-W	3.6	5.3	7.4	8.6	4.6	5.4	5.7	3.2
Operatives - non-durable goods	W	5.1	4.2	7.6	8.8	5.9	3.9	7.1	7.0
	NW	8.4	8.6	13.7	15.0	12.3	12.8	10.7	10.1
	NW-W	3.3	4.4	6.1	6.2	6.4	8.9	3.6	3.1

Appendix Table 6 (continued)

Intermediate occupation group	Series	Peak 1957		Trough 1958		Peak 1960		Trough 1961	
		1	2	1	2	1	2	1	2
Operatives - non-manufacturing	W	5. 7	5. 0	9. 6	10. 3	6. 5	5. 7	8. 8	9. 6
	NW	7. 6	7. 6	12. 7	12. 7	8. 3	8. 3	11.1	11. 1
	NW-W	1. 9	2. 6	3. 1	2. 4	1. 8	2. 6	2. 3	1. 5
Private household workers	W	1. 9		4. 3		3.		3. 2	
	NW	13. 3	12. 2	4. 0	3. 0	0. 0	0. 0	3. 1	4. 1
	NW-W	11. 4		-0. 3		-3. 0		-0. 1	
Protective service workers	W	1. 8	1. 4	3. 6	4. 8	1. 8	0. 4	3. 6	4. 1
	NW	3. 7	2. 8	17. 3	13. 8	5. 1	5. 9	11. 3	9. 2
	NW-W	1. 9	1. 4	13. 7	9. 0	3. 3	5. 5	7. 7	5. 1
Waiters, cooks, and bartenders	W	5. 1	3. 9	9. 6	11. 4	8. 8	7. 7	11. 2	12. 1
	NW	5. 2	4. 3	13. 3	12. 3	6. 3	5. 9	12. 8	16. 3
	NW-W	0. 1	0. 4	3. 7	0. 9	-2. 5	-1. 8	1. 6	4. 2
Other service workers	W	4. 7	3. 9	6. 9	7. 2	5. 7	3. 6	5. 6	8. 5
	NW	7. 2	7. 2	12. 6	12. 6	9. 7	9. 7	9. 8	9. 8
	NW-W	2. 5	3. 3	5. 7	5. 4	4. 0	6. 1	4. 2	1. 3
Paid farm laborers	W	8. 0	6. 7	11. 4	12. 3	7. 3	5. 8	10. 4	9. 5
	NW	7. 5	6. 7	13. 4	11. 3	12. 7	11. 0	10. 0	8. 5
	NW-W	-0. 5	0. 0	2. 0	-1. 0	5. 4	5. 2	-0. 4	-1. 0
Construction laborers	W	17. 3	13. 2	22. 4	21. 6	19. 3	14. 3	25. 4	25. 9
	NW	18. 2	16. 3	25. 3	23. 6	25. 1	23. 1	27. 1	25. 0
	NW-W	0. 9	3. 1	2. 9	2. 0	5. 8	8. 8	1. 7	-0. 9
Manufacturing laborers	W	8. 1	6. 9	17. 6	21. 8	11. 3	7. 4	17. 1	17. 3
	NW	10. 9	11. 4	21. 7	22. 8	14. 3	13. 1	19. 9	19. 5
	NW-W	2. 8	4. 5	4. 1	1. 0	˙3. 0	5. 7	2. 8	2. 2
Other laborers	W	7. 1	6. 1	9. 2	10. 6	8. 2	7. 1	10. 6	10. 8
	NW	9. 3	8. 8	13. 8	13. 8	12. 2	11. 0	15. 6	13. 7
	NW-W	2. 2	2. 7	4. 6	3. 2	4. 0	3. 9	5. 0	2. 9

[1] Based on three-point averages.

[2] Based on differential smoothing of white and non-white series.

Comments on Part II

COMMENT by Albert Rees

On Korbel

I should like to devote most of my attention to the paper by Korbel. The recent upsurge of interest in the determinants of labor force participation has carried our knowledge of this area far beyond the point to which it was taken by the pioneering contributions of Douglas and Woytinsky. Mincer has shown that the participation of married white women is positively related to the wages offered to women and negatively related to husband's income, with the relation being stronger with transitory components of the husband's income than with the permanent components. Preliminary results from the current work of Glen Cain at the University of Chicago suggest that the same variables explain, in roughly the same manner, cross-sectional differences in the labor force participation rates of married non-white women, though Cain has not yet been able to settle on any one explanation of why the rates for non-white women are higher than for white after many kinds of standardization. The work of Long, Mincer, Katz, and others has indicated, as Korbel points out, that both the "added worker" effect and the "discouraged worker" effect of unemployment on participation rates are present for different groups in the labor force, with the "discouraged worker" effect showing some tendency to dominate the aggregates.

I approach Korbel's contribution to this literature with some diffidence, since I am not familiar with the methods used in simulation studies. One point stands out at the beginning. His task would have been much easier if the gross-flow data from the Current Population Survey had been continuously available since 1949, and were cross-classified with more additional variables. In this connection we may note that the President's Committee to Appraise Employment and Unemployment Statistics in its recent report recommended that a program of research be undertaken to reduce defects in the gross flow data and that their publication be resumed as soon as possible. My understanding of the defects in the gross flow data is somewhat different from Korbel's. The principal difficulty seems to be the "first month bias" -- the systematic tendency of respondents to report their status differently when they first enter the panel and on later interviews. Gross flow data are much more sensitive to error of this kind than are estimates of levels, for in the latter but not in the former there is a tendency for errors to be offsetting.

One on the principal findings of Korbel's paper seems to be that the best predictor for the individual woman of labor force participation this year is participation last year. This is the kind of information that is useful to an employer who wants to minimize turnover, and I gather that it is useful in simulation studies because they too focus on individual behavior. From a broader point of view, however, this predictor is somewhat unsatisfactory, since it leaves unanswered the question why the woman was in the labor force last year. I am somewhat puzzled by the failure of education and husband's income to contribute much to the explanation of labor force participation in Korbel's sample. Education is a proxy for the income the woman could earn if she worked, and one would expect it to show a significant positive relation with participation when other factors are held constant. The general negative relation with husband's income is well known from other studies. Perhaps a different dependent variable, such as weeks worked in 1961, might have produced somewhat different results than the simple yes or no variable whether the woman worked during the year.

The finding that a small proportion of the women in the sample account for a large proportion of the total moves is consistent with the findings of many other mobility studies. It would be interesting to know much more than we do about this highly mobile fringe. For women in the child-bearing years, pregnancy can account for a number of changes in labor force status, but similar mobile fringes also occur in studies of male mobility.

On Marshall

Although I have focussed my discussion on Korbel's paper, I should like to add a word on employment opportunities for Negroes. Marshall's paper shows us how far we still have to go to attain equal job opportunities.

The problem of equal job opportunities is difficult enough so that no one approach to its solution is likely to take us very far. We need more FEPC laws and better administration of most of the ones we have. We also need better education for Negroes, in the North as well as in the South. But even given these, I am afraid that progress will remain slow in the absence of higher general levels of employment than we have experienced in recent years. When there are reserves of experienced unemployed white workers, some employers will continue to indulge their prejudices and some unions will cling to their restrictive practices, and even legal pressures may produce only token compliance. In a tight labor market, however, these employers may hire Negroes because they need to fill vacancies, and the unions concerned may relax barriers because the competition for jobs is less intense. The slow rate of improvement in job opportunities for minority groups is one of the many hidden costs we are paying for the low-pressure economy of the past five years. In time we may come to regard these costs as so high that we would prefer to suffer an upward creep of the price indexes or a modification of our rigid international monetary arrangements. However, it is not for us at this conference to discuss changes in national economic policy. Our purpose is the more limited one of better understanding urban labor markets, and the papers at this session have made an important contribution to it.

COMMENT by Jacob Mincer

On Korbel

The simulation experiment conducted by Orcutt and associates at Wisconsin has had its healthy share of followers, critics, and luke-warm sceptics. John Korbel's work suggests that no matter what simulation will or will not do ultimately, it is generating some worthwhile by-products in the meantime.

Period-to-period changes in numbers of women participating in the labor force are arithmetically equal to the difference between the numbers entering and the numbers leaving the labor force. The same net difference, that is, change in participation, can obtain with larger or smaller sizes of flows into and out of the labor force. Korbel stresses the importance of studying the sizes and distributions of these gross flows. It is clear, of course, that such information is indispensable for the purpose of the simulation project. It is equally clear that information on gross flows would be helpful in understanding many aspects of labor supply without resorting to simulation. Still, Korbel seems to overstate the case for gross flows (a) in intimating that without such information the determinants of labor supply cannot be properly assessed, and (b) in asserting that the "added worker hypothesis" remains untested for the same reason.

I am inclined to reject both allegations. On the matter of the "added worker," recent studies by Lee Hansen and by Arnold Katz have rather convincingly confirmed Clarence Long's view of the cancellation of "added" and "discouraged" workers, leaving a very slight, if any, elasticity of the labor force with respect to the business cycle. True, the studies shed no light on what happens during severe depressions, but neither would any current information on gross flows do that. Let us hope, at any rate, that this particular aspect of the question will continue to remain in the realm of metaphysics. As to the study of factors affecting labor force participation and its changes, a great deal has been done and more can be accomplished with sufficiently detailed information on the net magnitudes -- labor force rates -- classified by various population characteristics.

In a very real sense, the proper comparison of female labor supply is not with supplies of commodities but with effects of such factors as changes in inter-industry wage differentials or inter-industry changes in employment. There is no question that in studying effects of wages on inter-industry changes in employment we are interested in the net employment effects. In that context as well as in the context of movements of women between "home" and "market industries," the argument for studying gross rather than net movements must hinge on the evidence that better empirical relationships (explanations) are obtained between the causal factors and the gross movements. Perhaps this is a reasonable expectation. It should be tested.

Turning to Korbel's sample study, I notice that some of the statistics in his relatively small sample are quite similar to those in much larger surveys. Thus 42 per cent of his women worked sometime during the year, and about 63 per cent worked sometime during the twelve-year period. Given the degree of representativeness of the sample, and the data on job and labor force moves supplied by Korbel, some interesting comparisons between his findings and other data could be attempted. For example, various studies in the past showed that labor force mobility of women is much higher than that of men and that job mobility of men is higher than that of women. But if labor force mobility of women is viewed mainly as an inter-industry move, that is, a job move, one should add job moves and labor force moves for purposes of comparison. Based on a cursory examination, my impression is that remaining differences, if any, would be small. I also find that the distribution of "voluntary" and "involuntary" moves, given by Korbel in Table 11, is very similar for women and for men, once account is taken of the fact that the woman is self-employed in the home, and that a certain fraction of moves from home to market are equivalent to being "fired" from the job at home. Applying the male lay-off rate (1961, Monthly Labor Review) to Korbel's data on home-to-market moves, about 56 per cent of all the moves turn out to be voluntary and 44 per cent involuntary. This is precisely the same as the distribution of quits and accessions on the one hand, and lay-offs on the other, among males.

Tables 4 and 7 also bring out another phenomenon common to both sexes: mobility decreases with advancing age. However, the effect of age is probably exaggerated by the cross-section. The older women have less mobility not only because they are older, but also because they belong to a cohort which probably had less mobility at all comparable ages.

An issue which is repeatedly encountered in cross-sectional studies and from which wrong conclusions are often drawn emerges here too in Tables 9 and 10. No more than 16 per cent of total variance of the dependent variable (labor force status of women) is explainable by as many as seven demographic and economic variables, but about 43 per cent is explained simply by the lagged value of the dependent variable. I find these results neither surprising nor disturbing. Numerous variables enter the category of differences in tastes among families in a cross-section and show up in a huge residual variance. When aggregated to market levels, these idiosyncracies cancel out for the most part, the systematic variables do not. As to the lagged dependent variable, it is clearly of dubious explanatory value, being circular. But is it useful as a predictor? An analogy may help: supposing one studied cross-sectionally factors influencing ownership of a durable good which lasts four or five years and which is owned by about a half of the population in a given year. No doubt that last year's ownership status would be the most powerful "predictor" on the individual level. But would it predict changes in the aggregate ownership rate better than such cross-sectional cinderellas as income, relative prices, etc? The answer is no. Any half-way respectable estimating relation should have, and does have by definition, a higher predictive efficiency than a simple extrapolation from today to tomorrow.

On Gilman

I consider this work to be a serious effort in an area in which strong and popular beliefs tend to obscure facts and hamper interpretations. Gilman proposes to put to a test the notion that Negro workers are "the last to be hired and the first to be fired." He interprets this to be a statement about current discriminatory employment practices, not about consequences of a lopsided occupational and educational distribution of Negroes which is, in large part, a product of past discrimination. Thus, says Gilman, the validity of the "l-f" proposition would be established if discriminatory hiring and firing could be observed among workers of the same grade of occupational skill. Empirically, this means that unemployment differentials between the races must be standardized as closely as possible for the level of skill.

Before proceeding to the empirical tests, Gilman asks a theoretical question: under what conditions could one reasonably expect such discriminatory behavior to take place? His conclusion is that this could happen only when racial prejudices of employers and/or co-workers have not been fully paid off, or compensated in the form of lower wages paid to Negroes. He then considers a number of possible conditions under which discrimination might not be fully bribed away by a suitable wage differential. Gilman is inclined to feel that none of these conditions is plausible, but he admits that on a priori grounds one can neither definitely reject nor accept the "last hired and first fired" hypothesis. With this conclusion I fully concur, while disagreeing at several points with Gilman's views about theoretical plausibilities. However, for the empirical question at hand, such speculation is redundant.

In this paper Gilman singles out the empirical implications of the "1-f" hypothesis for time series behavior of employment and unemployment. One implication is that unemployment of Negroes should lead that of whites at peaks of the business cycle (that is, troughs of unemployment) and lag at the opposite turning points. No such relationships are apparent in the data, without or with standardization by major occupation. But the perception of leads and lags at cyclical turning points is perhaps too severe a demand on time series which are dominated by large random components, as is particularly true of the Negro unemployment series.

Another implication of the hypothesis is that amplitudes of fluctuations in unemployment should be greater for non-whites than for whites, after standardization for skill. The relevant measures, as Gilman points out, are absolute (not relative) differentials in disemployment rates. These are approximated, though with a very slight bias, by unemployment rates of experienced workers. The existence of differential amplitudes is tested, therefore, by a comparison of standard deviations (coefficients of variation are clearly inapplicable), and by a comparison of peak-to-trough differentials in unemployment rates.

The comparison of standard deviations, after an adjustment for random components due to differential sample sizes, is pushed to its most detailed level which the data permit in column 3 of Table 5. As in the case of rougher standardization for major occupation, here too the differences disappear at low skill levels, and are positive, but small, at the higher levels. Whether such differentials as remain would disappear with even more detailed, but unavailable standardization, remains an open question. The comparisons of peak-to-trough differentials, based on special smoothing procedures and shown in Table 10 seem to erase the differentials entirely across the whole occupational range. However, because of the biases in the procedure in the direction of oversmoothing the differentials, I give little weight to these particular results. They serve, roughly speaking, as one extreme and of a confidence interval. It remains true, however, that standardization from broader to more detailed groups reduces substantially the differences in variability. The factual validity of the "1-f" proposition turns out to be in grave doubt, indeed. Without any doubt, the proposition appears to be untrue when it comes to unskilled and semi-skilled workers, the bulk of the Negro labor force.

Is there anything left out of consideration that could rescue the "1-f" proposition? As the proposition refers to layoffs and not to quits, one possibility is that in good times temporary unemployment associated with quits is a larger component of white than of Negro unemployment. If true, equal amplitudes of total unemployment would mean larger amplitudes of involuntary unemployment among non-whites. The conjecture is plausible, but it needs to be investigated.

Finally, Gilman should emphasize more strongly that it is incorrect to infer from his results that the observable racial differentials in hiring and firing are due solely to historical factors affecting the educational and occupational distribution of Negroes. The occupational downgrading of currently qualified Negro workers and the discrimination in the provision of on-the-job training to them are current practices with direct consequences for current hiring and firing decisions. The patterns of unemployment differentials shown by Gilman are consistent with the existence of these phenomena.

COMMENT by Herbert Northrup

On Marshall

Professor Marshall has written a sound paper which brings together
a lot of information but which does not add very much to our knowledge of
the subject. This is not to disparage the work because an up-to-date
codification of these facts is valuable and timely. Until Professor
Marshall began his work, we had no studies of general significance in this
field since 1944.

Probably the most significant point made by Professor Marshall is
the need for more qualified Negroes. At the University of Pennsylvania,
for example, there is a great need for more Negroes to apply in the
graduate professional levels. Discrimination in these areas cannot be
eased until we get applicants and this is not easy even though an active
search is made for them. The same thing is true in the skilled trades
that is true in the professions. What some have called the "ghetto"
mentality prevents Negroes from aspiring to high levels where the opportunity to break through existing barriers would be welcome and commended. If discrimination is to be curbed further, we must devise
methods of getting into the Negro's home and family and build hope so
that those who have the ability will aspire to better jobs and better
positions throughout industry and the professions.

Let me add a final note about future research. In a trip through the
West this summer, I was impressed with the need to know what kind of
jobs American Indians hold and are prevented from holding. There was
great evidence in some of the Rocky Mountain states that the Indian there
is playing the role that the Negro does in many other states. I think it
would be significant if studies such as those before us today would turn
their focus also on the minority who preceded all of us here.

*PART III. SOME ECONOMIC IMPLICATIONS OF
HOUSEHOLD CONSUMPTION PATTERNS*

The Significance of Residential Preferences in Urban Areas

by John M. Richards

Why consumption rather than location theory is more applicable - Residential market is economically imperfect - Earlier work stresses locational aspects of spatial structures but are not useful for prediction - And consumption studies are not much better - Role of subjective (psychological) values - A new model explained and stated in formal terms - Important role of the spectrum of residential amenities at the margin - Empirical evidence found in Baton Rouge, La. - Inferences.

This paper is essentially an essay on consumer economics. More particularly it is concerned with one aspect of consumer behavior—that of residential consumption or selection. Whenever a consumer selects a housing unit, he consumes a special resource that has little, if any, reproducibility. This resource is urban land.

The paucity of urban sites is of vital significance for all human resources, since space for location becomes a prerequisite for their utilization as well as their very existence. It is true that urban land may be used more intensively by vertical expansion, or may be enlarged extensively by increasing the horizontal limitations of travel. Nevertheless, human pressure upon urban land has become so great that many foresee the threat of some urban Malthusian doctrine. Whether or not this dilemma has already appeared is incidental to the vital need for the better utilization of space within urban areas.

In considering the locational aspect of urban problems it is imperative that attention be given to consumer demands for residential location. It would seem legitimate to argue that the economic city, like other aspects of a capitalistic economy, should reflect consumer demands. While there are many competitive goals at work in urban locational situations, there are also adequate reasons for believing that consumer locational aims will tend to be subverted in the competitive market.

One of the primary reasons location decisions are not left solely to the competitive market is the degree of externalities involved. [1] External diseconomies arise from many alternative uses of land sites and cause a divergence of private and social interests. In such cases the competitive location of activities, without any attempt at control, would result in a highly unsatisfactory areal pattern.

Still another problem of urban location evolves from the permanency of past decisions. Unlike the market for commodities, the market for locational sites is not cleared for a fresh start with each market period. Urban location is a highly durable good and a city must live with solutions arrived at in the past. Urban land use patterns have been constituted through a series of short-run optimizations, each of which structures and limits the market results that follow.

There are adequate expressions of urban distrust of the purely competitive market system. Zoning regulations are the most prominent examples of interference in the desire to control and limit the externalities of individual locational decisions. In this sense, government intervention is active to prevent a social misallocation of spatial resources. Urban planning and urban renewal are also examples of similar restrictions. This use of "organized foresight plus corrective hindsight"[2] seeks to direct the future market results as well as to mitigate the effects of past results.

It must also be remembered that an important participant in most urban location markets is government itself. Armed with the power of eminent domain, society is constantly purchasing sites for the location of schools, streets, parks, and other public facilities. Correct location is a vital factor in the ultimate success and daily efficiency of these activities. The existence of the condemnation power yields the right to coerce the private market, but equally implies the obligation to arrive at optimum locational solutions for society.

The existing scarcity of urban sites and the demands raised by competing uses would be sufficient reason for casting attention upon the forces that act to structure one sector of those demands. However, when these facts are coupled with the government commitment to interfere for the benefit of the consumer, it becomes mandatory for greater attention to be focused upon the forces determining the residential decisions of individuals.

TWO CATEGORIES OF STUDIES

Past efforts concerned with residential consumption may be placed into two broad categories: 1) those studies seeking general models based on generic explanations governing residential distribution relative to other activity centers within the city, and 2) those studies concerned with geographical mobility and individual or family selection of housing units.[3]

The first of these will be referred to here as "locational" studies. While not necessarily based on economic location theory, they are involved with a space economy, explaining residential location as part of the spatial structure of urban areas. For the most part such studies use distance as a primary variable in explaining residential location, with accent on the distance separating home and work place. According to the conceptual framework of the researcher, distance may be used as a measure of total money costs (including rent and travel expense),[4] of the time and effort involved in travel,[5] or of some unexplained composite of these.[6]

The primary contribution of the locational studies is the provision of more powerful analytical tools for predicting general patterns of residential location in urban areas. Whether viewed from the point of travel demands or residential demands, the conclusion generally reached is that the bulk of workers live close to their work and, beyond a certain point, the proportion of workers decreases as the distance increases.[7]

The primary limitation or criticism of the locational approach to residential selection is the inability to explain deviations from a central tendency, or the scatter of population away from central work places. The degree of population dispersion and other evidence concerning residential selection indicate a more complex spatial structure of residences than often assumed.[8]

The nature of area complexity and the conflict of evidence becomes more apparent when attention is given to studies of household residential decisions. This approach may be referred to as "consumer" oriented, since primary emphasis is given to the physical and psychological needs that structure demands for housing.[9]

The primary contribution of consumer studies is to provide insight into the process of individual or household decision-making with regard to residence. This process includes both the manner in which possible choices come to the attention of families as well as deliberation among alternative housing opportunities. Such deliberation does involve relative location, but only along with many other attributes of housing units. In addition to housing features per se, almost equal concern is given to rent or housing costs, or the amount of income that will remain for non-residential consumption. One general conclusion of such studies is the relatively small significance given to the location of work place in final residential location.[10]

There are many limitations as to the use of consumer studies. While they provide generalizations as to motivational drives and final demands, they yield limited predictive value of what the final choice will be. What little predictive value is provided is primarily concerned with the timing of consumption relative to either family cycles[11] or economic conditions.[12] Information concerning spatial use and structuring is limited to general information concerning suburban development and dispersion.[13]

The dual approaches to residential selection, and the apparent conflicts involved, do pose an interesting field for the economist because of the interplay of two established areas of economic analysis. Both locational analysis and consumer analysis must be taken into consideration. Since all residences are locked into a spatial continuum, the individual decision to consume any housing unit co-determines his location in space.

Residential location certainly must be viewed within the functional specialization of the city as determined by optimal solutions between marginal rent and travel costs. However, to extend this more typical locational approach to the individual fails to provide complete analysis of all major factors involved. Site selection certainly involves the money costs of rent and transportation, but individuals also are concerned with the attributes of sites and the time and discomfort expended in approach to space, different from that of traditional theory. The need to travel involves not only the costs of travel, but also the effort of moving to other activity areas within the urban complex. To the extent that individuals desire to avoid discomfort and effort, spatial factors move beyond ordinary concern with the money cost of transportation.

In addition to desires to minimize effort, no approach is complete without some consideration being given to the desires for residential amenities that do vary among alternative residential locations. As a result, the totality of the individual's social and cultural environment becomes as important as the economic and spatial environment. The importance of the physical and social attributes of sites may be seen in residences located miles from other activities, perhaps on steep slopes or poorly drained swamps and lakes. The psychological productivity of such areas and the high value placed on "view" or social prestige certainly cannot be explained by typical location theory nor attempts to reduce travel time.

RESIDENTIAL PREFERENCES AND RESIDENTIAL SELECTION

The specific purpose of this paper is to attempt to provide some sort of synthesis in an approach to residential selection. The general thesis is as follows: the consumer in selecting a location seeks to maximize his utility; this involves balancing expenditures for space and amenities, expenditures for all other commodities, and the cost and effort of transportation at the margin.

The concept of spatially-fixed attractive forces influencing individual behavior is certainly not new to the literature. The idea that some people would prefer to carry on their activities in certain places is too obvious to be ignored. Goldner used the idea of "normal preference areas" which were "gleaned from spatial preferences of working men" as a theoretical tool for examining the geographical mobility of labor and labor supply areas.[14] Other studies concerned with either labor mobility or residential selection also point to such a phenomenon as playing a vital role in structuring the location of daily activity.[15]

The broader concept of residential preference is a mental by-product of Isard's original discussion of individual space preference.[16] Space preference is based upon the propensity of an individual to maintain some level of social contact through his location in space. Isard supports this assertion by referring to studies by psychologists and sociologists who, "...whether speaking of a gregarious instinct or acquired behavior patterns or of both, have emphasized the social nature of man and his propensity to associate with groups of various sorts."[17]

The concept of residential preference varies from that of space preference. Residential preference assumes a degree of space preference and is not directly concerned with broad social contact. Residential preference is concerned with attraction to characteristics located at points within a region -- to a residence and not a space. Residential preference is a limited aspect of Isard's hypothesis, taking the general areas of production, employment, and individual location as given. In order to satisfy his residential desires, the individual must still bear the burden of transportation. Not only must he bear the broader transport rate to his community, but also the additional costs of local movement in everyday activities of shopping and going to work. The costs and the attraction of specific residential amenities aid in structuring final location. The vital forces of residential selection involve all of the social and physical characteristics around a point. It is the sum of these immediate and surrounding forces that provides the net satisfaction from the occupancy of a residence.

Viewed from the vantage point of residential preferences (though it is not necessary to do so), the following picture develops: In a series of existing, or potential housing units each provides a different degree of utility according to the attributes it derives from the physical and cultural conditions involved. Individuals strive for that unit which will provide them with the largest degree of net utility or satisfaction, with the net figure arrived at according to value judgments concerning residential amenities and the costs necessary for occupancy.

A MODEL OF THE ALTERNATIVES IN RESIDENTIAL SELECTION

A model of the choices confronting an individual may be demonstrated by the use of indifference surfaces, as in the accompanying diagram. The shaded areas indicate two of a series of indifference surfaces indicating equal areas of satisfaction. Point O represents the most desired place of residence as chosen according to the individual's evaluation of residential amenities.

The vertical, OY axis indicates the level of the individual's net income, where net income is that remaining for the purchase of non-spatial goods and services after paying residential rentals and travel expenses. Income level Yb is the base income or the minimum subsistence that will make existence at point O tolerable. With net income of Yb the individual would still enjoy some satisfaction from the residential amenities at O, but all remaining income would be used for mere physical subsistence. As net income rose from Yb, the total satisfaction would increase as more goods and services could be consumed along with residential amenities.

If an individual lives at his point O, but is forced into frequent travel to other human activities, his total satisfaction would diminish due to the expenditure of his time and effort. The horizontal OA axis indicates the use of time and effort in travel. The slope of the edges of the indifference surfaces directly above the OA axis indicate the marginal rate of substitution of income for accessibility (defined as the amount of income that would be sacrificed to avoid one unit of travel, or to attain that degree of accessibility indicated by a unit of travel). If an individual were forced to travel a distance (weighted by time and effort) indicated by OX',

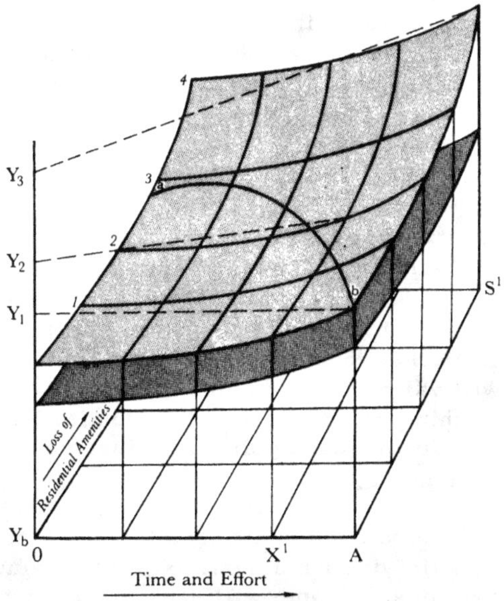

Figure 1. A residential indifference surface.

he could still enjoy the residential amenities at point O, but his income would have to increase if he were to remain on the same indifference surface and be no worse off than before. In the same manner, the distance along all axes parallel to OA represent the expenditure of time and effort made while living at locations other than O.

The adjacent, diagonal axis OS (parallel to AS') indicates the loss of residential amenities if location is changed to sites less preferred than O. As one moves from his most desired point, his satisfaction decreases with the loss of residential amenities and his income must necessarily rise if he is to receive the same total enjoyment from the over-all consumption package. The slope of the edges of the indifference surface above the OS axis indicates the marginal rate of substitution of income for residential amenities (defined as the amount of income that would be sacrificed to gain one degree of residential amenities as measured along the OS axis).

The indifference surfaces are extensions of the two sets of indifference curves and allow for substitution among income, residential amenities, and accessibility (or avoidance of using time and effort). The individual would always prefer a higher surface, but he would be indifferent as to the position he occupies on any one surface.

For example: living at point O with an income of Y_1 and traveling to use time and effort indicated by OA would be equal to living at point 2 with an income of Y_2 and traveling OX', or living at 4 with an income of Y_3 and traveling OA. Any of these combinations would provide equal satisfaction and would be superior to any combination lying on a lower surface.

If an individual is to change his location in space, and not vary his level of satisfaction, some substitution must take place between at least two of the factors involved. Thus, any movement along, or parallel to, the intersecting grid line would involve a substitution between income and accessibility or income and residential amenities. Any movement along a line such as AS, which denotes a constant level of income, would involve a substitution between residential amenities and accessibility. Other locational adjustments that intersect one of an infinite series of these lines and axes would involve a substitution among all three factors involved. In this way, a consumer faces a series of alternative substitutions in adjusting his spatial location to reach an optimum position.

THE COMPOSITE FORCES OF RESIDENTIAL SELECTION

The model is not introduced with the intention of yielding a specific equilibrium solution. Rather, it does call attention to significant factors that are involved in residential location. Three areas seem to stand out. First, there is a need to identify those residential amenities that structure residential preferences. Second, it would seem that the marginal rate of substitution will vary among various groups of population. Third, the principle of diminishing rates of marginal substitution implies that significant variation will occur among different income conditions, value judgments, and distance factors.

Complete and systematic understanding of those residential amenities that influence decisions could only be attained by following a difficult, if not impossible road. A list of such amenities would include both natural as well as social attractions. People will seek natural views,[18] spacious yards,[19] lakes and other water bodies.[20] In addition, people seek social attributes that lend prestige and provide other satisfactions. Just as "nature filled" space may distort the agricultural location model, so may "cultural filled" space distort the urban location model. The primary factors of residential location may well fall into Isard's second group of locational factors.[21] Within this group is included the "social and political milieu." If significant factors do fall into this group they still may not be ignored, but should be examined for process and effect.

When attention is given to the implications of differing marginal rates of substitution and the accompanying principle of diminishing rates of substitution, more definite statements might be made. These principles would imply that different groups would behave differently. This is to say, people would segregate themselves spatially according to a series of groups with which they either directly identify themselves or indirectly share similar residential values.

Such group segregation might stem from people in similar income circumstances who evaluate the alternatives of amenities, accessibility, and income in a like manner to some and in a different manner than others. An example of varying intensities of demand (or differing marginal rates of substitution) among residential amenities, accessibility, and income may be seen by viewing the residential locations of certain occupational groups. There are certain occupations that are less exacting in their time demands. Fixed working hours and a lower emphasis upon punctuality may allow proprietors, managerial employees, and other day and hourly workers to have greater freedom in their choice of residential selection. Other occupational positions place a greater stress upon the limited time of the individuals involved. The latter groups might or might not include college professors, but would most certainly include physicians. In the case of physicians, frequent off-hour calls and the emergency nature of such calls make accessibility to hospitals and offices a primary concern in selecting a home.

In the city of Baton Rouge, Louisiana, the residences and work places of physicians, college faculty, attorneys, and managers and supervisors of one large corporation were obtained from telephone directories. The average airline distance between homes and offices were as follows:

Physicians	1.8 miles
College faculty	2.1 miles
Attorneys	3.0 miles
Managers and supervisors	5.2 miles

A second feature accounting for group segregation would seem to stem from the fact that people in different income circumstances would be more willing to sacrifice income for additional residential amenities or accessibility. Numerous examples demonstrate the tendency for high income groups to segregate themselves. Some evidence is available to indicate that high income groups typically not only sacrifice income for residential amenities but sacrifice accessibility as well.[22]

There are other data indicating that in some instances high income groups are purchasing accessibility to central work places, which would appear to come at a high price if no great sacrifice is to be made in housing quality or residential amenities. One would anticipate that accessibility desires would vary over time in large, growing urban areas. As the population increases and the city grows outward by accretion, the problem of accessibility becomes more important to urban inhabitants.

130

Table 1. Coefficients of Localization by Classification and Groups in Census Tracts, Baton Rouge, Louisiana, 1960

Classification base	Group	Coefficient of localization
I. Family income (Total families)		
	a. $4,000-$5,999	8.1
	b. $6,000-$7,999	13.9
	c. $8,000-$9,999	17.6
	d. $10,000-$14,999	23.3
	e. $2,000-$3,999	23.5
	f. less than $2,000	28.8
	g. $15,000-$24,999	43.4
	h. over $25,000	51.9
II. Industry (Total employed)		
	a. Retail trade	8.2
	b. Transportation and utilities	11.4
	c. Mining and construction	11.5
	d. Wholesale trade	13.7
	e. Personal and business services	16.7
	f. Manufacturing	18.1
III. Occupation (Total employed)	a. Clerical workers	14.6
	b. Sales workers	18.1
	c. Craftsmen and foremen	18.3
	d. Operative workers	19.7
	e. Occupation not reported	22.3
	f. Managers and proprietors	25.1
	g. Professional and technical	28.4
	h. Service (excluding private households)	29.5
	i. Laborers	38.9

Source: U. S., Bureau of the Census, 1960.

Table 2. Concentration Index for Family Income Group, Baton Rouge, Louisiana, 1960

Income group	Centralization index	Job availability index	Low rent index
Less than $2,000	13.7	18.6	57.6
2,000 - 3,999	13.4	17.6	23.1
4,000 - 5,999	7.3	9.9	10.1
6,000 - 7,999	1.5	0.1	-4.2
8,000 - 9,999	-1.1	-12.1	-18.9
10,000 - 14,999	-1.8	-21.8	-22.2
15,000 - 24,999	-3.1	-40.5	-41.6
over 25,000	-3.6	-50.5	-46.3

Source: U. S., Bureau of the Census, 1960.

As greater distances must be covered in reaching primary activity centers, certain individuals become willing to substitute amenities on the
outer fringe to satisfy their desire for accessibility to the inner core.[23]

Table 1 shows coefficients of localization or segregation when individuals are placed in various socio-economic groupings. The coefficient is computed by adding the positive differences obtained by subtracting the percentage of the base classification in each census tract from
the percentage of the group in each tract. The possible range of the coefficient is from 0 to 100.0; the higher the value of the coefficient the
more the group is localized or concentrated relative to the base classification.[24]

Table 2 presents indexes of concentration of income groups in areas
of the city distinguished by a) distance from the central city (Centralization Index), b) estimated jobs available (Job Availability Index) and c)
median rentals of occupied dwellings. The indexes were obtained by
classifying tracts according to distance from the central city, the number of estimated jobs available and low-to-high areas, respectively;
computing by intervals the percentage distribution of all families and
each family income group; cumulating the distributions; and calculating
$\Sigma(X_{i-1}Y_i) - \Sigma(X_iY_{i-1})$, where X_i is the cumulated percentage of each
income group to each interval and Y_i is the cumulated percentage of all
families. The index varies from 100 to -100. Positive values indicate
concentration towards the central city, job availability, or low-rent
areas; and negative values, concentration away from such areas.[25]

Table 3 is a distribution, or association, index obtained in a manner
similar to that used in Table 1. The index is the sum of the positive
differences obtained in each census tract by subtracting the percentage
of each income group from the percentage of each of the other classifications. This value is subtracted from 100 to give values from 0 to 100,
with the higher value indicating equal percentage distribution of the two
compared groups in the tracts and 0 indicating no association in any
tract.

One of the problems involved in many studies relating residential
location to distance from work is that people are grouped by place of
work, or industry of employment. This classification arises from the
alternative difficulties of obtaining a general measure of accessibility
for each and all urban locations, or soliciting the distance to work on an
individual basis.

There is, however, ample evidence that people are more prone to
localize or segregate themselves by other factors than their industry or
place of employment. The position they hold would appear to be more
significant, since classification of workers by occupational types show
greater tendencies of residential segregation. Group classifications by
income, age, race, and other socio-economic characteristics will, on
occasion, also display high degrees of residential segregation.[26]

A greater tendency to segregate according to occupation as opposed
to industry is shown in Table 1. In Baton Rouge the two most dispersed
occupational groups approximate the degree of localization of the most
localized industrial classes. Industrial groups range in coefficients
from 8.2 to 18.1. Occupational groupings have a higher and broader
range from 18.1 to 38.9. The latter also have a standard deviation of
7.2, compared to that for industry groups of 3.4, indicating a greater
variation from the established distribution of employees.

The classification base with the most striking example of localization is family income groups. A range from 8.1 to 51.9 and a standard deviation of 13.9 indicate a pattern of varying degrees of localization. Segregation, or localization, is highest for the high (over $15,000) income groups and least evident in the middle ($4,000 to $9,999) income families. Such a pattern indicates the significance of income limitations and economic forces in residential selection.

In general, each individual will be more concerned with economizing that resource of time or money which is dearest to him. In this respect he may well be on a threshold of indifference concerning remaining alternatives.

Table 3. Distribution (Association) Index for Family Income Groups and Socio-economic Groups, Baton Rouge, Louisiana, 1960

Family income group	Age			
	20-34	35-49	50-64	Over 65
Less than $2,000	67.4	75.7	70.5	78.5
$2,000 - 3,999	72.0	69.1	44.0	78.3
4,000 - 5,999	84.4*	84.3*	73.2*	78.4
6,000 - 7,999	62.0	82.5	62.4	80.0*
8,000 - 9,999	63.1	81.8	59.1	77.7
10,000 - 14,999	66.9	76.8	57.7	74.2
15,000 - 24,999	47.0	56.8	46.9	55.6
over 25,000	38.6	46.6	39.6	46.7

Family income group	Owner-occupied housing units	Children under 6 years	Children under 18 years	People walking to work
Less than $2,000	62.1	65.1	66.9	72.8*
$2,000 - 3,999	64.9	76.5	69.2	67.5
4,000 - 5,999	82.6	85.9	84.1	60.7
6,000 - 7,999	87.7*	88.2*	88.5*	48.3
8,000 - 9,999	84.1	83.2	84.3	42.9
10,000 - 14,999	79.2	75.0	76.0	40.9
15,000 - 24,999	67.2	53.7	55.2	26.1
over 25,000	47.4	54.3	46.3	26.7

Source: U. S., Bureau of the Census, 1960

*Indicates highest column figure.

The Concentration Indexes show low-income groups primarily located in low-rent districts, somewhat near the central city, but with equal emphasis on job availability. In addition, Table 3 indicates a greater tendency to walk to work as income decreases. The explanation for these patterns is probably found in data indicating that these income groups are primarily laborers engaged in service industries as well as in their lower income. This data would seem to indicate that low-income groups will attempt to conserve on residential amenities in order to obtain a sufficient supply of non-residential necessities. Such groups would thus tend to be highly concentrated in low-cost housing, generally found in either the belt of transition from intensive land use near the central city and industrial work places or along the outer fringe in the transition from urban to agriculture.[27] The extent to which individuals would locate in each area would depend on relative values of residential amenities and accessibility, but many buyers among the low-income group see no great distinction among houses within their income range. "When you see one house, you've seen them all," becomes a common complaint.[28] One would expect that the larger the city and the greater the distance to activity centers the more low-income groups would select housing near job opportunities. Money travel costs would also induce these groups to select places accessible to work.

Median income groups would have more freedom to seek a wider variety of residential amenities in a mix of some luxuries in their non-residential consumption. Katona has pointed out the increasing importance of "better homes" and "better locations" in consumer desires that is of particular importance in families having incomes above $3,000, younger members, and holding expectations of being in an improved income position in the future.[29]

In seeking important, but not highly luxurious, residential amenities, such individuals would be more likely to attach themselves to other groups seeking similar residential attributes. If single, they may seek temporary quarter with others in similar circumstances. Married couples with young children may seek spacious and desirable neighborhoods with good schools.

Though there is no little significant variation in Table 3, it is still indicative that middle income groups (particularly $4,000 to $9,999) have a higher degree of association with all the compared classifications. Such data are an indication of more homogeneous neighborhoods sought out for their residential amenities.

The Concentration Indexes for middle-income groups indicate that they are less concerned with accessibility than are lower income families. The Coefficients of Localization also indicate a pattern of relative dispersal throughout the city. Such families are able to afford some degree of residential amenities, but are not likely to purchase the added luxury of locating these amenities on land surrounding activity centers. Thus, they are more likely to locate in a random pattern seeking residential amenities available within their income range.

High income groups have the ability to attain luxury items which would include high-valued residential amenities. To the extent that high quality housing is one of the luxuries of consumers, a high degree of residential concentration would occur as is indicated by the Coefficient of Localization.[30] A second luxury of high incomes could be increased accessibility.[31] This becomes a luxury item for such groups because it is often necessary to clear valuable land to combine access with quality housing and neighborhoods. While such behavior is not shown in either the Centralization Index or the Job Availability Index, it is hinted at in the low association between the highest income group and owner occupied homes. This drop in the index is primarily due to high value apartments, often located near the central city and other activity centers. The inclusion of many professions within this group would also allow for accessibility to be attained by the location of offices more adjacent to residences. High income groups would thus appear to have highly localized areas of residential location, but the degrees of accessibility would depend upon the separate use evaluations among the various members.

The above analysis of income group behavior would also explain other classifications of residential localizations. Occupational categories would be high because income restraints are more often determined by an individual's occupation rather than the industry in which he is either employed or engaged. In addition, similar occupational categories are more likely to share similar value judgments concerning residential amenities as well as accessibility and non-residential consumption. Finally, it would seem that among similar occupations certain social and economic values would arise from a closer residential association.

The use of industrial group localization coefficients would seem to indicate the importance of accessibility in residential location. The coefficients are highest for manufacturing and wholesale workers and lowest for retail workers. By the same token, manufacturing and, to a lesser extent, wholesale houses are concentrated in particular areas of the city. Retailing, however, is carried on in the central city and in highly dispersed centers. To the extent that accessibility is paramount in location, retail workers would naturally be more scattered in all areas. Analysis would thus appear to vary according to the method of classification, but a composite would seem to indicate the complexity of forces at work in residential selection.

SUMMARY

The concept of residential preferences only serves as a base on which to build a framework for analyzing residential location. The use of a consumer theory would, however, seem to provide a better background for understanding than the locational type of analysis. The evidence would seem to validate the supposition that consumers are seeking the three goals of maximum residential amenities, accessibility, and non-residential income. More significantly, evidence indicates that residential spatial structures are more complex than is often assumed. Future efforts need to be made along paths that explore residential adaptation to the social environment as well as to the economic and spatial environment.

5

NOTES

[1]Charles M. Tiebout, "Intra-urban Location Problems: An Evaluation," American Economic Review, LI (1961), pp. 271-78.

[2]George B. Galloway, Planning for America (New York: Henry Holt and Co., Inc., 1941), p. 3.

[3]One might include as a third category, or as some combination of the above, cartographic generalizations of typical urban land-use patterns. For examples, see: Robert E. Park and Ernest W. Burgess, The City (Chicago: University of Chicago Press, 1925), Chapter 2, pp. 47-62; Chauncy D. Harris and Edward L. Ullman, "The Nature of Cities," Annals of the Academy of Political and Social Science, CCXLII (1945), pp. 7-17; Homer Hoyt, The Structure and Growth of Residential Neighborhoods (Washington: Federal Housing Administration, 1939).

[4]Leo F. Schnore, "The Separation of Home and Work: A Problem for Human Ecology," Social Forces, XXXIII (1954), pp. 336-43.

[5]J. Douglas Carroll, Jr., "Home-Work Relationships of Industrial Employees" (Harvard University, 1950); "Some Aspects of Home-Work Relationships of Industrial Workers," Land Economics, XXIV (1949), pp. 414-22; "The Relation of Homes to Work Places and the Spatial Pattern of Cities," Social Forces, XXX (1952), pp. 271-82.

[6]Edgar M. Hoover and Raymond Vernon, The Anatomy of a Metropolis (Cambridge: Harvard University Press, 1959), pp. 127-52.

[7]Carroll, op. cit.; Hoover and Vernon, op. cit.; Thurley Bostick, et al., "Motor Vehicle-Use Studies in Six States," Public Roads, XXVIII, pp. 99-125; Leonard P. Adams and Thomas Mackesey, Commuting Patterns of Industrial Areas (Ithaca: Cornell University, 1955).

[8]Schnore, op. cit.; Adams and Mackesey, op. cit.; John M. Richards, "Spatial and Nonspatial Factors Affecting Residential Location" (El Paso: an unpublished paper, 1962).

[9]See: Peter H. Rossi, Why Families Move (Glencoe, Ill.: The Free Press, 1955); Edward T. Paxton, What People Want When They Buy A House, U.S. Department of Commerce (Washington: U.S. Government Printing Office, 1955); Richard Dewey, "Peripheral Expansion in Milwaukee County," American Journal of Sociology, LIV (1948), pp. 417-22; Theodore Caplow, "Home Ownership and Locational Preferences in a Minneapolis Sample," American Sociological Review, XIII (1948), pp. 725-30; George Katona and Eva Mueller, Consumer Expectations: 1953-1956 (Ann Arbor: University of Michigan, 1956), pp. 76-90.

[10]See: Adams and Mackesey, op. cit., p. 66; Caplow, op. cit., p. 727; Rossi, op. cit., pp. 82 and 85; Katona, op. cit., pp. 125-26 and 130-37.

[11]Rossi, op. cit., p. 9.

[12]Katona, op. cit., p. 125.

[13]Caplow, op. cit., p. 727; Dewey, op. cit., p. 422.

[14]William Goldner, "Spatial and Locational Aspect of Metropolitan Labor Markets," American Economic Review, XLV (1955), pp. 113-28.

[15]For some examples, see: Lloyd G. Reynolds, The Structure of Labor Markets (New York: Harper and Brothers, 1951), p. 52; Indiana Employment Security Division, Indiana State Employment Service, "Survey of Commuting Patterns in Selected Indiana Areas," (a mimeographed report, Indianapolis, Indiana); Roy Gerard, "Commuting and the Labor Market," Journal of Regional Science, I (1958), pp. 124-125; Richard Dewey, op. cit., p. 422; Peter H. Rossi, op. cit.

[16]Walter Isard, Location and Space-Economy (New York: John Wiley & Sons, 1956), pp. 22-23; 84-85; 287.

[17]Ibid., p. 84.

[18]Leonard M. Cowley, "The Value of View," The Appraisal Journal, 1951, pp. 235-242.

136

[19]Paxton, op. cit., p. 16: Hoover and Vernon, op. cit., pp. 220-221.

[20]Hoyt, op. cit., p. 114.

[21]Isard, op. cit., p. 138.

[22]Herbert S. Parnes, A Study in the Dynamics of Labor Force Expansion (Columbus: Ohio State University Research Foundation, 1951), p. 175. Parnes' study shows that, with two exceptions, as the average wage rose from less than 80 cents per hour to above $1.70, the median distance from plant to home rose from 3.5 miles to 8.78 miles.

A rough random sample in Baton Rouge of 450 of 2,171 (20.7%) city census blocks gave the following information concerning the median value of homes within the block and the average distance from the central business district:

less than $10,000	1.2 miles
over $10,000, but less than $15,000	1.4 miles
over $15,000, but less than $20,000	1.7 miles
over $20,000, but less than $25,000	2.2 miles
over $25,000	3.3 miles

See, also, the discussion in Hoover and Vernon, op. cit., pp. 163-176.

[23]Hoover and Vernon, op. cit., p. 176, point out that the population of the city's inner core is primarily composed of both high and low income groups. Also, see the general discussion in: Chester Rapkin and William Grigsby, Residential Renewal in the Urban Core (Philadelphia: University of Pennsylvania Press, 1960); and John M. Richards, "Residential Preference, Residential Location and Home-Work Separation," an unpublished dissertation, Louisiana State University, 1961, pp. 231-235; L. G. Reeder, "Social Differentials in Mode of Travel, Time and Cost in the Journey to Work," American Sociological Review, XXL (1956), p. 61.

[24]See: Walter Isard, ed., Methods of Regional Analysis (New York: John Wiley & Sons, 1960), pp. 249-262; Otis D. Duncan and Beverly Duncan, "Residential Distribution and Occupational Stratification," American Journal of Sociology, LX (1955), pp. 493-503.

[25]Duncan and Duncan, ibid.

[26]Ibid.; Hoover and Vernon, op. cit., pp. 154-155.

[27]Walter Firey, "Ecological Considerations in Planning for Urban Fringes," Cities and Society, Paul Hatt and Albert Reiss, editors (Glencoe, Ill.: The Free Press, 1957), pp. 793-794.

[28]Paxton, op. cit., p. 12.

[29]Katona, op. cit., pp. 121-126.

[30]Margaret G. Reid, "Effect of Income Concept Upon Expenditures Curves of Farm Families," Studies in Income and Wealth: Vol. 15 (New York: National Bureau of Economic Research, 1952), pp. 133-174.

[31]Reder, op. cit., p. 61; Hoover and Vernon, op. cit., pp. 163-176.

Differences Among Consumers Attributable to Locations

What are unique characteristics of urban spending? - Data from Michigan Survey Research Center arranged by degree of urbanization - The variables - Analysis of housing status and mobility, travel experience, and other variables including consumer debt payments and expenditure of public funds - Conclusions.

Increasing interest in urban economic problems raises the general question: Which problems are uniquely identified with urban living? Differences in many characteristics between urban and rural families have been observed over long periods, but it does not necessarily follow that the observed differences can be attributed to differences in location. This paper attempts to identify some observed differences between consumers in different locations as differences caused by location or attributable to other variables which affect consumer behavior. The results of the analysis suggest that location may be directly associated with differences among consumers with respect to home ownership, residential mobility, and some attitudes toward the use of tax revenues. Apparent differences between consumers with respect to some other variables can be attributed to characteristics other than location.

THE INDEPENDENT VARIABLES

The analysis is based on data from a cross-section sample survey conducted by the Survey Research Center in October, 1960. One of the location variables available in the survey results was developed by Bernard Lazerwitz and is based upon a set of "belts" surrounding the central city of a Standard Metropolitan Area.[1]

Author's note: The preparation of this paper has been supported by a research grant from Resources for the Future, Inc. The Survey Research Center, University of Michigan, provided access to data and assistance in processing the data. The University of Kentucky Computing Center assisted in calculation. I am grateful to these three institutions for their generous aid.

137

The areas, or "belts," are defined as follows:

1. <u>Largest central cities (LCC)</u> includes residents of cities of 250,000 or more which are contained within the twelve largest Standard Metropolitan Areas.

2. <u>Other central cities (OCC)</u> includes residents of cities of 250,000 or more which are contained within Standard Metropolitan Areas other than the twelve largest.

3. <u>Suburbs of largest central cities (SLC)</u> includes residents of counties which contain largest central cities but who are located outside the city limits. When the central city is its own county, the suburbs include the adjoining counties.

4. <u>Suburbs of other central cities (SOC)</u> includes residents of counties which contain other central cities but who are located outside the city limits.

5. <u>Adjacent areas (AA)</u> includes residents of counties the geographical centers of which are within fifty miles of the center of a central city.

6. <u>Outlying areas (OA)</u> includes residents of counties which have geographical centers more than fifty miles away from the centers of central cities.

The above definition of a locational variable may have operational significance if the variable assists in distinguishing between families who differ in their characteristics or behavior. Since it has been well established, however, that differences among consumers can frequently be associated with differences in income, age, education, and family size, these four variables have been chosen as independent variables along with location. The four variables have been defined as follows:

<u>Income (Y)</u> is measured as it was coded in the survey data and the code approximates a transformation of the logarithm of dollar income. The coded dollar values are:

10.	Under $1,000	16.	6,000 - 7,499
11.	$1,000 - 1,999	17.	7,500 - 9,999
12.	2,000 - 2,999	18.	10,000 - 14,999
13.	3,000 - 3,999	19.	15,000 - 19,999
14.	4,000 - 4,999	20.	20,000 and over
15.	5,000 - 5,999		

<u>Age (A)</u> is measured approximately in decades as follows:

1.	18 - 24	4.	45 - 54
2.	25 - 34	5.	55 - 64
3.	35 - 44	6.	65 and over

<u>Education (E)</u> attempts to combine the values of grades completed and non-academic training in the following arbitrary form:

0. Grade school, none
2. Some high school
3. Some high school plus non-academic
5. Completed high school
6. Completed high school plus non-academic
7. Some college
9. College degree

<u>Number of children (C)</u> is measured in terms of the actual number reported, under age 18, and ranging from zero to 8 for eight or more children.

Table 1 shows the means of coded values of income, age, education, and number of children for each of the six areas. As expected, incomes are higher in the suburbs of large cities than in all other areas and lower in outlying areas than in all other areas. The average level of education is higher among families living in suburbs of large central cities than among families in other areas. The families living in suburbs of other central cities tend to be younger and have more children compared to families in other areas.

It should be noted that Table 1 and all following tables must be inter-

Table 1. Differences between Means of Independent Variables for Pairs of Areas*

Income

Areas	Mean	LCC	OCC	SLC	SOC	AA	OA
Largest central cities (LCC)	14.44	x					*
Other central cities (OCC)	14.50		x				*
Suburbs of largest central cities (SLC)	15.97	*	*	x	*	*	*
Suburbs of other central cities (SOC)	15.02	*	*		x	*	*
Adjacent areas (AA)	14.24					x	*
Outlying areas (OA)	13.39						x

Age

Areas	Mean	LCC	OCC	SLC	SOC	AA	OA
Largest central cities (LCC)	3.67	x			*		
Other central cities (OCC)	3.88		x		*		
Suburbs of largest central cities (SLC)	3.74			x	*		
Suburbs of other central cities (SOC)	3.33				x		
Adjacent areas (AA)	3.85				*	x	
Outlying areas (OA)	3.78				*		x

Education

Areas	Mean	LCC	OCC	SLC	SOC	AA	OA
Largest central cities (LCC)	3.37	x					
Other central cities (OCC)	3.77		x			*	*
Suburbs of largest central cities (SLC)	5.00	*	*	x	*	*	*
Suburbs of other central cities (SOC)	4.26	*			x	*	*
Adjacent areas (AA)	2.88					x	
Outlying areas (OA)	2.97						x

Number of Children

Areas	Mean	LCC	OCC	SLC	SOC	AA	OA
Largest central cities (LCC)	.94	x					
Other central cities (OCC)	1.00		x				
Suburbs of largest central cities (SLC)	1.21			x			
Suburbs of other central cities (SOC)	1.70	*	*	*	x	*	*
Adjacent areas (AA)	1.38	*	*			x	
Outlying areas (OA)	1.28	*	*				x

*Asterisks in this and succeeding tables indicate a significant difference, at the 5% level, between the mean for the area indicated by the row and the mean for the area indicated by the column. The asterisks are placed so that the row mean is the higher. For example, under income, the mean for SLC is significantly higher than all other means, and the mean for OA is significantly lower than all other means. The "pooled variance" method for testing differences between means has been used in all tests.

preted with some caution. Although formal procedures for testing statistical hypotheses have been followed for analytical purposes, some limitations must also be considered. First, the testing procedures are based on simple random samples; no correction has been made for the

140

effect of stratification and clustering in the Survey Research Center's sampling procedures. Second, a large number of "t" tests have been performed in the analysis, and errors of both Type I and Type II are surely present.

The differences between areas noted in Table 1 are to be used in interpreting differences in other dependent variables in the following sections. Table 2 summarizes Table 1 in schematic form indicating the variables and areas for which differences were observed. The representation of differences between areas with respect to a given independent variable by a letter (Y, A, E, or C) is carried forward to other tables below.

Table 2. Composite Differences between Areas on Income (Y), Age (A), Education (E), Number of Children (C)

Areas	LCC	OCC	SLC	SOC	AA	OA
Largest central cities (LCC)	X			A		Y
Other central cities (OCC)		X		A	E	YE
Suburbs of largest central cities (SLC)	YE	YE	X	YAE	YE	YE
Suburbs of other central cities (SOC)	YEC	YC	C	X	YEC	YEC
Adjacent areas (AA)	C	C		A	X	Y
Outlying areas (OA)	C	C		A		X

Table 3. Home Ownership*

Area	Relative Frequency	LCC	OCC	SLC	SOC	AA	OA
Largest central cities (LCC)	.32	x					
Other central cities (OCC)	.61	*	x				
Suburbs of largest central cities (SLC)	.78	*YE	*YE	x			*YE
Suburbs of other central cities (SOC)	.75	*YEC	*YC		x		*YEC
Adjacent areas (AA)	.76	*C	*C			x	*Y
Outlying areas (OA)	.62	*C					x

Area	R²	Intercept Frequency	Change in Frequency per Unit of:			
			Income	Age	Education	No. of Children
LCC	.16	-.930	.055	.093	.021	.042
OCC	.16	-.711	.068	.073	.000	.058
SLC	.20	-.563	.055	.121	-.009	.050
SOC	.17	-.548	.065	.100	-.007	.005
AA	.11	-.233	.048	.079	-.001	.008
OA	.19	-.722	.062	.121	-.002	.045

*See Table 1 for interpretation of asterisks. Underlined coefficients are significantly different from zero at the 5 per cent level.

HOUSING STATUS AND MOBILITY

Home ownership is much lower in frequency in the largest central cities than in other areas. The frequency is high in the suburban and adjacent areas. Table 3 shows the differences in frequency of home ownership among the six areas and also carries forward, from Table 2,

notation on pairs of areas which differ with respect to income, age,
education, and number of children. The lower part of Table 3 reports
the results of regression equations relating home ownership to the four
independent variables in each of the six areas. Home ownership is correlated with income and age in all areas and with the number of children
in some areas. The low frequency of ownership in large central cities
relative to the suburban areas could be attributed to the higher levels of
income in suburban areas. When the frequencies are compared between
large central cities and other areas, however, no alternative explanation is obvious and it is concluded that there may be some intrinsic
characteristics of housing in large central cities which result in low frequency of home ownership.

This result is surely not surprising to the casual observer of very
large cities. Thus the only value claimed for the finding is that it tends
to give some credibility to the analytical procedure.

The number of years' residence in the currently occupied dwelling
increases with age in all areas, increases with income in large central
cities, and decreases with family size in suburbs of other central cities.
Given that the residents of suburbs of other central cities are younger
families, it is perhaps surprising to find, in Table 4, no significant differences among areas with respect to length of residence.

Table 4. Length of Residence*

Area	Means	LCC	OCC	SLC	SOC	AA	OA
Largest central cities (LCC)	11.10						
Other central cities (OCC)	11.50						
Suburbs of largest central cities (SLC)	11.51			No			
Suburbs of other central cities (SOC)	10.64			differences			
Adjacent areas (AA)	12.22						
Outlying areas (OA)	12.39						

				Change in Residence Length per Unit of:		
Area	R^2	Intercept	Income	Age	Education	No. of Children
LCC	.25	- 19.575	.883	4.555	.074	.987
OCC	.20	- 14.287	.541	4.377	.376	- .461
SLC	.21	1.074	.052	3.394	- .520	- .365
SOC	.36	- 1.633	-.110	4.679	.077	-1.157
AA	.22	- 8.926	.489	4.038	- .355	- .233
OA	.28	- .830	-.109	4.116	- .191	- .238

*Underlined coefficients are significantly different from zero at the
5 per cent level.

The data on tendency to move, reported in Table 5, are based on
replies to several different questions in the survey. Respondents were
asked whether they expected to buy or build a house in the next year or
in the following year. If the replies to these questions were negative,
they were asked if they expected to be living in the same place a year
later. If the reply was "Yes," they were asked if they would like to stay
in their current residence, no time period specified. The responses to
these questions were combined for this analysis to differentiate between
those families with some tendency to move and those with no tendency to
move.

The residents of large central cities show a greater tendency to
move than families in adjacent or outlying areas. Age is the principal
explanatory variable and there are no differences between residents of

Table 5. Tendency to Move*

Area	Relative Frequency	LCC	OCC	SLC	SOC	AA	OA
Largest central cities (LCC)	.50	x				*	*Y
Other central cities (OCC)	.44		x				
Suburbs of largest central cities (SLC)	.40			x			
Suburbs of other central cities (SOC)	.46				x		
Adjacent areas (AA)	.37					x	
Outlying areas (OA)	.38						x

Area	R^2	Intercept Frequency	Change in Frequency per Unit of:			
			Income	Age	Education	No. of Children
LCC	.06	.955	- .012	- .072	- .013	.028
OCC	.12	1.186	- .014	- .132	.001	- .039
SLC	.09	1.250	- .047	- .064	.022	.023
SOC	.16	1.168	- .030	- .113	.013	.033
AA	.10	.668	- .007	- .075	.023	.022
OA	.11	.923	- .012	- .097	.014	- .041

*See Table 1 for interpretation of asterisks. Underlined coefficients are significantly different from zero at the 5 per cent level.

these three areas with respect to age. The greater mobility among families in large central cities is therefore an apparently unique characteristic of these areas.

TRAVEL EXPERIENCE

Other aspects of mobility are covered in several questions included in the survey and relating to the family's experience with various types of travel. Questions were asked about the head of the household (the husband in most cases) concerning travel abroad and travel by air. Other questions were asked about the respondent (in some cases the husband) concerning recent travel by train or bus and long travel by auto.

The relative frequency of having traveled abroad is high in the largest central cities and their suburbs and it is low in adjacent and outlying areas. The regression analysis shows mixed results. The frequency of foreign travel increases inconsistently among areas with income, age, and education. The residents of large central cities and their suburbs, when compared with areas in which foreign travel is relatively infrequent, tend to have higher incomes, and/or tend to be older, and/or tend to have a higher education. Thus there seems to be no intrinsic characteristic of location which is associated with a higher frequency of foreign travel.

Some experience with air travel is reported more frequently by residents of central cities and suburban areas than by families in adjacent and outlying areas. Again, however, this apparent difference cannot be attributed to location. The regression coefficients indicate that air travel is related to income and education, and the outlying and adjacent areas tend to have lower levels of income and education than the central cities and suburbs.

The residents of large central cities and their suburbs report a recent train trip of more than 100 miles more frequently than residents

Table 6. Travel Overseas*

Area	Relative Frequency	LCC	OCC	SLC	SOC	AA	OA
Largest central cities (LCC)	.11	x				*	*Y
Other central cities (OCC)	.06		x				
Suburbs of largest central cities (SLC)	.17	*YE		x	*YAE	*YE	*YE
Suburbs of other central cities (SOC)	.07				x		
Adjacent areas (AA)	.04					x	
Outlying areas (OA)	.04						x

Area	R^2	Intercept Frequency	Change in Frequency per Unit of: Income	Age	Education	No. of Children
LCC	.07	- .066	.002	.035	.014	- .031
OCC	.03	- .194	.007	<u>.033</u>	.008	.000
SLC	.06	- .357	.022	<u>.043</u>	.008	- .021
SOC	.14	- .355	.013	<u>.054</u>	<u>.015</u>	- .011
AA	.10	- .367	<u>.019</u>	<u>.025</u>	.009	.006
OA	.04	- .178	<u>.012</u>	.008	.005	.009

*See Table 1 for interpretation of asterisks. Underlined coefficients are significantly different from zero at the 5 per cent level.

Table 7. Travel by Air*

Area	Relative Frequency	LCC	OCC	SLC	SOC	AA	OA
Largest central cities (LCC)	.29	x					*Y
Other central cities (OCC)	.30		x			*E	*YE
Suburbs of largest central cities (SLC)	.49	*YE	*YE	x	*YAE	*YE	*YE
Suburbs of other central cities (SOC)	.35				x	*YEC	*YEC
Adjacent areas (AA)	.21					x	
Outlying areas (OA)	.19						x

Area	R^2	Intercept Frequency	Change in Frequency per Unit of: Income	Age	Education	No. of Children
LCC	.20	.038	.013	- .026	<u>.061</u>	- .048
OCC	.19	- .429	<u>.044</u>	- .005	<u>.038</u>	- .031
SLC	.25	- .904	<u>.063</u>	.029	<u>.056</u>	- .005
SOC	.28	- .871	<u>.057</u>	.048	<u>.053</u>	- .008
AA	.19	- .406	<u>.037</u>	.003	<u>.045</u>	- .030
OA	.19	- .378	<u>.037</u>	- .002	<u>.035</u>	- .020

*See Table 1 for interpretation of asterisks. Underlined coefficients are significantly different from zero at the 5 per cent level.

144

Table 8. Recent Train Travel*

Area	Relative Frequency	LCC	OCC	SLC	SOC	AA	OA
Largest central cities (LCC)	.13	x				*	Y*
Other central cities (OCC)	.08		x				
Suburbs of largest central cities (SLC)	.11			x			YE*
Suburbs of other central cities (SOC)	.08				x		
Adjacent areas (AA)	.06					x	
Outlying areas (OA)	.06						x

Area	R^2	Intercept Frequency	Change in Frequency per Unit of:			
			Income	Age	Education	No. of Children
LCC	.06	.344	- .018	- .011	.028	- .015
OCC	.06	- .386	.017	.043	.010	.009
SLC	.06	- .179	.014	- .011	.019	- .001
SOC	.05	- .011	.013	- .012	.000	- .034
AA	.07	- .290	.025	.003	.001	- .015
OA	.04	.025	- .002	.003	.017	- .001

*See Table 1 for interpretation of asterisks. Underlined coefficients are significantly different from zero at the 5 per cent level.

Table 9. Recent Bus Travel*

Area	Relative Frequency	LCC	OCC	SLC	SOC	AA	OA
Largest central cities (LCC)	.08	x					
Other central cities (OCC)	.09		x	*			
Suburbs of largest central cities (SLC)	.04			x			
Suburbs of other central cities (SOC)	.06				x		
Adjacent areas (AA)	.05					x	
Outlying areas (OA)	.09			*			x

Area	R^2	Intercept Frequency	Change in Frequency per Unit of:			
			Income	Age	Education	No. of Children
LCC	.03	.377	- .014	- .024	.004	- .022
OCC	.01	- .023	.003	.014	- .002	.014
SLC	.05	- .064	- .004	.028	.012	.004
SOC	.03	.282	- .018	.009	.010	- .007
AA	.01	.066	- .005	.011	.000	.004
OA	.01	.013	.006	.006	- .002	- .016

*See Table 1 for interpretation of asterisks. Underlined coefficients are significantly different from zero at the 5 per cent level.

Table 10. Long Auto Travel*

Area	Relative Frequency	LCC	OCC	SLC	SOC	AA	OA
Largest central cities (LCC)	.48	x					
Other central cities (OCC)	.67	*	x				
Suburbs of largest central cities (SLC)	.72	*YE		x		*YE	*YE
Suburbs of other central cities (SOC)	.66	*YEC			x		
Adjacent areas (AA)	.63	*C				x	
Outlying areas (OA)	.60	*C					x

Area	R^2	Intercept Frequency	Change in Frequency per Unit of: Income	Age	Education	No. of Children
LCC	.12	.012	.034	- .041	.031	.018
OCC	.19	- .294	.050	.025	.032	.022
SLC	.04	.503	.010	- .005	.022	- .031
SOC	.21	- .098	.038	.012	.048	- .030
AA	.13	.416	.021	- .039	.046	- .041
OA	.16	.153	.041	- .036	.031	- .049

*See Table 1 for interpretation of asterisks. Underlined coefficients are significantly different from zero at the 5 per cent level.

of adjacent and outlying areas. There are widespread interaction effects of income, age, education, and number of children among the six areas. One of the four independent variables is correlated with reported travel by train in each area. There is no firm basis for concluding that a higher propensity to travel by train is or is not a unique characteristic of the largest Standard Metropolitan Areas.

A recent bus trip of 100 miles or more is reported most frequently by residents of central cities and residents of outlying areas, two extremes on the "belt" system of location. The regression analysis shows some tendency for bus travel among suburban residents to decline with income and increase with age and education. The observed differences in frequency of bus travel between areas cannot be explained by the other independent variables, but neither is there an obvious rationale for the differences observed between areas. It is possible that none of the variables included in the analysis is closely related to travel by bus.

Travel by automobile on trips of 500 miles or more is significantly less frequent among residents of large central cities than among residents of all other areas. The frequency is largest among residents of the suburbs of the same largest central cities. The regression analysis shows a positive relationship between auto travel and income and education in most of the areas. The greater propensity to make long auto trips among families living in the suburbs of the largest central cities can therefore be explained by their higher levels of income and education relative to other areas. The low frequency of auto travel from the largest central cities cannot be explained in the same way and seems to be a unique phenomenon of location. Comparable data for analysis of ownership of automobiles have not been included in this study. Other survey data as well as casual observation suggest that there is an intrinsic locational effect on automobile ownership. The automobile is simply not "a way of life" in larger cities.

OTHER ECONOMIC VARIABLES

The amount of monthly payments, in dollars, for consumer debt does not differ between areas. Table 11 also indicates, however, that there are interaction effects among the six areas in the effect of income, age, and number of children on the amount of debt payments. Thus, while the amounts of debt may not differ among areas, there may be intrinsic dif-

Table 11. Amount of Monthly Debt Payments*

Area	Means	LCC	OCC	SLC	SOC	AA	OA
Largest central cities (LCC)	33.65	x					
Other central cities (OCC)	33.03		x				
Suburbs of largest central cities (SLC)	28.61			x			
Suburbs of other central cities (SOC)	39.45	No			x		
Adjacent areas (AA)	32.47	differences				x	
Outlying areas (OA)	34.65						x

				Change in Frequency per Unit of:		
Area	R^2	Intercept	Income	Age	Education	No. of Children
LCC	.07	11.042	5.788	-12.487	- 4.492	.198
OCC	.13	119.948	.361	-20.868	- .055	- .493
SLC	.10	57.716	.167	- 8.166	- 2.815	10.622
SOC	.08	31.489	4.606	-14.887	- 1.960	-1.815
AA	.10	- 13.759	2.953	- 2.924	- .420	11.985
OA	.10	- 43.731	5.343	- 1.555	- .423	10.822

*See Table 1 for interpretation of asterisks. Underlined coefficients are significantly different from zero at the 5 per cent level.

Table 12. "Good" Re-employment Chances*

Area	Relative Frequency	LCC	OCC	SLC	SOC	AA	OA
Largest central cities (LCC)	.44	x					
Other central cities (OCC)	.46		x				
Suburbs of largest central cities (SLC)	.53			x		YE*	
Suburbs of other central cities (SOC)	.50				x		
Adjacent areas (AA)	.41					x	
Outlying areas (OA)	.56	C*				*	x

		Intercept		Change in Frequency per Unit of:		
Area	R^2	Frequency	Income	Age	Education	No. of Children
LCC	.03	.579	.010	- .066	- .002	- .050
OCC	.06	.335	.013	- .051	.013	.038
SLC	.09	.272	.024	- .080	.034	- .012
SOC	.20	.900	- .020	- .108	.050	.002
AA	.07	.321	- .002	- .020	.026	.059
OA	.09	.813	- .008	- .076	.027	- .003

*See Table 1 for interpretation of asterisks. Underlined coefficients are significantly different from zero at the 5 per cent level.

ferences between areas with respect to the kinds of families who have
different amounts of debt.

One of the questions asked in the survey was, "Suppose that for some
reason (the head of the family) loses his job during the next few months.
What do you think are the chances of getting another job that pays about
the same?" The original detailed coding of the responses have been
collapsed for this study into, "Good," and, "Not good." The relative
frequencies of "Good" responses are shown in Table 12. Families who
live in outlying areas are more confident of being re-employed favorably
than those who live in large central cities or adjacent areas. The regression analysis does not support an alternative explanation and the
observed difference may be an intrinsic characteristic of location. The
greater job confidence among residents of the suburbs of larger central
cities compared with residents of adjacent areas can probably be attributed to higher levels of education in the suburbs.

ATTITUDES TOWARD USE OF PUBLIC FUNDS

Two questions included in the survey provide the data for this section. First the respondents were asked, "Some people say that there
will be some disarmament and therefore our government will spend less
on arms and defense. Suppose this is the case, what would you say
should be done with the money saved?" After obtaining an answer to
this open question, the interviewer continued, "Here are some suggestions
that have been made (a card is handed to the respondent). Please tell
me which use of the money appears best to you, which is second best,
third, etc." The cards listed the following uses:
  A. Should be used to increase financial help to other countries.
  B. Should be used to reduce government debt.
  C. Should be used to reduce income taxes.
  D. Should be used to build schools, highways, and the like.
  E. Should be used for public welfare programs - to help needy peo-
      ple in the U.S.
The ordering of the uses causes a reversal of the usual numerical con-
ventions and the reader is reminded that a low rank number indicates
relatively high favor in Tables 13-18.

Table 13 indicates that families living in central cities favor foreign
aid more than residents of suburbs of other central cities, adjacent areas
and outlying areas. Moreover, the tendency for rank to decline as education increases offers no alternative explanation and one must conclude
that the difference in ranks arises from an intrinsic locational characteristic. One readily available hypothesis is that differences in ethnic
origin may be a concealed correlate for which location serves as a proxy.

Residents of larger central cities are clearly less concerned about
the national debt than residents of other areas. The results of the regression analysis suggest that rank declines with level of education, but
this and other results do not explain the differences in means of ranks.
Again, there is apparently a difference in attitudes which might be attributed to location.

An unusual configuration of differences between areas in rank
assigned to tax reduction appears in Table 15. Those most in favor of
tax reduction live in larger central cities, suburbs of large central
cities, and adjacent areas. Those least in favor of tax reduction live in
other central cities and outlying areas. The regression analysis does

Table 13. Rank Assigned Foreign Aid*

Area	Means	LCC	OCC	SLC	SOC	AA	OA
Largest central cities (LCC)	4.03	x					
Other central cities (OCC)	4.17		x				
Suburbs of largest central cities (SLC)	4.26			x			
Suburbs of other central cities (SOC)	4.47	$\overset{*}{Y}$EC	$\overset{*}{Y}$C		x		
Adjacent areas (AA)	4.45	$\overset{*}{C}$	$\overset{*}{C}$			x	
Outlying areas (OA)	4.37	$\overset{*}{C}$					x

Area	R^2	Intercept	Change in Rank per Unit of:			
			Income	Age	Education	No. of Children
LCC	.03	5.191	- .064	- .026	- .045	.045
OCC	.07	3.873	.046	.012	- .105	- .001
SLC	.14	3.316	.058	.165	- .132	.048
SOC	.10	5.634	- .083	.067	- .046	.047
AA	.02	3.969	.010	.098	- .007	- .012
OA	.06	4.149	.038	.013	- .078	- .078

*See Table 1 for interpretation of asterisks. Underlined coefficients are significantly different from zero at the 5 per cent level.

Table 14. Rank Assigned Government Debt Reduction*

Area	Means	LCC	OCC	SLC	SOC	AA	OA
Largest central cities (LCC)	3.51	x	*	*	$\overset{*}{A}$	*	$\overset{*}{Y}$
Other central cities (OCC)	2.98		x				
Suburbs of largest central cities (SLC)	2.89			x			
Suburbs of other central cities (SOC)	2.87				x		
Adjacent areas (AA)	2.76					x	
Outlying areas (OA)	2.76						x

Area	R^2	Intercept	Change in Rank per Unit of:			
			Income	Age	Education	No. of Children
LCC	.02	3.723	.028	- .090	- .071	- .028
OCC	.10	4.761	- .018	- .254	- .132	- .025
SLC	.05	3.821	.001	- .152	- .081	.010
SOC	.09	3.674	.010	- .079	- .136	- .057
AA	.01	2.692	.005	- .021	- .009	.072
OA	.11	5.249	- .137	- .132	- .070	.064

*See Table 1 for interpretation of asterisks. Underlined coefficients are significantly different from zero at the 5 per cent level.

Table 15. Rank Assigned Tax Reduction*

Area	Means	LCC	OCC	SLC	SOC	AA	OA
Largest central cities (LCC)	2.67	x					
Other central cities (OCC)	3.12	*	x	*		*E	
Suburbs of largest central cities (SLC)	2.71			x			
Suburbs of other central cities (SOC)	2.90				x		
Adjacent areas (AA)	2.73					x	
Outlying areas (OA)	3.00	*C		*		*	x

				Change in Rank per Unit of:		
Area	R^2	Intercept	Income	Age	Education	No. of Children
LCC	.05	3.873	- .113	.063	.065	- .021
OCC	.04	3.984	- .040	- .114	.056	- .060
SLC	.07	3.490	- .040	- .155	.075	.050
SOC	.05	2.969	.029	- .162	.004	.014
AA	.02	3.440	- .040	- .055	- .008	.071
OA	.02	3.992	- .057	- .077	.023	- .012

*See Table 1 for interpretation of asterisks. Underlined coefficients are significantly different from zero at the 5 per cent level.

Table 16. Rank Assigned Public Assets*

Area	Means	LCC	OCC	SLC	SOC	AA	OA
Largest central cities (LCC)	2.34	x					
Other central cities (OCC)	2.34		x				
Suburbs of largest central cities (SLC)	2.27			x			
Suburbs of other central cities (SOC)	2.25		No		x		
Adjacent areas (AA)	2.33		differences			x	
Outlying areas (OA)	2.26						x

				Change in Rank per Unit of:		
Area	R^2	Intercept	Income	Age	Education	No. of Children
LCC	.01	2.658	- .004	- .073	.014	- .046
OCC	.02	2.456	- .031	.080	.021	- .040
SLC	.02	1.915	.002	.061	.031	- .051
SOC	.04	1.800	- .014	.151	.039	- .007
AA	.07	3.013	.001	- .085	- .048	- .171
OA	.02	1.900	- .002	.088	.019	.003

*See Table 1 for interpretation of asterisks. Underlined coefficients are significantly different from zero at the 5 per cent level.

Table 17. Rank Assigned Public Welfare*

Area	Means	LCC	OCC	SLC	SOC	AA	OA
Largest central cities (LCC)	2.12	x					
Other central cities (OCC)	2.20		x				
Suburbs of largest central cities (SLC)	2.55	*YE	*YE	x	*YAE		*YE
Suburbs of other central cities (SOC)	2.29				x		
Adjacent areas (AA)	2.46	*C	*C			x	
Outlying areas (OA)	2.31						x

			Change in Rank per Unit of:			
Area	R^2	Intercept	Income	Age	Education	No. of Children
LCC	.12	- .252	.124	.099	.076	- .058
OCC	.16	- .085	.061	.202	.140	.071
SLC	.02	2.502	- .012	.037	.040	- .067
SOC	.18	.629	.055	.067	.136	.019
AA	.01	1.781	.026	.043	.040	.023
OA	.16	.072	.143	.027	.090	- .031

*See Table 1 for interpretation of asterisks. Underlined coefficients are significantly different from zero at the 5 per cent level.

Table 18. Summary of Means of Ranks*

Area	Foreign Aid	Reduce Debt	Reduce Taxes	Public Assets	Public Welfare
LCC	4.03	3.51	2.67	2.34	2.12
OCC	4.17	2.98	3.12	2.34	2.20
SLC	4.26	2.89	2.71	2.27	2.55
SOC	4.47	2.87	2.90	2.25	2.29
AA	4.45	2.76	2.73	2.33	2.46
OA	4.37	2.76	3.00	2.26	2.31

*Underlined means are the lowest among the alternative uses of funds for a given area.

not yield alternative explanations, and the observed differences may be true locational differences in attitudes.

There are no differences in mean ranks between areas for construction of public assets as a desirable use of public funds. The regression analysis provides almost negligible explanation of differences in rank. Perhaps the most interesting insight is gained from Table 18 which shows that in four areas the mean rank for public assets is the lowest of means for alternative use of funds.

The use of funds for welfare purposes is least favored by residents of the suburbs of larger central cities and adjacent areas. This alternative is most favored by residents of central cities. Favoring this option tends to decline as education increases, however, and this relationship explains the high rank assignments in the suburbs of larger central cities. No alternative explanation is suggested for adjacent areas and the relatively high rank reported for the residents of these areas may reflect a true locational difference.

CONCLUSION

Many differences among consumers which are apparently attributable to differences in location can easily be explained by differences in alternative variables. This ambiguity in data describing people in different locations serves to emphasize the importance of carefully identifying "urban problems." Surely the long-run impact of increasing research on urban economic problems will be a better description of urban life and policies aimed at improving urban life. It is possible that the search for a satisfactory description and satisfactory policies may benefit from careful identification of characteristics unique to urban life because it is these characteristics which will point out the necessity of new theories of economic behavior and unusual policies.

152

NOTES

[1] Bernard Lazerwitz, "Metropolitan Community Residential Belts, 1950 and 1956," American Sociological Review, April, 1960; also, "Some Characteristics of Residential Belts in the Metropolitan Community, 1950-1956," unpublished Ph.D. thesis, University of Michigan, 1958.

Expenditures on Vacations and Pleasure Travel in New Hampshire

by Richard L. Pfister

Ways of making expenditure estimates and the wide range of results - Identification of type of expenditures to be included - The two New Hampshire methods of estimation include analysis of Census of Business data for New Hampshire firms and a comparison of New Hampshire per capita sales with New Hampshire per capita consumption - Conclusions.

"When the U.S. Government conducted its first census, in 1790, it found that only 4 per cent of Americans lived in cities. For the 96 per cent who did not, the idea of a grand holiday was a trip to town to enjoy the noise and excitement of commerce and crowds. Now, some hundred and seventy years later, two-thirds of Americans live in cities and suburbs, and in their leisure time they have a mighty urge to get away from the city to sea and sun and unspoiled countryside."[1] Most of these trips to "get away" are day trips or weekend trips that end within a fifty mile radius of the traveler's home. The greatest congestion thus occurs in the recreation areas surrounding major urban centers. New Hampshire feels the impact of this urge to "get away" because many people drive farther than fifty miles on their trips, particularly when it is a vacation of a few days or more. The bulk of the out-of-state vacationers in New Hampshire comes from the urban centers of the New England and Middle Atlantic states.

This paper is concerned with what people spend when they "get away" for vacations and pleasure travel. The number of these brief escapes from urban living has increased with the growth of population, income, leisure, and mobility. Attendance figures for various tourist attractions (parks, recreation areas, resorts, etc.), the number of motels and other tourist accommodations, and sales of recreation equipment all indicate that expenditures on vacations and pleasure travel have grown rapidly in the last several years. Just how rapidly this spending has grown and just how great it is have never been answered satisfactorily, despite the attention given the matter by state promotional agencies, travel organizations, and tourist and vacation businesses.

EXPENDITURE ESTIMATES AND THEIR DIFFERENCE

Various agencies and organizations have prepared estimates for individual states and for the nation. The estimates sometimes vary widely and show some obvious inconsistencies. For instance, Fortune magazine estimated that Americans spent $10.4 billion in 1958 on domestic vacations and pleasure travel while the National Association of Travel Organizations (NATO) and the American Automobile Association (AAA) estimated that the total was around $19 or $20 billion.[2] These estimates were intended to cover roughly the same types of expenditures. A comparison of state estimates indicates that some states not particularly known as tourist attractions have been excessively generous in estimating their tourist expenditures.

How is it possible that there can be such great disagreement and inconsistencies in the estimates? Clearly the situation can exist only because of the unavailability of reliable information about tourist expenditures. The unavailability in turn stems from the difficulty of obtaining the required information at a justifiable cost. This fact probably explains why there are no official estimates by federal agencies of tourist spending. To a minor extent, the discrepancies may be due to the use of different definitions, a point that will be discussed later.

Most statistics of consumption expenditures come either from the U.S. Department of Commerce or from studies using the Department's consumption categories. Whereas the Department of Commerce estimates personal consumption expenditures for a variety of goods and services, it is, unfortunately, not possible to identify expenditures for vacations and pleasure travel from their data. Recreation is one of the categories, but it includes such items as reading matter, toys, sporting goods, radio and television, and amusement admissions. Many of these items are not related to vacations and pleasure travel. On the other hand, numerous items that are a part of vacation and travel spending are included in other categories. The amounts spent for hotel, motel, and cabin accommodations are included in housing; purchased meals and beverages are included in the food and tobacco category; and amounts spent on transportation for pleasure purposes are included in the transportation category which includes travel for all purposes. Other studies of family expenditures have generally used the expenditure categories of the Department of Commerce, so they provide no additional information on vacation-travel spending.

The lack of benchmarks means that the various estimates of tourist expenditures must be judged primarily on the methods used. From this standpoint, the Fortune estimate for 1958 must be considered the most reliable. Their expenditure figures for vacations and pleasure travel were derived from a study of expenditures for all types of leisure and recreational activities. The study appears to be a careful attempt to determine the amounts of these various types of expenditures. The authors relied on a variety of data, including personal consumption expenditures of the Department of Commerce, advice from persons in charge of compiling the Commerce data, and intermittent surveys of travel expenditures of subscribers to various periodicals. Several categories of personal consumption spending as estimated by the Department of Commerce were used as controls to prevent the over-all estimates from getting out of bounds.

NATO and AAA both depended primarily upon state agencies for estimates of tourist expenditures in individual states. NATO reported that the state estimates were adjusted "where necessary to comply with regional trends."[3] State promotional agencies who supplied the figures for NATO and AAA are generally overly optimistic in their estimates of tourist spending.

Examples of exaggerated estimates are provided by comparing the receipts of all lodging places as reported in the 1958 Census of Business with the estimates of tourist expenditures for lodging as made by various state agencies.[4] In order to obtain the estimated lodging expenditures of tourists, it was assumed that 21 per cent of total state tourist expenditures was for lodging. Numerous studies have shown that approximately this percentage is spent on lodging. Of thirty-three states for which data were available, eighteen showed estimated lodging expenditures of tourists to be in excess of the receipts of lodging places as reported in the U.S. Census. Kentucky, Mississippi, and New Jersey are the extreme cases as their estimated lodging expenditures by tourists exceeded the Census receipts figure in 1958 by 279 per cent, 295 per cent, and 331 per cent, respectively. Theoretically, a slight overstatement would be possible as the Census does not include tourist homes, but their receipts would be small in comparison with those of the covered lodging establishments. In addition, a significant share of lodging receipts in any state would be from business, commercial, and residential clientele.

Most reports on tourist spending continue to cite the high figures of NATO and AAA rather than the sounder and more reliable figures of Fortune. The reason is apparent: if the lower Fortune estimate were accepted, most state figures would be revealed as obvious overstatements. Reliable and authoritative national figures would serve as a check on the state estimates. It is not easy to determine each state's share of the U.S. total, but a firm figure for the latter would do much to stimulate better estimating on the part of interested state agencies.

It was noted above that some variation in estimates of tourist expenditures could be the result of using different definitions. Despite the prominence of the tourist industry, it is hard to define and it is statistically slippery. Different types of businesses provide a wide variety of goods and services to vacationers and pleasure travelers. Non-travelers frequently engage in the same activities as travelers do, and many businesses serve both non-travelers and travelers. The same can be said for seasonal residents and permanent residents in any locality.

The terms and concepts in tourist research have not yet been standardized. Tourists have been defined as visitors from other states, persons traveling for pleasure or recreation, persons making overnight stops away from home, etc. In my opinion, however, the different definitions do not affect the estimates of tourist expenditures significantly. A major discrepancy could arise in the case of state estimates if some studies are restricted to expenditures by out-of-state visitors while others include, in addition, vacation spending within a state by its own residents.

TYPES OF EXPENDITURES INCLUDED IN THE STUDY

At the beginning of a study of this sort, it is advisable to set down conceptually what expenditures are to be included in the estimates even though the available data will not permit accurate calculations of the desired measures. It is easiest to list first what was not included in the New Hampshire study. Expenditures for recreational and leisure activities not connected with pleasure traveling or vacationing, such as that for sports, movies, radio, reading, television, and occasional dining out were excluded. It must be admitted that tourists and especially seasonal residents spend money on these items. The reason for excluding such expenditures was that the bulk of them are made by persons who are not engaged in vacationing or pleasure traveling. Also excluded were expenditures made in New Hampshire by residents of border states unless made in the course of a vacation or pleasure trip.

The expenditures of all vacationers and pleasure travelers were included regardless of state of origin and regardless of length of time spent in the state. The pleasure traveler thus included those persons making day trips into and within the state even though they did not stay overnight and even though they were New Hampshire residents. Conceptually, expenditures made on purely business trips were excluded. Finally, important components were the expenditures made by persons occupying seasonal homes and by those attending boys' and girls' camps. Thus, a tourist or pleasure traveler was actually defined to be anyone who traveled away from home in whole or in part for pleasure purposes.

METHODS OF CONSTRUCTING EXPENDITURE ESTIMATES

Obviously the available information was insufficient to permit an accurate calculation of the desired measures. Many studies have relied on the interview method of obtaining expenditure data directly from travelers. We did not use this approach. The funds available for our study were inadequate to permit an extensive interview program in addition to the other work in the project. We felt that another interview study of tourist spending patterns would add little to existing knowledge of this matter. To be sure, the collection of information from a carefully designed sample would have facilitated the estimation of total tourist expenditures in the state. However, there is reason to question the accuracy of responses given by travelers to interviewers. People on vacation frequently have an aversion to being interviewed. When being interviewed, they are likely to be in a hurry and otherwise occupied so that their answers to questions concerning expenditures may be rough guesses. Many vacationers do not keep detailed records of their expenses. If questionnaires are mailed to tourists some time after their vacations, they may well have forgotten the details of their expenditures. A poor response to mail questionnaires may cause the sample to be biased.

As a result of these considerations, we decided to explore two different methods of estimating vacation-travel expenditures in an indirect manner. The first involved an intensive study of the receipts of those businesses that provide the bulk of the goods and services purchased by vacationers and pleasure travelers. The second consisted of comparing the actual receipts of selected retail and service establishments in New Hampshire with the hypothetical receipts to be expected from purchases solely by the state's year-round residents. The hypothetical receipts were calculated from the assumed consumption patterns and the per capita incomes of New Hampshire residents. The results of the two methods were compared and evaluated and the final estimates were determined by using certain techniques from both methods.

First Method of Estimation.

As stated above, the first method involved a careful study of those categories of business which initially receive the bulk of the expenditures of vacationers and pleasure travelers. Numerous tourist studies have shown that roughly 80 per cent of total expenditures by tourists are received by businesses in the following categories: Commercial Lodging Establishments; Eating and Drinking places (including liquor stores); Gasoline Service Stations; Tire, Battery and Accessory Stores; Auto Repair Services and Garages; and Amusements, Except Motion Pictures. The receipts of these businesses were obtained from the 1958 Census of Business. Each category was examined in detail and an estimate was made for the proportion of total receipts that came from vacationers and pleasure travelers. The sum of tourist spending in each of the above business classes was assumed to constitute 80 per cent of the total, the remaining 20 per cent being widely scattered among various other retail and service establishments.

This procedure does not include most of the spending by the occupants of seasonal homes in the state. An estimate was made for the total spending of these persons, and the amount not included by the above calculations was added to give a total for all vacationers as well as pleasure travelers. The occupants of seasonal homes constitute a major part of the vacation-travel market for New Hampshire as there were nearly 30,000 such homes in the state in 1958.

Table 1. Total Receipts of Travel-Serving Businesses and the Allocation to Vacationers and Pleasure Travelers, New Hampshire, 1958

Establishment class	Total receipts ($1,000)	Percentage for vacationers and pleasure travelers	Amount for vacationers and pleasure travelers ($1,000)
Seasonal hotels	$ 12,286	93	$11,426
Year-round hotels	7,509	60	4,505
Motels, tourist courts	6,843	94	6,432
Trailer parks	260	20	52
Sporting, recreational camps	[1]6,695	100	6,695
Eating places	36,059	35	12,621
Drinking places	2,182	35	764
Liquor stores	23,042	23	5,300
Gasoline service stations	42,922	25	10,731
Tire, battery, accessory stores	6,217	25	1,554
Auto repairs and services	12,494	25	3,124
Amusements, except movies	10,820	35	3,787
Total	$166,213	40	$66,991

Source: Total receipts from U.S., Bureau of the Census, 1958 Census of Business Retail Trade and Selected Services, New Hampshire area reports.

[1]The Census figure was adjusted upward. For explanation, see text.

Table 1 contains the 1958 receipts of the travel-serving businesses and the percentage and dollar amount allotted to tourist spending. Five separate components of the lodging category were treated separately. The percentage allotments for the first three components were based on estimates made by the operators during a field survey of all lodging establishments. For seasonal hotels, 93 per cent of receipts was attributed to vacationers and pleasure travelers. Business conventions account for the remainder and even this derives largely from the recreational opportunities afforded by these hotels. The figure for year-round hotels was 60 per cent while for motels and tourist courts it was 94 per cent. These percentages may seem high, but travel for strictly business and trade purposes is not great in New Hampshire with its relatively small population and few concentrations of population that serve as major commercial centers.

The 20 per cent allotment for trailer parks is a guess. It is known, however, that the bulk of trailers in the parks are there on a semipermanent basis. Thus, for the most part, those who dwell in trailers are not pleasure travelers. The receipts of trailer parks are small, so even a large error in the percentage figure would not significantly affect the total.

All of the receipts of sporting and recreational camps were assigned to vacationing. The receipts figure in Table 1 is 20 per cent larger than the amount reported in the Census of Business. The upward adjustment was made because numerous boys' and girls' camps were outside the scope of the census. Actually the number of boys' and girls' camps registered with the State Department of Health in 1958 was 30 per cent greater than the census total for the entire category of sporting and recreational camps. The census excluded day camps and non-commercial establishments. Average receipts for these excluded camps are generally somewhat less than for those included in the census. For this reason, the census figure was increased by only 20 per cent even though the number of camps was at least 30 per cent greater.

For both eating places and drinking places, 35 per cent of receipts was attributed to vacationers and pleasure travelers. This figure is largely an informed guess although it appeared reasonable in the light of several relevant factors and discussions with informed persons in this field.

In order to estimate the share of liquor store sales resulting from the state's vacation and pleasure-travel business, the average monthly sales for the first four months were calculated for each year from 1955 to 1960. This figure was assumed to be the normal monthly consumption by New Hampshire residents and regular customers from border states for each year. It was multiplied by seven to give the expected consumption for the months May through November provided there was no additional consumption by regular customers and provided there were no purchases by visitors. It was assumed that December purchases, which are high because of the holiday season, were made entirely by regular customers. The excess of the actual purchases for May through November above the "expected" amount, obtained by the previous calculation, was attributed to visitors and increased purchases for recreational purposes by the regular customers. This excess was then expressed as a percentage of total annual sales for each year from 1955 to 1960; the annual average was 20 per cent. On the assumption that some part of the purchases during the first four months of a year was due to vacations and pleasure travel, we felt that probably around 23 per cent of annual sales was attributable to such activities.

The procedure in allocating gasoline sales was similar to that used for liquor sales. Average sales per month were computed for the first four months of each year from 1955 to 1960. This average was then multiplied by twelve to obtain the annual total that would have existed if travel in the rest of the year had been at the same monthly rate. The resulting figure was assumed to be the amount that would be consumed by New Hampshire residents for their non-pleasure driving. The actual consumption for one year exceeded this hypothetical figure, by an average of 20 per cent. Since skiing activities account for a substantial amount of travel during the first four months of the year, we felt that at least 25 per cent of gasoline sales should be allotted to pleasure travel.

The 25 per cent estimate for gasoline sales was also used for Tire, Battery and Accessory Stores and for Auto Repair Shops and Garages. The figure may be too high for these businesses as out-of-state travelers might reasonably be expected to have had their autos serviced for their trips. Hence, their expenditures for repairs and accessories might not be in proportion to the amount of traveling done.

The receipts for amusements was difficult to handle. Probably three-fourths of the total of $10.8 million was accounted for by race tracks. Even though the bulk of the betting was undoubtedly done by out-of-state residents, primarily persons from Massachusetts, the race track receipts could not be considered in the same category as most vacation-travel expenditures. Going to the races is a form of recreation for those attending, but most persons attend just as they might go to a movie (although the distance involved probably would be greater). The main point is that attending the races is not generally a part of a vacation trip or pleasure travel in the usual meaning of those terms. At most, 10 per cent of race track receipts might be assigned to vacationers and pleasure travelers. For the remaining establishments in the amusement category, it was assumed that two-thirds of the receipts were due to vacation-travel spending. The tourist share of receipts of all amusement enterprises in New Hampshire was thus approximately 35 per cent.

Adding the amounts thus allocated for vacations and pleasure traveling gives a total of $66,991,000 for 1958. An adjustment is needed, however, for some of the figures in the last column of Table 1. The receipts of eating and drinking places do not include tips which are counted by tourists as part of their expenditures. The same is true in the case of hotels for both food and lodging service. Tips were assumed to be 15 per cent of receipts of eating and drinking places and 10 per cent of receipts of seasonal and year-round hotels. The total addition for tipping was $3,601,000 which brings the expenditure total for the business groups in Table 1 to $70,592,000. On the assumption that the latter figure is 80 per cent of total expenditures by tourists, the total becomes $88,240,000. The 20 per cent of expenditures not caught in Table 1 includes those for various personal and professional services; gifts and curios; fees for toll roads, state parks, fishing and hunting; clothes; sporting equipment; and food purchased at grocery or general stores.

The total of $88,240,000 fails to include most of the spending by residents of seasonal homes in the state. A study in Wisconsin provided the guide for estimating the expenditures by residents of private cottages or seasonal homes.[5] In that study, 1959 expenditures were obtained for

both resident and non-resident owners of seasonal homes through use of
a mail questionnaire. It was assumed that the Wisconsin figures were
valid for New Hampshire. The method of allocation used in Table 1 has
already included the expenditures of occupants of seasonal homes for
purchased meals and part of that for amusements and transportation. It
was estimated that 15 per cent of the total spending per cottage was included in Table 1. The remainder came to $1,180 per cottage for 1958
at which time New Hampshire had approximately 29,560 seasonal homes
or cottages. Spending of cottage residents in addition to that caught in
Table 1 was thus estimated to be $34,881,000. The grand total for all
vacationers and pleasure travelers amounted to $123,121,000 as indi-
cated in Table 2.

Table 2. Vacation-Travel Expenditures in New Hampshire, 1958

Amount for vacationers and pleasure travelers, from Table 1	$66,991,000
Allowance for tips	3,601,000
	$70,592,000
If above is 80%, then 100% is	$ 88,240,000
Additional spending by seasonal residents	34,881,000
	$123,121,000

Second Method of Estimation

The second method of estimating vacation travel expenditures begins
with a comparison of the per capita sales of a variety of goods and services in New Hampshire with the national averages. Per capital retail
sales in New Hampshire exceeded the national average by about 6 per
cent in 1958 despite the fact that the state's per capita income was approximately 8 per cent below the U.S. average. The New Hampshire
per capita receipts also exceeded the nation's for several types of service expenditures, including the following: lodging; auto repair services
and garages; and amusements and recreational services except motion
pictures.

These consumption figures provide clues to the spending by visitors
because the per capita figures were based on a population count that excluded visitors and temporary residents.[6] There are many factors to
consider, however, before the expenditures of visitors can be estimated
from the receipts of retail and service establishments. Visitor expenditures would not significantly affect the receipts or sales of establishments in several different retail and service categories. For some
categories there may be a significant effect, but the New Hampshire per
capita figure may be above or below the national average for other reasons. Consumption patterns may vary from one state or region to another because of differences in income, climate, degree of urbanization,
and family characteristics.

Adjustments must be made in the data on sales and receipts in an
effort to allow for the effects of these differences. The goal, of course,
is to determine the expected consumption of various goods and services
by the year-round New Hampshire residents. The difference between
this expected amount and the actual amount would be attributed to visitors.

A detailed study was made for each kind-of-business group in the
1958 Census of Business for retail trade and selected services. Several
groups and individual business classes were eliminated from the compu-

tations on the grounds that their sales would not be significantly affected by expenditures of visitors or that other factors had far greater effect on sales. Sales of the following were quickly excluded: farm equipment dealers; hay, grain, and feed stores; and other farm and garden supply stores; fuel dealers; and automotive dealers.

Whether to include or exclude various other categories of retail and service trades was less clear-cut. The general merchandise group of stores showed substantially smaller per capita sales in New Hampshire than in the nation. This group includes department, dry goods, limited price variety, and general stores. Visitors should have an effect on their sales. The New Hampshire per capita figure was much above the national average for each sub-category except department stores. The primary reason is that New Hampshire has few major population centers but many towns and villages. General stores are quite common in towns and resort areas while large department stores are found primarily in large cities. In New Hampshire, department stores accounted for only one-third of the sales of the entire group as compared with nearly two-thirds for the nation. The impact of spending by vacationers in New Hampshire would undoubtedly be reflected in the sales of general stores rather than department stores. However, even in the absence of the spending by vacationers, the per capita sales of general stores in New Hampshire could be expected to exceed the national figure because this is the state's dominant type of store for the group. As a result of these considerations department store sales were excluded, and only half of the New Hampshire excess for the other general stores in the group was attributed to vacation spending.

New Hampshire per capita sales of furniture, home furnishings, and home equipment stores were somewhat below the national figure. The per capita figure would certainly be further below that of the entire country if it were not for the vacation business in the state. Apparently many of the commercial establishments, especially the larger ones that cater to tourists, buy their furnishings and equipment from wholesale stores in metropolitan areas outside the state. This factor apparently outweighs the stimulus of visitor spending on New Hampshire sales. Therefore, the sales of this group were omitted in the computations.

Per capita sales of drug and proprietary stores were also below the U.S. average. Here again vacationers should add to the sales of drug stores, and certainly the sales would have been even lower had it not been for the spending by visitors. The lower New Hampshire figure was probably due to differences in types of drug stores. Chain and cut-rate drug stores are common in large cities. These stores are really variety stores selling many products not ordinarily found in the typical small-town drug store. This factor alone might explain the higher per capita sales of drug stores in the more urbanized states, and it justifies elimination of drug stores.

Among the major categories of services covered in the Census of Business, personal services, miscellaneous business services, miscellaneous repair services, and motion pictures were excluded from the computations. The vacation business of New Hampshire could be expected to have some impact on receipts of these establishments, but per capita receipts in the state were substantially below the national figure for each category. An examination of data for several different states revealed that per capita sales for these categories were closely associated with the extent of urbanization. The fact that New Hampshire has a relatively small population and a relatively low degree of urbanization overwhelms any effect that visitor spending might have on such service receipts.

Table 3 shows the business groups that were retained for use in the computations. A few comments are necessary concerning the figures for some of the groups. The receipts of eating and drinking places as reported in the Census of Business yielded a per capita figure for New Hampshire that was surprisingly below the national average. Spending by vacationers should have given a strong boost to the receipts of such establishments. An explanation is readily at hand, however. Per capita sales in New Hampshire were high for liquor stores but low for drinking places because of the state's laws pertaining to drinking places. By combining the two, the New Hampshire figure is about the national average. Per capita receipts of eating places were low apparently because such a large amount of spending on meals by vacationers was included in the receipts of lodging places. According to the 1954 Census of Business, 40.0 per cent of the receipts of year-round hotels (with payroll and 25 or more rooms) in the United States came from the sale of meals and beverages; in New Hampshire the corresponding figure was 51.2 per cent. Motels and tourist courts with payroll and ten or more rental units obtained 9.5 per cent of their receipts from the sale of goods and beverages in the entire United States; for New Hampshire the figure was 21.9 per cent.[7] If a part of the New Hampshire hotel and motel receipts is transferred to eating places so that the ratio of lodging receipts derived from the sale of food and beverages is equal to the national figure, then the New Hampshire per capita sales by eating places exceeded the national average.

The per capita receipts or sales for the selected groups of businesses are contained in Table 3. The expected per capita figures for New Hampshire's permanent residents would be lower than the national average because the state's per capita income is approximately 8 per cent lower. However, the permanent residents would not necessarily spend 8 per cent less on purchases from each business group. The effect of the difference in income would vary from one type of purchase to another, depending upon the income elasticity of demand.

Income elasticities for the United States were computed from data in the 1950 study of consumer expenditures.[8] Some adjustments were made, as explained later, to allow for the exclusion of the rural population in that study. A few simple calculations yielded the average coefficients of income elasticity for each of seventeen different expenditure categories. As the expenditure categories did not correspond exactly with the business groups in Table 3, it was necessary to decide what elasticity coefficients (or combination of them) were appropriate for each business group. Table 4 contains the elasticity coefficients which were used. Although these coefficients were based on 1950 family budget data, they were assumed to apply in 1958.

Earlier it was stated that the per capita personal income in New Hampshire as estimated by the U.S. Department of Commerce was about 8 per cent below the U.S. average. However, another factor must be considered in the comparison of income levels. Because of the colder climate in New Hampshire, the state's residents spend more per person on fuel than the national average. At any given level of family income, New Hampshire residents would therefore have less money left to spend for other things. In effect, the higher fuel expenditures cause

Table 3. United States and New Hampshire Per Capita Receipts for Various Businesses and Estimated Amount Attributable to Visitors, 1958

Establishment	Per capita receipts U.S.	Per capita receipts N.H.	Estimated consumption by N.H. residents Per capita	Estimated consumption by N.H. residents Total	Actual N.H. consumption	Excess attributed to visitors
Lodging places	$ 22.3	$ 55.9	$ 19.8	$ 11,524	$ 32,477	$ 20,953
Eating and drinking places	87.4	65.8	81.3	[3] 45,897	38,241	-7,656
Liquor store sales	24.1	39.6	21.4	12,455	23,042	10,587
Food stores	281.7	345.5	264.8	[3] 149,491	200,860	51,369
General mdse. stores, except dept. stores	49.0	61.5	45.1	26,248	35,772	[4] 4,762
Lumber, bldg. material, hardware stores[1]	63.9	64.4	59.4	34,571	37,417	2,846
Gasoline service stations	81.5	83.6	72.5	42,195	48,624	6,429
Auto repair, auto service, garages	22.1	24.2	19.7	11,465	14,098	2,633
Amusements, recreational services, except motion pictures	15.3	18.6	13.6	7,915	10,820	2,905
Apparel, accessory stores	72.0	77.6	64.8	37,714	45,126	7,412
Aircraft, boats, motorcycles, trailers	5.8	9.0	5.2	3,026	5,223	2,197
Other retail stores[2]	38.8	42.4	34.5	20,079	24,673	4,594
						$109,031

Source: Retail sales and service receipts from U.S., Bureau of the Census, 1958 Census of Business; population data obtained from 1961 Statistical Abstract of the United States.

[1] Excludes farm equipment dealers.
[2] Excludes liquor stores: hay, grain, and feed stores; and fuel and ice dealers.
[3] Reduced by 3 per cent for reasons described in the text.
[4] Only one-half of the excess was included. See text for explanation.

real incomes in the state to be lower. The 1950 study of consumer expenditures included three cities in northern New England. Families in those cities spent on the average about 3 per cent more of their incomes on fuel than did all urban families in the United States. An allowance was made for this factor by assuming that New Hampshire per capita incomes were 11 per cent lower than the national average instead of just 8 per cent lower.[9]

Table 4. Coefficients of Income Elasticity of Demand for Selected Expenditure Categories, Based on 1950 Data

Establishment	Coefficient
Lodging places	1.04
Eating and drinking places	.89
Liquor stores	1.00
Food stores	.51
General mdse. stores, except department stores	.70
Lumber, bldg. material, hardware stores	.66
Gasoline service stations	1.00
Auto repair, auto service, garages	1.00
Amusements, recreational services, except motion pictures	1.04
Apparel, accessory stores	.95
Aircraft, boats, motorcycles, trailers	1.00
Other retail stores	1.04

Source: Calculations based on data in U.S., Bureau of the Census and the Wharton School, University of Pennsylvania, Study of Consumer Expenditures, Income, and Savings, University of Pennsylvania, 1957, Vol. XVIII.

The national per capita consumption figures were next adjusted downward in accordance with the lower incomes in New Hampshire and with the appropriate income elasticities for the expenditure categories. The results gave the expected per capita purchases by New Hampshire residents in view of their lower incomes. One additional adjustment was made to allow for the fact that New Hampshire had 12 per cent more of its population living in rural communities than the nation in 1960. It was assumed that per capita sales by both food stores and eating places in New Hampshire would be approximately 3 per cent less than the U.S. average for each income class in the absence of the effects of tourist spending. Family budget studies have consistently shown that rural families, both farm and non-farm, produce a substantial amount of the food they consume and thus purchase less than urban families.

The third column of Table 3 contains the estimated per capita purchases by New Hampshire year-round residents, adjusted for the income difference and for the effect of the high percentage of rural population on food purchases. The multiplication of the adjusted per capita figures by the population of the state gave the total purchases expected to be made by the permanent residents. The excess of the actual over the expected purchases was attributed to spending by visitors with the exception noted in the table.

The excess comes to $109 million for 1958 as shown in Table 3. This figure would be a valid estimate only if no New Hampshire residents left the state during the year. Since a substantial number of the state's residents go to other states and countries for vacations and pleasure trips, an allowance must be made for this factor. The excess of sales attributed to visitors would have been somewhat greater had the permanent residents remained in the state the entire year.

Most studies of state tourist expenditures have not attempted to determine the out-of-state expenditures of their own residents. Only one study, the Kansas Tourist Survey, 1952, collected information on out-of-state spending by the state's residents.[10] It was estimated that Kansans spent $90 million on pleasure travel outside the state. If spending for out-of-state pleasure travel by New Hampshire residents were in the same ratio to that of Kansas as the ratio of the populations of the two states, and if an allowance be made for the rise in consumer prices, the New Hampshire figure would be approximately $25 million for 1958.

Even if $25 million is accepted as the amount spent on out-of-state pleasure travel by New Hampshire residents, this amount could not be added to the $109 million in Table 3. These persons would probably have spent less than $25 million in New Hampshire had they remained in the state. New Hampshire sales of the business groups in Table 1 might well have been $15 million greater had the permanent residents remained in the state for the entire year. Using this figure, the total attributable to tourists would then be around $124 million.

A major shortcoming of the second method is that it simply does not catch vacation-travel spending by residents of the state. The estimate of what residents would have spent had they not left the state for vacations elsewhere is not at all reliable. This technique also misses, for the most part, the in-state vacation-travel expenditures of New Hampshire residents.

An Evaluation and a Final Estimate

The two methods give estimates that are rather close together. At various stages in both procedures, other assumptions could reasonably have been made and the resulting estimates would have been changed somewhat. Many assumptions involved a matter of judgment, and the path followed was that which seemed most justified to us in light of the relevant considerations.

It does appear, however, that the first method is the sounder in that fewer assumptions and adjustments had to be made and more adequate information was available to guide the estimating. The second method required more arbitrary and questionable assumptions. Many variables can affect the per capita receipts of the various business groups, and available information was inadequate to permit accurate adjustments for their effects. Despite the belief that the first method represents the sounder approach, the second method offers some useful information to supplement the first.

A comparison of individual categories for Tables 1 and 3 permits a more detailed evaluation of the two methods. In Table 1 total receipts of lodging establishments attributed to vacationers and pleasure travelers amounted to $31,120,000 which was the sum of the first five figures in the last column. An allowance for tips boosts the total higher. The lodging figure in Table 3 was $20,953,000. In our view, the former amount is the more accurate. It was based on the receipts reported in the Census of Business and on the operators' estimates of the share of their total business due to vacationers and pleasure travelers. The use of the U.S. per capita figure for New Hampshire residents, despite the adjustment for income differences in Table 3, certainly overstates lodging expenditures by the state's residents. The national figure is heavily influenced by the business and commercial trade of lodging places in urban centers while this type of trade is much less important relatively in New Hampshire.

In the case of eating and drinking establishments, the figure of $12,621,000 in Table 1 is surely closer to the correct amount than the negative sum of $7,656,000 shown in Table 3. Without doubt a large share of the receipts of the state's restaurants and cocktail lounges is attributable to tourists. The amount for liquor stores in Table 1 ($5,300,000) is also more accurate than that listed in Table 3 ($10,587,000). The latter figure is pushed up by the across-the-border purchases which should not be considered sales to vacationers and pleasure travelers. Similar comments could be made for the other categories of the two tables that are comparable. In virtually all cases there are strong reasons for feeling that Table 1 gives more accurate estimates.

The excess of $51,369,000 for food stores in Table 3 is significant, however. What are the reasons for this great excess in food store sales? Some of the smaller camps, restaurants, and lodging places that serve meals buy part of their food supplies from local food stores. However, it is known that the large establishments buy virtually all their supplies directly from distributors or wholesalers, so this factor would not account for much of the excess. It would appear that the large excess is due primarily to two factors. First, the commercial cabin and motel establishments of the state have a higher proportion of units of the housekeeping variety than is true for the nation, so presumably more of their guests prepare their own meals. Second, informed persons believe that a high (higher than average for the United States) proportion of visitors stay with friends and relatives, especially those with summer homes. The figure of $34,881,000 in Table 2 for seasonal resident expenditures includes slightly less than $8,000,000 for purchase from food stores. The latter amount is less than one-sixth the total excess for food stores given in Table 3.

The large excess of food store sales in Table 3 leads to the conclusion that the $123 million estimate in the first method is too low. That method accounts at most for $15 to $20 million of the $51 million surplus of food sales. The high per capita food store sales in New Hampshire was not a fluke event of 1958; it is a firmly established fact supported by the data from earlier volumes of the Census of Business. In view of this, it appears that as a conservative estimate, total vacation-travel spending in New Hampshire in 1958 was at least in the neighborhood of $135 million.

There are other reasons for believing that the 20 per cent is insufficient allowance for expenditures in addition to those in Table 1. Hunting and skiing are important activities for vacationers and pleasure travelers in New Hampshire. Equipment expenditures are high for these activities. Most studies of tourist spending have not explicitly covered hunters and skiers, so equipment expenditures are probably not adequately reflected in the data from which the 20 per cent was derived. Without doubt, many non-resident hunters and skiers purchase their equipment from stores in their home communities. However, it is to be expected that a considerable number of them would purchase equipment in New Hampshire.

Assuming an expenditure figure of $135 million, the New Hampshire share of the Fortune estimate of U.S. domestic tourist spending in 1958 was 1.30 per cent. By comparison, the New Hampshire shares of U.S. totals were 0.34 per cent for population, 0.35 per cent for retail sales, and 0.31 per cent for personal income. The state's share of national tourist spending was thus about four times as great as its share of population, income and retail sales, a relationship that appears reasonable.

INCOME ORIGINATED AND THE MULTIPLIER EFFECT

Not all of the $135 million spent in New Hampshire by vacationers and pleasure travelers in 1958 represented net income to the persons and businesses serving this market. A large part of these receipts was spent for the purchase of materials, supplies, and services from other businesses. An additional portion was allotted for depreciation charges and state and local taxes. The remainder constituted the net income derived from the initial expenditures. It consisted of payments to the factors of production and took the form of wages and salaries, rent, interest, and profits. These factor payments directly traceable to tourist expenditures constituted the income originated in New Hampshire by the tourist industry. Income originated, or a similar measure-value added, is the best measure for evaluating the economic importance of an industry to a local area.[11]

The first step in estimating the income originated was to distribute total tourist expenditures among the various types of businesses that initially received the money. The second step was to break down these business receipts to show what they were allotted to.

Table 5. Income Originated by the Vacation-Travel Spending in New Hampshire, 1958

Income source	Thousands of dollars
Wages and salaries	$24,801
Profits	11,775
Rents	2,810
Interest	1,001
Total	$40,387

Source: See text for discussion of the numerous sources drawn upon.

It was arbitrarily assumed that an establishment's factor payments attributable to its vacation-travel receipts were the same percentage of total factor payments as vacation-travel receipts were of total receipts. Table 5 shows the aggregate amounts that were allotted to wages, salaries, profits, rents, and interest. The sum of $40 million is the income originated directly by the $135 million of vacation-travel expenditures. The remaining $95 million was used for the purchase of materials, supplies, equipment, and services; for depreciation allowances; for payment of state and local taxes; and for other miscellaneous expenses.

This estimate of factor payments was based on information from several sources. The 1947 Interindustry Relations Study of the U.S. Bureau of Labor Statistics was very useful for some of the business groups. The Census of Business provided much helpful information, particularly for the relationship of payrolls to receipts. Additional information was obtained from Statistics of Income published by the U.S. Internal Revenue Service and from special tabulations made by the New Hampshire State Tax Commission.

By employing the economic base theory and assuming the vacation-travel industry to be a basic industry, the initial receipts are assumed to generate additional income through the multiplier process. Some extremely rough calculations based on assumptions concerning the New Hampshire propensity to import, the propensity to save, and tax leakages resulted in the conclusion that the gross vacation-travel expenditures of $135 million in 1958 generated an estimated $220 million of the state's income.[12] In other words, the state's income would have been $220 million less if nothing had been spent in the state on vacations and pleasure travel and if no part of the $135 million had been spent within New Hampshire for other goods or services. Since these estimates involved many rough calculations based on personal judgment, they should be considered indicators of the order of magnitude rather than refined and accurate measures.

Obviously it requires time for these successive rounds of spending to take place. The income generated in 1958 actually resulted in part from the carry-over effects of tourist expenditures of the preceding year or two and in part to the expenditures made in that year. Likewise, the effects of tourist spending in 1958 carried over into the next couple of years. Provided that there were no sharp year-to-year changes in the various measures discussed above, the estimate of $220 million for 1958 would not be thrown off significantly by neglecting the time factor.

CONCLUSION

The methodology of this study of New Hampshire is relevant to studies of metropolitan regions which attempt to determine the economic base. Such studies try to estimate the receipts from the sale of recreation and vacation services to people from outside the metropolitan region. The conclusions concerning the two methods used in New Hampshire are reinforced when considering their applicability to metropolitan regions. In particular, the second method is less applicable in a metropolitan region where a larger share of expenditures is made by local inhabitants. This method would be most applicable to resort areas consisting of individual counties or groups of counties where most of the spending stems from visitors. The second method holds the most promise for large metropolitan areas. With more time and money, the data can be improved over that used in the New Hampshire study. It is certainly less difficult to collect the desired data from a sample of recipients of vacation-travel expenditures than it is to collect them from the much larger number of actual vacationers and pleasure travelers.

NOTES

[1]Gilbert Cross, "The Costly Crush to Get Outdoors," *Fortune*, July, 1962, p. 157.

[2]Charles E. Silberman, "The Money Left Over for the Good Life," *Fortune*, November, 1959, p. 134ff; American Automobile Association, Profile of the American Tourist, 1960-1961 edition, Washington, p. 3; National Association of Travel Organizations, "Expenditures by U.S. Citizens at Home and Abroad---1935-1957" (mimeographed), June, 1958. The AAA estimated the total to be "something like $25 billion" currently, so their estimate for 1958 presumably would have been at least $20 billion.

[3]National Association of Travel Organizations, op. cit.

[4]This paragraph is based on Tourist Travel Trends, Bureau of Business Research, University of Colorado, 1961, pp. 38-39.

[5]I. V. Fine and E. E. Werner, Private Cottages in Wisconsin, Wisconsin Vacation-Recreation Papers, Vol. 1, No. 4, Bureau of Business Research & Service, University of Wisconsin, April, 1960.

[6]College students are an exception. They are included in the population of the communities where the colleges are located.

[7]Comparable figures for 1958 were not yet published when these computations were made.

[8]U. S., Bureau of Labor Statistics, and the Wharton School, University of Pennsylvania, Study of Consumer Expenditures, Income, and Savings, University of Pennsylvania, 1957, 18 volumes.

[9]It was assumed that there was no "money illusion."

[10]Horace W. Harding, The Kansas Tourist Survey, 1952, Bureau of Business Research, University of Kansas, October, 1953.

[11]Income originated is similar to the concept of value added in manufacturing. Value added is obtained by subtracting from the value of shipments the cost of materials, supplies, containers, fuel, purchased electric energy, and contract work. Income originated is more net in that additional costs are subtracted from value of shipments. Among them are depreciation charges, state and local taxes, allowances for bad debts, telephone service, and a few other items. Income originated is simply the sum of factor payments--wages and salaries, profits, rents and interest.

[12]For an explanation of the calculations, see New Hampshire Department of Resources and Economic Development, Vacation Travel Business in New Hampshire, 1962, pp. 144-49.

Comments on Part III

COMMENT by Jack L. Knetsch

On Pfister and Richards

I would like to address my comments in the main to issues touched
on or raised in the Pfister paper on recreation expenditure in New Hampshire and somewhat to those of the Richards paper on residential preferences.

I might begin by noting that, having had occasion to look through a
number of tourist and recreation expenditure studies, people attempting
to do work in this general area certainly welcome this kind of attempt to
put meaningful numbers together. For we see here the many ways in
which such estimates can take on varying complexions depending on
choices of assumptions and other conditions. The need for objective
analysis of this type of data is considerable.

Beyond this, however, it seems that there is a real need to structure
this whole question of vacation trips, recreation areas, and patterns of
behavior. And I would suggest that in large part this means looking at
some demand relationships or household consumption patterns of urban
people. Getting information together and deriving estimates of expenditures is no small matter, but it seems that this kind of work really begins to pay off when it is put into a framework of people, particularly
those in urban centers; their consumption patterns; the supplies of various resources, in this case generally recreation resources; the values
that are attached to these; and the resulting policy and planning implications.

By themselves, data on the size of expenditure for this given class
of goods and services is useful. This can be used by people interested in
such things as the flow of expenditure into New Hampshire, in the area's
economic well-being, and for possible promotion of policies to enlarge
certain aspects of this kind of expenditure. But even these people could
use more information than provided by these totals.

If we want to gain a better understanding of this general area of
household expenditure, we need to know about the relationships between
travel, recreation, and population characteristics. We may want to
know, for example, in the case of New Hampshire, why given numbers of
people make certain expenditures; where they come from; the effect of
other similar areas on the drawing power of this area; the effects of such
things as transportation improvements on numbers going to this and other
regions; the relation of improved recreation areas close to urban centers
on travel in more remote areas, and possibly on the implications these
may have on residential preferences. These are only some of the issues
that can be involved; I only want here to indicate the types that hinge on
these relationships.

But being able to say very much meaningful about these issues almost necessitates that we know something fairly specific about these
functional relations. For example, we know almost intuitively that, all
else equal, fewer people are going to travel from an urban community
to a more remote area than to one less remote. However, only rarely
do we have any quantitative ideas of this, and we know very little about
how this travel relationship and the expenditure relationship is altered
with: different degrees of attractiveness of the two areas; or with

changes in the characteristics of the tributary zone; or with different alternatives available to potential users of these areas. Nor do we really have good notions about how this pattern is changing over time. We know that travel and recreation are becoming more important and in some cases the rate of growth in demand for resources is not only very high but showing few signs of decreasing, but again we know little about how the different determinants of this demand affect different aspects of the travel and recreation situation.

These things clearly have implications for not only certain portions of consumer expenditure but also for the economic well-being of certain areas; for land-use decisions, pertaining particularly to recreation developments; and for residential preferences of urban people, as such preferences are in part a function of access to recreation areas both within the urban area and outside of it.

Given that these things have some importance, and that it is helpful to know about the important patterns that affect parts of the urban environment that we may be interested in, we might then say something about ways in which we can acquire some insight into these relationships. This, it seems to me, involves very much the observation and analysis of current activity. In large part this is not too much different from many market-type studies of other goods and services which are generally available. In a sense our economy or society produces parks and recreation areas and enjoys their benefits in much the same way as it produces and consumes many or most other goods and services, and I think there is much to be gained by addressing questions in this area in much the way we address those involving these other sectors. While there are some added problems stemming from such things as the necessity of often imputing values to resources, and of the immobility of resources, these I do not believe are as difficult as some seem to imply. The economic significance here is considerable, even if our social calculus is possibly incomplete.

If our interests are in New Hampshire, for example, we might determine how many people visit the area, their expenditures, their characteristics as to incomes, ages, etc., and where they come from. And we would want to know how this use pattern is changing over time. We may want to do the same for areas such as the Maine woods, the Green Mountains, Cape Cod, the Catskills, and possibly others to determine not only the degree of competitiveness and complementarity existing between such areas, but also which segments of the various urban populations are involved or might be expected to become involved.

And much the same thing can be said for determining some of the important supply and demand relationships of areas within and relatively close to urban areas. Here again we are interested in who and how many people make use of varying kinds or types of outdoor areas, and how these depend on distances involved, population characteristics of the area, the characteristics of the open areas, the alternatives open to people, the size and incidence of expenditures for these activities, and importantly how these relationships are changing over time.

With this type of information it seems that we can then begin to impute some values to various policy or program alternatives that might be considered. These would include the cost of doing different things; and also what we may expect to be gained from them. This last, I would think, would include possibly some income flows that might be desirable; the values that people, through their behavior, imply that they attach to various resource uses, which we might call user benefits; and how vari-

ous arrangements may help meet other regional objectives such as
through preservation of strategic open land so as to preclude undesirable
development for one reason or another. All of this would thus seem to
lead to useful conclusions which would have considerable economic consequences and would help guide private and public entities in making decisions that could result in better use of resources.

COMMENT by J. B. Lansing

On Guthrie

Of the three papers presented in this group the one which comes
closest to my own research interests is that by Professor Guthrie. The
problem to which his paper is addressed is clear-cut: are there differences in consumer attitudes and behavior which result from the location
of the residence of consumer units? Do these differences persist when
other independent variables are taken into account?

These questions are of interest to the student of consumer behavior
who is primarily concerned with other independent variables but does not
wish to be misled by confusing the effect of these other variables with the
effect of location. It may be, for example, that high income people who
live near the center of large cities have quite different patterns of expenditure than high income people in suburbs. In addition, students of
urban areas will be interested directly in the effect of the location variables. For example, there is interest in differences in residential
mobility associated with differences in location.

Are Guthrie's statistical methods optimal? He begins by testing for
the significance of differences in the means of dependent variables between areas. Unfortunately, as is mentioned in the paper, the test as
made assumes a simple random sample whereas the actual sample on
which the data are based is a multi-stage, stratified, clustered sample.
Under these circumstances it would have been helpful if we had been
given the actual estimate of the standard error of the difference between
the means instead of the statement that the difference was significant at
the 5 per cent level. When the extensive calculations needed for exact
estimates of sampling errors from a survey have not been carried out,
it is conservative practice to use the three sigma level rather than the
two sigma level of confidence and interpret it in terms of the .05 probability.

A possible second stage of statistical sophistication would have been
to test whether the location variable has any incremental effect on each
dependent variable when the other four variables which Guthrie mentions,
income, education, age, and number of children, have been taken into
account. Instead of testing this type of hypothesis Guthrie goes to a
third level of sophistication immediately. He tests whether the effects
of the four socio-economic variables mentioned differ from one location
to another. He does find interesting interaction effects by this method,
but we are sometimes left without a direct test of the hypothesis that
location would make a difference in addition to the linear effect of the
four other independent variables. One might think of testing this hypothesis by comparing the constant terms in the equations that Guthrie
calculates separately for the several locations. But there are difficulties
with these constant terms. Many of them have negative values in a context where a negative value makes no sense, as in the data for home

ownership. It would seem, therefore, that there is still work to be done either in improving the fit and obtaining reliable estimates of the constant terms or in reverting to what I have called the second stage of sophistication.

What are the main results of the calculations? For some of the dependent variables Guthrie finds clear evidence that the location variables have an effect after allowing for the linear effect of income, age, education, and number of children. For home ownership, tendency to move, and attitudes toward government spending, there is clearly a location effect. There is also evidence that the people who live in different areas differ in their feelings about the chances of getting another job if they lose their job during the next few months.

With regard to travel data, Guthrie finds evidence of differences in automobile travel which he explains, I think correctly, in terms of the differences in automobile ownership between central cities and other areas. He also finds differences in bus travel which he is at a loss to explain. I would suggest that differences in the availability of bus compared to other common carriers may be at the heart of the matter. For example, people in outlying areas ordinarily have good bus service but poor plane service. Hence, they travel more by bus. Special studies which we have done at the Survey Research Center of non-business air, rail, and bus travel all tend to confirm that there are differences among locations in the frequency with which people take trips by the different modes allowing for the effect of the usual socio-economic variables.

Has Guthrie also proved that there are interaction effects, that is, differences in the effect of the other independent variables which depend upon location? Lacking the actual standard errors of the regression coefficients or any test of the differences between coefficients from one equation to another, one is often left with some uncertainty on this point. There are indications of such interactions in the travel data, but the evidence is strongest in the analysis of consumer debt. Even here, however, the evidence is not conclusive. The coefficient of multiple determination never rises above .13. We are not offered an interpretation of the reasons for the existence of such differences as are observed and I, at least, am not able to supply an explanation. One is left wondering what might be done with more careful attention to the shape of the relations of the several independent variables to this particular dependent variable -- are they linear? Might the departures from linearity have different importance in different locations? Over-all, I think, we must conclude that the evidence for the existence of the interaction effects is less persuasive than the evidence for the existence of direct locational effects.

These results raise some interesting theoretical questions. What causes the observed differences in consumer behavior from one location to another? That is, what are the characteristics of urban areas which cause differences in peoples' attitudes and behavior? Can we improve on the classification of locations presented here? What long-range shifts may we expect in the proportion of the population living in different locations? What changes does this suggest may take place in consumer behavior in the future? This paper should cause us to turn to these questions with renewed interest.

On Pfister

I welcome the paper by Richard L. Pfister as a contribution to a field in which there has been too little sober economic analysis. The initial discussion of existing estimates of expenditures on vacations and pleasure travel seems to me particularly helpful. It is hard to disagree with Pfister's conclusion that the main difficulty is the lack of reliable information about tourist expenditures.

How, then, can better estimates be constructed? Pfister makes use of two methods, but expresses a preference for the first. I share his preference. I note, however, that a major component of the total, spending by owners and residents of seasonal homes and cottages, had to be estimated on the shaky basis of a single survey in Wisconsin using a mail questionnaire. There is a need for better studies of expenditures in connection with vacation homes. But there is no reason why such surveys cannot be made. Similarly, there is no basic obstacle to improvements in the breakdown of the total receipts of various businesses between sales to vacationers and sales to others.

The second method Pfister uses, it will be recalled, rests on comparing estimates based on the 1958 Census of Business of sales of selected goods and services in New Hampshire with what one would predict from average national sales per capita, adjusted for differences in income level. While this method has something to be said for it in a state like New Hampshire where most vacation spending is by non-residential people, I do not believe it could be applied to larger states where people often spend their vacations in the state where they reside. The method leads to an estimate of the excess of spending in New Hampshire over what would be expected for a state like it on the basis of the national average. To complete his estimate Pfister needs to know, first, how much spending there is by New Hampshire residents out of the state. He is forced to rely on a single survey conducted in Kansas in 1952 concerning out-of-state spending by Kansans. The only adjustments made are for the populations of the two states and the change in prices from 1952 to 1958. Pfister comes up with an estimate of $25 million which he arbitrarily adjusts to $15 million for reasons not entirely clear to me. Second, he needs to know the amount of in-state vacation travel spending by residents of the state. This amount he is unable to estimate directly by his second method. The difficulty here is partly logical. It seems to me a dubious, even an impossible, procedure to estimate spending for non-vacation purposes by adjusting national averages of total outlay of a certain type for income elasticity. We know, on the contrary, that expenditures for vacations and pleasure travel are themselves elastic with respect to income. To measure this elasticity more exactly should be one of the objectives of the research itself. Where one is interested primarily in the amount of spending by people from outside of a state, as in New Hampshire, it is probably not too bad an approximation to make an estimate of total spending by New Hampshire residents and assume the rest is by people from other states.

It is interesting to speculate whether better results might not be obtained by carrying this method one step further and working with data at the county level. In some states the areas where people take vacations and the areas where they live the year around can be distinguished. There may well be groups of counties for which this type of estimate would make sense, that is, counties where the vacation spending by permanent residents is so small that it can be ignored. An additional possibility would be to interview a cross-section of the population of the state concerning how much they spent on vacations inside the state and outside the state. Let us hope that the necessary resources are made available so that data collection can proceed to the point where improved estimates of spending on vacation and pleasure travel become possible.

COMMENT by Richard F. Muth

On Richards

Many writers have considered the effect of distance and transport costs on the residential location decisions of urban households. Some of the implications of distance and transport cost theories, namely that population densities decline with distance from the center of a city and that workers tend to concentrate around their work places where the latter are outside the central business district, are consistent with behavior actually observed. In his paper, however, Professor Richards asserts that distance and transport cost theories are insufficient to explain residential location choices. He further suggests that the preferences of households for residential amenities may be of primary importance in determining patterns of urban residential location. I share with Professor Richards the belief that better theories of residential location would be desirable. But the hypothesis that households are influenced by the desire for residential amenities is so general that it adds little to our knowledge about residential location choice.

The only bit of evidence which Richards cites against distance and transport cost theories of residential location choice is the fact that high income households generally do not live in areas immediately adjacent to central work places. He argues that such groups ought to be as willing to pay the higher housing costs associated with central locations to avoid the costs of travel as lower income groups. Now certainly higher income persons generally have a higher opportunity for cost of time spent in travel because of their higher wage rates. These higher transport costs would tend to make central areas more attractive for them. However, higher income households also generally purchase more housing, including land, than lower income households. Since land tends to be less costly in the outer parts of cities than close to their centers, higher income households can save more on land costs by locating away from central work places than can lower income households. Whether this latter factor is sufficiently strong to outweigh the greater transport costs of higher income households is primarily an empirical question. But it is not necessarily inconsistent with distance and transport cost theories for higher income households to avoid location near city centers.

Nor do I find the evidence Professor Richards cites as favoring the amenities hypothesis very convincing. Two kinds of evidence are given. The first is that in Baton Rouge low income migrants from rural areas have tended to locate on the outer edges of the city, which is attributed to their desires for rural amenities. The fact that in larger cities low-income migrants tend to locate near the central part is attributed to their willingness to substitute accessibility to the city center for residential amenities. Now of course residential amenities have their price, the time and cost of transport to the city center in this case, and as city size increases the price of the rural amenities to be found on the urban fringe increases. But transport costs increase for middle and upper income residents as well, and there is no strong reason to believe that their demands for amenities are more inelastic than those of lower income persons.

The other piece of evidence is the fact observed by Professor Richards that in Baton Rouge physicians and professors live closer to their places of employment than do dentists, lawyers, and managers and proprietors. He suggests that this may be partly due to the residential amenities associated with a hospital or a university. The relations observed in Baton Rouge may be due to special circumstances of that city, since I am not at all certain that college professors generally tend to congregate in areas immediately surrounding a university. While I certainly find it believable that a higher proportion of professors than plant managers would find residences close to their place of work desirable, I am suspicious that physicians find hospitals more appealing than dentists find their offices or lawyers the court house. Actually, the behavior of physicians and professors seems quite consistent with the transport cost hypothesis. Both groups have very irregular working hours, and the latter require extra work trips. For this reason transport costs are greater for physicians and professors than for dentists and lawyers, whose toothaches and divorce cases can usually wait until morning.

But my major criticism of the amenities hypothesis as advanced by Professor Richards is that it is consistent with virtually any kind of behavior and, hence, practically useless for prediction. I would certainly not deny that people are influenced by their tastes and by quality differences in their residential location choices and that tastes may vary from one group to another. And I find it fairly easy to think of reasons why various persons might prefer to live where they do because of so-called residential amenities if I already know where they live. But unless one specifies something about the tastes and preferences of different groups and how the spatial pattern of quality differences in houses and neighborhoods are determined, the hypothesis that households are influenced by the desire for amenities is useless for predicting unknown location choices.

On Guthrie

I would also like to make one comment on Professor Guthrie's paper. He finds that for variables such as length of residence and monthly debt payments locational differences are accounted for by income, age, education, and number of children. For others, such as home-ownership, locational differences persist when the above noted independent variables are included in the analysis. I doubt, however, that location itself affects human behavior. Location in a cross-section regression analysis, like a time-trend in a time-series regression, is but a proxy for other variables, some of which we understand but find very difficult to measure and others which we do not understand at all. Home-ownership is a good example of the former. Owner-occupancy for most people is limited to houses, since the market for co-operative apartments is still very thin. Houses tend to use more land relative to other productive factors than apartments, and land tends to be relatively expensive in the central parts of large cities. For this reason people who prefer ownership to renting will tend to locate outside the central parts of large cities, while people who prefer a location near the city center will tend to find housing cheaper in apartments and become renters. In cases where the location variable is not included as a proxy for another variable, its inclusion is a good check for the omission of important variables from a cross-section analysis. Where significant locational differences are found it is but an indication that the analysis is incomplete. For, I would argue, location itself is nothing more than a particular combination of values of variables upon which human behavior depends.

PART IV.   *AREA WAGE DIFFERENTIALS AND MIGRATION*

The Impact of Puerto Rican Migration to the United States

by Belton M. Fleisher

Puerto Rican migration in the stream of American history - American internal migration as affected by the entry of Puerto Ricans - Impact of Puerto Rican migration on earnings and employment in New York City - The factors affecting these earnings and employment levels - Conclusion.

The Puerto Rican migration to the United States--principally to the New York metropolitan region (see Table 1)--is the most recent in a long series of similar population movements affecting the U.S. labor market. The purpose of this paper is to examine the effects of the Puerto Rican inflow upon: (1) population flows within the United States and (2) earnings[1] and industrial location. Since the migration has been predominantly (although not exclusively) directed toward the New York metropolitan region, the analysis will be conducted principally by comparing various aspects of New York's labor market behavior with those in other regions.

A discussion of the impact of Puerto Rican migration is important for two reasons: the first is that very little has been said heretofore regarding the impact of the migration, although a great deal has been said about the factors attracting Puerto Ricans to the United States;[2] the second is that little is known about the response of wages and internal population movements to an influx of new population from outside the continent.

The problem of separating causal relationships from historical sequences is always a difficult one, requiring essentially a statement about what would have happened had a particular event not occurred. In order to gain the proper perspective for an analysis of the effects of Puerto

Author's note. The author wishes to acknowledge the helpful advice of Donald V. T. Bear, W. Lee Hansen, H. G. Lewis, M. W. Reder, and Albert Rees.

Table 1. Net Migration: Puerto Rico to the United States

Decade totals[1] (in thousands)		Yearly totals[2]		
Period	Net migration	Year	Net departure	Percentage going to New York City[3]
1910-19	11	1947	23,955	95
1920-29	42	1948	32,661	90
1930-39	18	1949	29,182	90
1946-55	406	1950	38,867	85
		1951	49,346	80
		1952	61,729	77
		1953	69,382	75
		1954	22,871	75
		1955	42,350	70
		1956	48,833	65
		1957	36,631	n.a.
		1958	25,982	n.a.

[1]A. J. Jaffe, People, Jobs and Economic Development (Glencoe, Illinois, 1959), p. 65. Jaffe's figures refer to total net migration from Puerto Rico to the United States through 1939. The 1946-55 figures are for total net departures from Puerto Rico.

[2]Puerto Rico, Planning Board, Bureau of Economics and Statistics, Statistical Yearbook: Historical Series, pp. 218 ff.

[3]New York City, Department of City Planning, Puerto Rican Migration to New York City. According to this source, there were approximately 34,500 Puerto Ricans living in New Jersey in 1956. This was the largest group of Puerto Ricans living in the continental United States outside New York, and there is a presumption that a significant proportion of those Puerto Ricans not migrating to New York City nevertheless end up in the greater metropolitan region.

Rican immigration, it is necessary to inquire into the likely effects of the great European migrations to the United States. Consequently, a rather large portion of this essay is devoted to analyzing the impact of immigration from European countries on the assumption that the basic response mechanisms operating now have much in common with those of the past.[3]

Further, we shall explore the effects of Puerto Rican migration upon migration flows internal to the United States. This is done principally by trying to predict, on the basis of historical experience, what these internal flows would have been in the absence of Puerto Rican migration and then by comparing this prediction with the actual course of events.

And finally, we shall examine the impact upon the wage structure (in manufacturing industries) in a similar manner. Since constructing wage series for a large number of finely defined industries over any considerable length of time is nearly impossible due to the various definitional problems involved, I have chosen to concentrate attention upon the single industry most important for employing Puerto Ricans and previous immigrant groups--women's and children's garments. The behavior of wages in this industry is examined along with the group of industries which employ a few Puerto Ricans and the group of remaining industries which employ considerable numbers of Puerto Ricans. The results of these examinations are the basis of suggestive if not definitive conclusions regarding the impact of the migration upon wage structure--especially upon interregional wages.

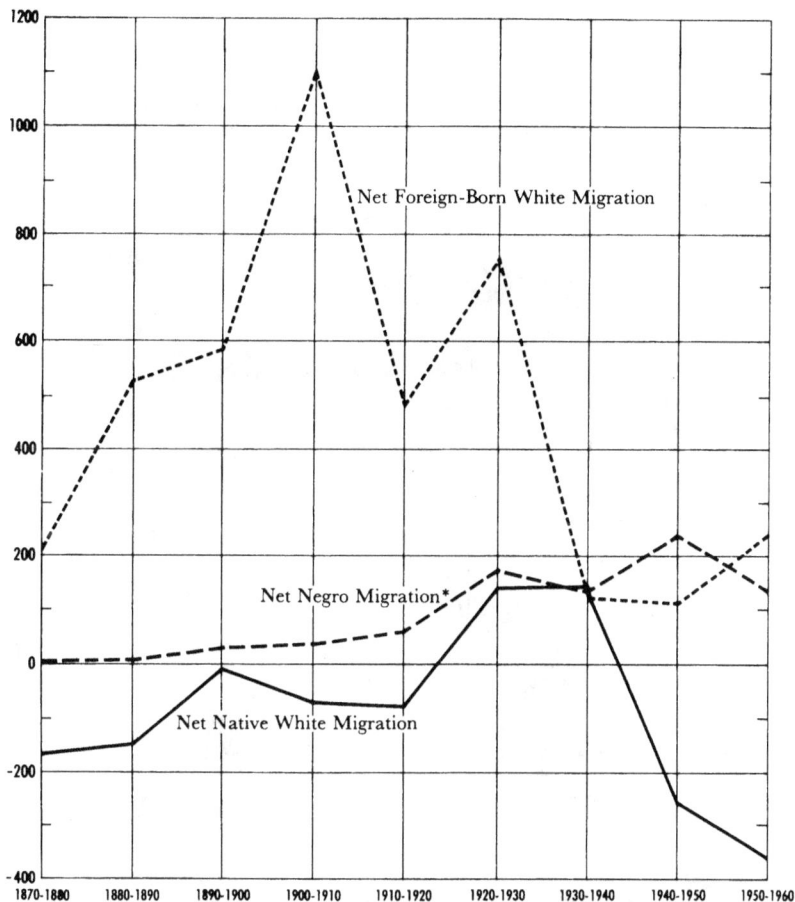

For 1950-1960 the figures are for net non-white migration

Figure 1. Net migration to New York State. (Note: The 1870-1950 estimates are from Everett S. Lee, et. al., Population Redistribution and Economic Growth in the United States, 1870-1950; those for 1950-60 are my own, derived insofar as possible in the same manner as those from the former source. Since complete and final 1960 Census data are not yet available in all cases, it has been necessary to estimate 1950-60 net migration for all age groups [Lee excludes ages 0-10] and to use actual number of births [U.S. Department of Health, Education, and Welfare, Vital Statistics of the United States] rather than birth rates by age groups in predicting 1960 population.)

THE IMPACT UPON MOVEMENTS INTERNAL TO THE CONTINENTAL UNITED STATES

It is impossible to understand what effect net Puerto Rican immigration may have had upon the United States labor market without placing it in its historical context. When one examines the history of immigration to the United States[4] and then notes that a large proportion of Puerto Rican migrants have been settling in the New York metropolitan region, it becomes obvious that a comparison with European immigration is appropriate. While the rate of inflow of Puerto Ricans has not reached the maximum ever attained by European immigrants, it does appear to be of the same order of magnitude. Migration from Europe reached its peak in the decade 1900-1910, when the net inflow of foreign-born whites to New York State amounted to 1,100,000 (see Figure 1). During the decade

Thousands

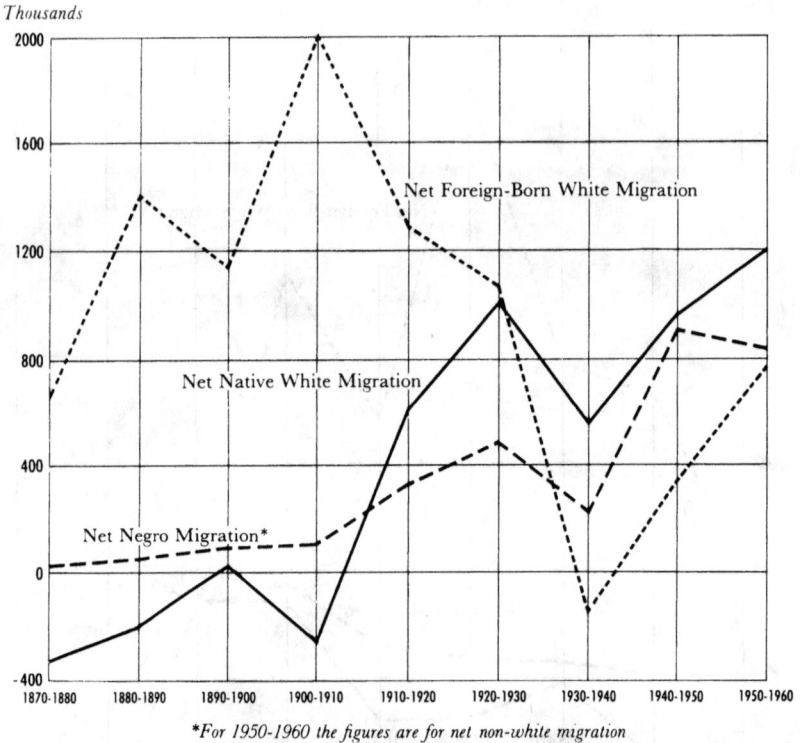

For 1950-1960 the figures are for net non-white migration

Figure 2. Net migration to California, Illinois, Massachusetts, Michigan, Missouri, Ohio, Pennsylvania, and the District of Columbia. Excluding New York, these areas contained the ten largest SMA's in 1950. (See note, Figure 1.)

following World War II, net Puerto Rican immigration to the United States was over 400,000 and most of this was directed toward the New York City region. In fact, since European immigration affected diverse regions of the United States (see Figure 2), whereas the Puerto Rican immigration has so far been much more localized in its impact, one suspects that the effect of Puerto Ricans upon the New York region's--relative to the nation's--economy could be substantial indeed by comparison.

An examination of Figure 1 brings out the following patterns. Net immigration of native-born whites to New York State has been negative in all periods except 1920-40. (A possible explanation of this will be presented below.) Net migration of Negroes has always been positive and has fallen in only two periods: from the decade of the 1920's to the decade of the 1930's and from the 1940's to the 1950's. However, the flow attained large proportions only after the curtailment of European immigration by World War I and the restrictive legislation of the 1920's. Historically, European immigration appears to have damped the migration of both native whites and Negroes to New York State. Following World War I both native white and Negro immigration rose substantially, but the increase in Negro immigration has been more consistent over the years.

Figure 2 shows the same information as Figure 1, but for the states which contained the ten largest standard metropolitan statistical areas (SMA's), except New York, in 1950. Here we may discern a similar set of relationships between net migration of foreign-born whites, native whites, and Negroes, except that the migration of native whites is less often negative. Later on, by comparing the two figures, we shall make inferences regarding what the behavior of migration to New York State might have been in the absence of a Puerto Rican influx.

On the basis of the relationships exhibited in Figures 1 and 2, it is tempting to hypothesize that several principal labor markets in the United States (and New York City's in particular) were characterized, especially before World War I, by a constant inflow of new labor, competitive with workers already in these areas. The new labor, having recently arrived in the United States and not being able to take advantage of many opportunities involving further migration, and probably being of somewhat lower quality than other workers due to difficulties involving language, education, etc., tended to keep wages low. Thus it seems likely that there was a constant labor turnover, with new immigrants replacing old immigrants and native whites. [5,6] When the supply of European immigrants was cut off, those labor markets which had become dependent upon it in order to maintain competitive positions within the United States' economy gradually found it necessary to raise wages relative to other areas in order to retain adequate labor supplies. For gradually the older migrants and their offspring were able to take advantage of better opportunities. Older immigrants were, of course, diminishing in number. This hypothesis can explain the rise in the net migration of both Negroes and native-born whites to the areas of Figures 1 and 2 during the period 1910-30.

The response of intra-continental population movements to the cessation of European immigration was not uniform. Total (intra-continental plus extra-continental) net immigration to New York State fell by over 50 per cent between 1900-10 and 1910-20, reaching the 1900-10 level once again only during the 1920's. On the other hand, the response of intra-continental flows to the states represented in Figure 2 was sufficient to keep total net immigration rising throughout this period.

One explanation of this differential response of intra-continental migration is that New York is situated rather far away from the principal rural southern sources of excess population in the United States. Thus, migrants must first pass through Saint Louis, Washington, D.C., Philadelphia, etc., before reaching New York City. [7] Massachusetts and California are the only states represented in Figure 2 which might be considered farther away from the principal sources of intra-continental migration than is New York. Between 1910-20 and 1920-30, Negro migration to these two states rose by 12 per cent and 60 per cent, respectively, compared with approximately 100 per cent for New York and 200 per cent for all the states represented in Figure 2. The behavior of native white migration is more difficult to describe, because the net flow was negative to most of the areas in 1900-10, but positive in 1910-20. The flow to New York increased less than to the other states, however.

The Great Depression of the 1930's naturally had a damping effect upon net immigration of all kinds, but it seems to have had smaller impact upon the flow to New York than to the states represented in Figure 2. The flows declined as follows between the two decades: foreign-born whites, 84 per cent to New York, from 1,077,000 to -146,700 to the other states; native whites, unchanged to New York, 45 per cent to the other states; Negro, 18 per cent to New York, 53 per cent to the other states. Interestingly, there may have been a redistribution of foreign-born whites toward New York during the 1930's, since net migration of this group was negative in Figure 2 but positive in Figure 1. Three explanations of this behavior present themselves a priori. One is that the depression did not affect New York State as severely as the others. Another is that the demand for labor did not fall as sharply in New York, because the cessation of European immigration had had a greater effect on the labor market there than elsewhere. A third, related, explanation is associated with the lagged response of intra-continental migration described above; the failure of New York to maintain its pre-World War I level of total immigration between 1910-1920 may have had long-run effects which moderated the degree of unemployment brought on by the depression of the 1930's. The third explanation has some bearing upon the behavior of interregional earnings, and is discussed in this context below.

During the 1940's and 1950's, the immigration of foreign-born whites continued at relatively low levels imposed by legal restrictions (although the inflow seems to have increased in the period following the depression and World War II). [8] Intra-continental movement of Negroes continued at a high level, tapering off slightly for the states in Figure 2 during 1950-60, and quite sharply for New York during the same decade. Between the 1930's and the 1940's, net immigration of Negroes rose over fourfold to the states in Figure 2, but less than twofold to New York. Net migration of native whites to the states in that Figure rose to new, high levels following the depression, but became negative to New York State. [9,10]

One is at a loss to explain the sharp decline in intra-continental migration to New York relative to other urban areas of the United States without resorting to the sharp upswing in Puerto Rican immigration. The apparent impact is indeed striking. It seems likely that Puerto Rican immigration curtailed the movement of both native whites and Negroes to New York, but the impact on the movement of native whites seems to have been greater.

The conclusions drawn at the end of the next section provide further evidence regarding the effect of Puerto Rican immigration on population flows within the United States. There we explore the immigration's impact upon earnings in New York relative to those in other parts of the United States. The depressing of relative earnings in New York is no doubt due at least in part to the response of the economic system to an inflow of population from outside the continent; thus the impact of immigration from other areas is spread throughout the economic system as internal population flows are redirected.

THE IMPACT OF PUERTO RICAN IMMIGRATION ON
EARNINGS AND EMPLOYMENT

In this section we investigate whether Puerto Rican immigration has affected noticeably interregional earnings and employment--especially in those industries where Puerto Ricans have most frequently been employed.

Precise knowledge of the relative quantitative importance of individual industries as sources of employment for Puerto Ricans and of Puerto Ricans as sources of labor to these industries is unavailable; but we do have knowledge of the names of the industries. According to Segal:

> Both Puerto Ricans and Negroes, male and female, are found
> in a variety of industries and services and they are becoming a
> larger proportion of the employees in some of these activities--the
> assembly of television sets being one striking example. Our inter-

Table 2. Net Migration of Whites and Non-Whites to Selected Standard Metropolitan Statistical Areas

SMA	1940-50[1]			1950-60[2]			White population 1950	Non-white population 1950
	White	Non-white	Total	White	Non-white	Total		
Chicago	-110,984	207,054	96,070	91,420	205,756	297,176	4,981,566	604,430
Cleveland	-12,682	54,654	41,792	34,280	57,702	91,982	1,311,632	153,879
Detroit	87,331	150,136	237,467	6,921	93,606	99,527	2,654,253	361,944
Los Angeles	934,196	123,593	1,057,789	1,375,753	270,050	1,595,803	4,090,233	275,178
Philadelphia	27,183	96,564	123,747	94,111	85,803	179,914	3,186,470	484,578
Pittsburgh	-126,110	9,620	-116,440	-106,670	5,988	-100,682	2,076,016	137,220
San Francisco	404,520	122,622	527,142	122,777	73,552	196,329	2,030,135	210,632
St. Louis	43,582	43,283	86,865	4,801	26,964	31,771	1,464,359	216,922
Washington, D.C.	227,083	67,044	294,127	145,473	65,290	210,763	1,121,493	342,596
Total	1,474,118	874,570	2,348,559	1,768,866	834,711	2,602,583	22,916,157	2,787,379
New York-Northeastern New Jersey[3]	-261,868	272,700	+10,832	-234,090	+291,032	+56,942	11,866,122	1,045,872

[1] Source: Donald J. Bogue, Components of Population Change: 1940-50 (Oxford, Ohio, 1957).

[2] The estimates for 1950-60 were made in a manner similar to Bogue's and are, therefore, comparable. The sources are U.S., Bureau of the Census, Census of Population: 1950, Vol. II, "Characteristics of the Population," and U.S. Department of Health, Education, and Welfare, Vital Statistics of the United States.

[3] The estimates of net migration to the New York-Northeastern New Jersey Consolidated Area exclude Puerto Ricans.

Note: The Standard Metropolitan Statistical Areas were the nine largest besides New York in 1950. Boston was excluded because birth data were not available according to color for the entire 1950-60 decade.

views show that the following industries [besides apparel] constitute
a substantial source of employment for both Puerto Ricans and
Negroes in the Region: Bakeries, Office building services, Confec-
tionery, Canning plants, Paint and varnish plants, Paper files and
envelopes, Paperboard boxes, Photographic supplies, Maritime
service, Television plants, Toilet preparations, Toys.

In addition, the following industries and services employ substantial numbers of Negroes, though not Puerto Ricans: Building
trades, Department stores, Trucking, Domestic help, Local transit.[11]

Because earnings and employment data are available from the Census
of Manufactures over a relatively long period of time, I have chosen to
confine this study to the impact on manufacturing industries. Furthermore, not all the manufacturing industries listed above can be included,
since statistics on what are now three-digit classifications are the most
finely detailed available over the span of time required. Women's and
Children's Garments[12] represent the apparel industry, and the other industries included in the study are: Paperboard containers and boxes,
Paints and allied products, Canning, preserving, and freezing, Confec-
tionery, Luggage, and Toys and sporting goods. (These industries will
be referred to from now on as the Puerto Rican industries.) Data for the
states of New York and New Jersey[13] are compared with those for the
entire United States.

Women's and Children's Garments accounts for by far the largest
share of employment in the Puerto Rican industries listed above, and this
industry will receive most of our attention; its production workers constituted 68 per cent of the employment in all these industries for New
York and New Jersey in 1958. (These and other references to employment and earnings in the industries studied here are derived from the
same sources as are Figures 3, 4, and 5.) According to an unofficial
ILGWU estimate, there were about 40,000 to 50,000 Puerto Rican union
members in 1961. On the basis of the 1958 level of employment in
Women's and Children's Garments for New York and New Jersey, and
assuming that all Puerto Rican garment workers found employment in
these two states, we can estimate that Puerto Ricans accounted for about
20 per cent of the New York-New Jersey employment. If we assume a
60 per cent labor force participation rate for Puerto Ricans (about that
reported in the 1950 Census),[14] it follows that about 20 per cent of employed Puerto Ricans are to be found in Women's and Children's Garments. These estimates are not violently inconsistent with, although
larger than, those in Herberg's study of the ethnic composition of the

Table 3. Percentage Composition of Local 22 Membership[1]

Year	Jewish	Latin-American	Negro	Other
1934	70.5	6.5	9.5	13.5
1945	63.4	8.1	14.5	14.0
1948	53.4	12.0	17.0	16.2
1953	51.0	16.2	15.4	7.4

Source: Will Herberg, "The Old-Timers and the Newcomers: Ethnic Group
Relations in a Needle Trades Union," Journal of Social Issues, 1953, No. 9,
pp. 12-13.

[1]Local 22 contained slightly less than 45 per cent of the members of the New
York dressmakers union--the Joint Board of Waist and Dressmakers Unions of
Greater New York--in 1953.

ILGWU's Local 22 (which includes all "non-Italian" dress workers except cutters and pressers). (See Table 3.) They would suggest that the trends he pointed out as of 1953 have continued to operate.

In order to give special attention to the behavior of employment and earnings in Women's and Children's Garments, we shall analyze separately three subgroups of industries: Women's and Children's Garments, the remaining Puerto Rican industries, and the non-Puerto Rican industries.

One needs merely to consult any one of a number of studies of the garment industry to be made aware of the crucial importance of European immigration for the original location of the industry principally in the New York region.[15] Figure 5 shows the relationship between New York-New Jersey's share of national employment and the level of average annual earnings per production worker in these two states divided by that in the entire nation. (These variables will hereafter be called relative employment and relative earnings, respectively.) The two series appear inversely related until after World War II, and both behave consistently with an hypothesis emphasizing the importance of immigration for the location of industry in the New York region. The employment share reaches a peak during the 1900-1910 period, levels off and then declines following the restriction of immigration by World War I and the restrictive legislation of the early 1920's. At the same time, relative earnings reaches a trough, levels off, and gradually rises, reaching a peak in the late 1920's and early 1930's.

The behavior of relative employment and earnings during the 1930's is more difficult to rationalize. Average annual earnings fell to about 60 per cent of pre-depression levels in both New York-New Jersey and throughout the nation during the early years of this period; they subsequently rose less rapidly in New York-New Jersey than elsewhere. On

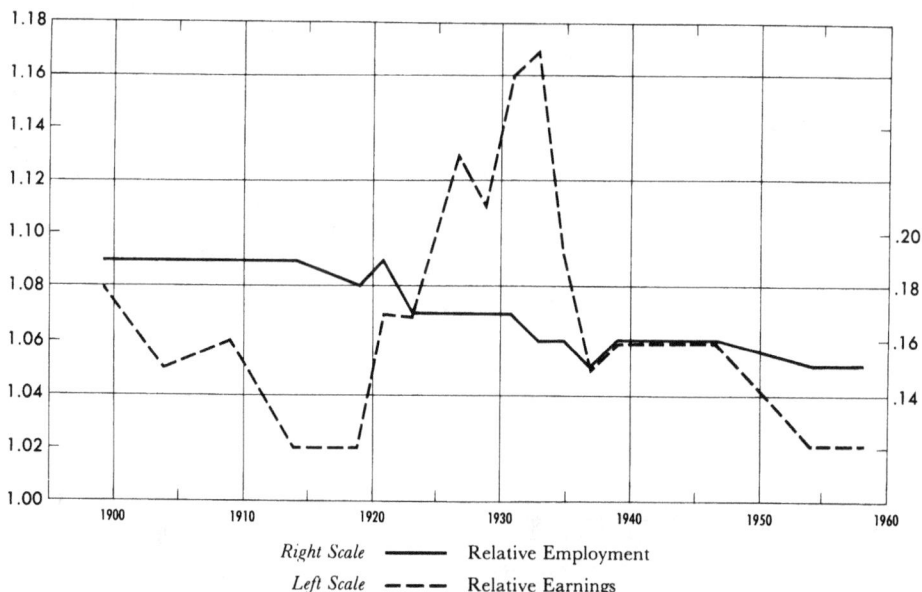

Figure 3. Non-"Puerto Rican" manufacturing industries: Relationship between the New York-New Jersey share of national employment and the level of average annual earnings per production worker in these two states divided by that in the entire nation. (Source: U.S. Census of Manufactures.)

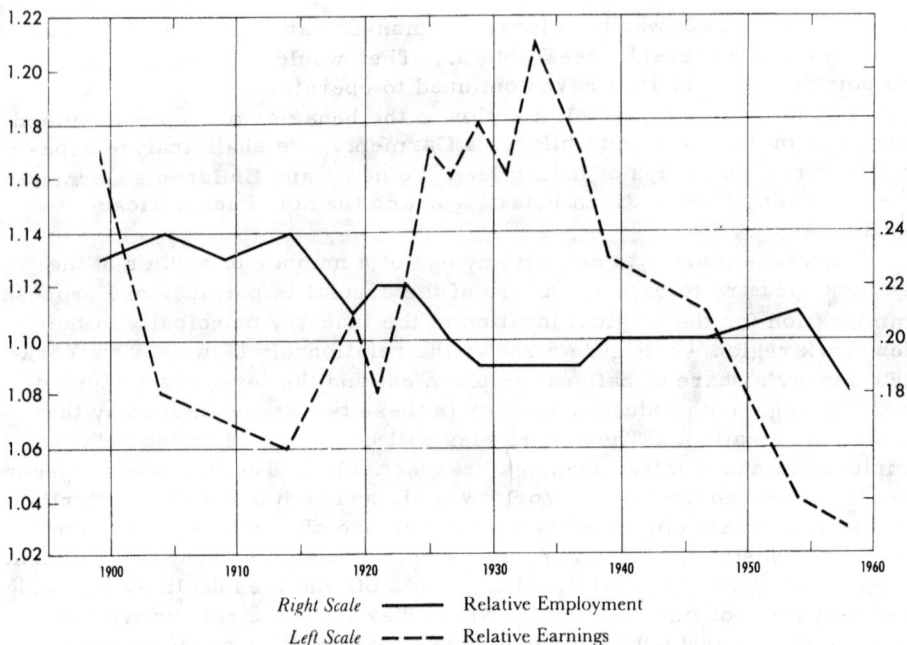

Figure 4. "Puerto Rican" manufacturing industries (excluding garments): Relationship between the New York-New Jersey share of national employment and the level of average annual earnings per production worker in these two states divided by that in the entire nation. (Source: U.S. Census of Manufactures.)

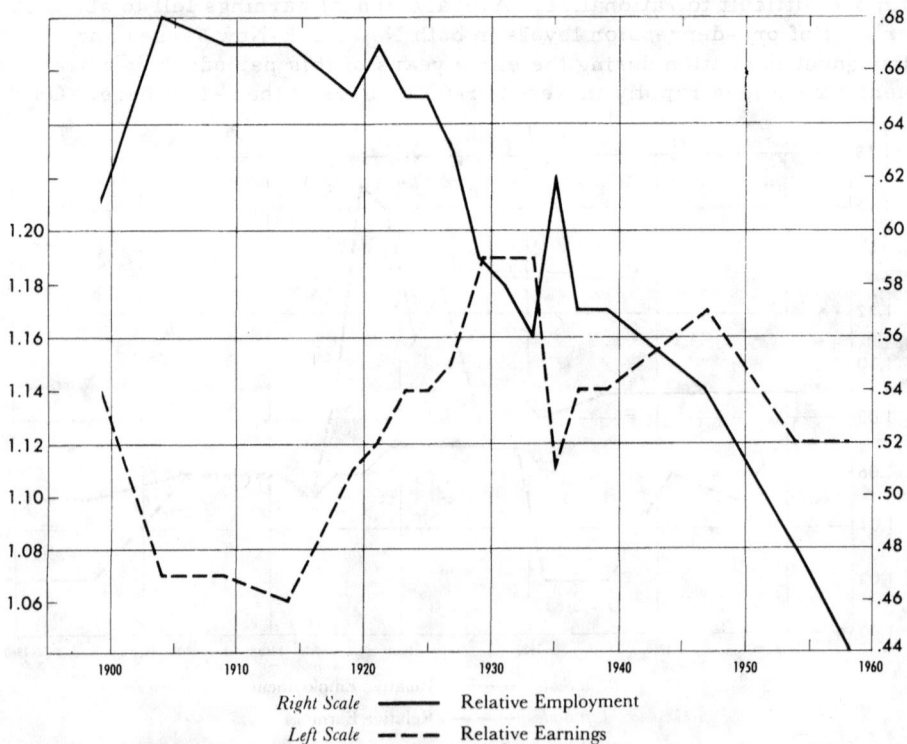

Figure 5. Women's and children's garments: Relationship between the New York-New Jersey share of national employment and the level of average annual earnings per production worker in these two states divided by that in the entire nation. (Source: U.S. Census of Manufactures.)

the basis of the behavior of European immigration, there is no reason to expect that New York-New Jersey's relative earnings should have fallen in the middle and late 1930's, but two other factors may have been responsible.

The first factor is a change in regional union strength. It is likely that union organization first became strong in the older garment center and later spread to the rest of the United States. Thus, various acts facilitating labor organization which were passed during the New Deal period may have resulted in an improvement in New York's competitive position as wages rose more rapidly elsewhere. In fact, the ILGWU had much ground to gain in the early 1930's due to intra-union strife which had sharply reduced membership during the late 1920's.[16] Between 1930 and 1934, union membership rose from 50,800 to 200,000[17] while employment of production workers for the United States rose from 173,000 to 220,000. Thus, by the latter year, a very high proportion of the industry had become unionized. In 1941, union membership was 280,000,[18] and the number of production workers was probably in the neighborhood of 300,000; union strength relative to the size of the industry changed little in the late 1930's.

It is quite likely that as union strength grew between 1930 and 1934, many areas outside New York were organized--distributing union strength more uniformly over the various garment producing centers than it had been previously. It is noteworthy in this context that New York-New Jersey's relative earnings fell sharply between 1933 and 1935, and then rose slightly to remain roughly constant for the remainder of the 1930's.

A second factor which may have contributed to the decline in New York-New Jersey's relative earnings during the 1930's is the apparent lag in the response of net intra-continental migration to the New York region during the 1910-1920 period. The lag may have contributed to the concurrent rise in New York-New Jersey's relative earnings which was followed by a smaller curtailment of net intra-continental migration to New York than to other regions of the United States during the 1930's.

It is entirely possible that changes in regional union strength and the behavior of intra-continental migration both influenced the behavior of relative earnings. The present state of our knowledge unfortunately does not provide enough information to determine their relative importances. One bit of information which tends to cast doubt upon the importance of unionism is the similar behavior of relative earnings for the industries represented in Figures 3, 4, and 5. While the regional distribution of unionism may have been changing for these industries as well as for Women's and Misses' Garments during the 1930's, the amount of such change was probably much smaller. Ten and nine-tenths per cent of the workers in all manufacturing were union members in 1929,[19] and 22.8 per cent in 1939.[20] The rate of increase in the proportion of workers unionized was somewhat smaller than for Women's and Children's Garments, but much more important, the highest degree of unionism attained was much lower. The number of workers directly affected by union organization was much smaller relative to total employment for all manufacturing than it was for Women's and Children's Garments alone, yet all three groups of industries analyzed here exhibited similar behavior of relative earnings during the 1930's.

The behavior of relative earnings in Women's and Children's Garments after 1939 suggests that the immigration of Puerto Ricans following World War II had an impact similar to that of previous migrations.

However, throughout the period 1939-1958, the New York-New Jersey region's share of national employment was falling (absolute employment in Women's and Children's Garments actually declined between 1954 and 1958), and it is difficult to separate the effects of immigration and the relatively slow growth of demand for labor. At least one student of this subject suggests that competition in the garment industry from other regions of the United States can explain practically all of the postwar decline in relative earnings.[21] It is my opinion, however, that the evidence indicates Puerto Rican immigration has probably had a significant impact upon wages and earnings in the New York-New Jersey region.

We should note, first of all, that relative earnings in the New York-New Jersey region have not yet fallen below their average level for the 1910-1930 period, throughout which the region's share of national employment was declining.[22] Neither have they reached the historical low attained in the years just prior to World War I; nor has net immigration since the war been as great as it was prior to 1914.

If immigration of labor to the New York region has in the past enabled the area to compete with lower cost labor markets in other parts of the United States, and if other factors affecting the garment industry, such as those described in the following paragraphs, have not changed, then we should not expect the competition of labor-plentiful regions to affect relative earnings significantly. As immigration dwindles, we should expect relative earnings in New York to begin to rise; marginal firms would then begin transferring operations (or new firms would begin operations in low cost regions, perhaps forcing old, marginal firms out of business). Thus, New York would lose employment relatively to the United States in the presence of rising, not falling relative earnings. We must resort to some technological change or a change in the structure of demand to rationalize a falling employment share together with falling relative earnings.

In fact, some aspects of the garment industry have been changing in recent years. One change has been in fashions; casual and other types of clothing which demand less the kinds of special artistic and production skills that New York has to offer and more the plentiful labor of other regions have become increasingly in demand.[23] In a sense, New York has been losing its comparative advantage in the production of clothing due both to a lack of immigration and to a change in the structure of demand which has made a highly skilled labor supply less important to the garment trades.[24]

Equally and perhaps more important has been the impact of modern highways and trucking services. Now, almost any small eastern town where labor costs are low due to, say, shut down mines, can serve as a production area for New York garment manufacturers. Helfgott[25] estimates the cost of shipping a garment from New York to Wilkes-Barre, Pennsylvania, and back as $.065. Thus, he estimates that a dress which costs $2.50 to manufacture completely in New York (and which wholesales for $6.75) could be manufactured for $2.06 including transportation if shipped out.

It is perhaps a moot point whether technological and demand changes have been as important as immigration in determining relative earnings in Women's and Children's Garments since World War II, but the behavior of relative earnings and employment between 1954 and 1958 leads one to believe that immigration has been the more important factor. Note that relative earnings declined between 1947 and 1954, but remained unchanged during the next four years. On the other hand, New York-New

Jersey's employment share dropped steadily throughout the entire period; the region actually lost 26,000 production workers in this industry between 1954 and 1958 while the remainder of the country gained 8,000. This was the first time since 1931-33 that New York-New Jersey failed to increase its absolute employment in the industry. If changes in technology and the structure of demand had been important in causing the 1947-54 decline in New York-New Jersey's relative earnings and employment share, why did they not have the same effect in 1954-58? One concludes that net immigration, which was larger on the average during 1947-53 than during 1954-58, was the more important factor affecting earnings.

The remaining Puerto Rican industries do not exhibit individually as well-defined a pattern of relative employment and earnings as does Women's and Children's Garments. However, when they are aggregated as in Figure 4, a similar pattern emerges.[26] The principal difference between Figures 4 and 5 in the behavior of the earnings series is that in Figure 4 it declines continuously after 1947. Relative employment, however, does not display the sensitivity to relative earnings (and, presumably, to migration) in the remaining Puerto Rican industries as in garments. Note also that New York-New Jersey's employment share is considerably lower for these industries than for garments. One suspects that labor market factors have been more important in determining the location of Women's and Children's Garments than of the remaining industries.

The behavior of the non-Puerto Rican manufacturing industries is similar to that of the non-garment Puerto Rican industries. The response of the employment share to immigration and to relative earnings is small, and the level of the employment share is slightly below that of the non-garment Puerto Rican industries. Relative earnings declined throughout the 1947-58 period.[27]

CONCLUSION

The New York area has been the recipient of many kinds of immigration throughout its history. The evidence at our disposal indicates that large numbers of people were continually entering while others were leaving the labor force of the area. It seems fruitful to view the region as having acted as a processing center for many groups of people entering the United States labor market for the first time, and to view the behavior of population movements, location of employment, and levels of earnings as showing the effects of this. Therefore it is not surprising that we observe responses to Puerto Rican immigration which are similar in many respects to the responses of the economy to European immigration in previous years.

We explored first the behavior of intra-continental population movements in response to European and Puerto Rican immigration and saw that the Puerto Rican influx had an effect on intra-continental movements not dissimilar to that of European immigration. Then we presented evidence which shows why changes in the direction of intra-continental migration have taken place. That is, earnings in New York were depressed relative to other regions during periods of rapid immigration from outside the continent. While complicating factors prevent us from inferring with all desirable certainty that the Puerto Rican immigration has affected relative earnings in exactly the same way as did European immigration, we can conclude with considerable confidence that the effect has been significant.

NOTES

[1]It is beyond the scope of this paper to discuss the appropriate variable (e.g., annual or weekly earnings, or hourly wage rates) which should be used to measure the impact of migration upon the price of labor. Let it suffice to justify the use here of average annual earnings that this is the only variable for which data are available throughout the period of time covered by the present discussion.

[2]A fairly extensive bibliography on Puerto Rican migration to the United States can be found in the author's unpublished Ph.D. dissertation, "Some Economic Aspects of Puerto Rican Migration to the United States" (Stanford University, 1961).

[3]Essentially the problem of assessing the impact of immigration upon internal population flows, earnings, and the location of industry requires an elaborate and well-tested model of location. Since none is available the methods of inference and prediction used in this paper are necessarily impressionistic.

In the cases of both European and Puerto Rican migration, the empirical analysis really refers to a "reduced form" of some (undefined) location model; it is because certain of the conclusions refer to "structure" that our results cannot be definitive. But at a minimum, they should be suggestive.

[4]See, for instance, Simon Kuznets and Ernest Rubin, Immigration and the Foreign Born, National Bureau of Economic Research Occasional Paper No. 46 (New York, 1954); and Everett S. Lee, et al., Population Redistribution and Economic Growth in the United States, 1870-1950 (Philadelphia, 1957).

[5]An interesting discussion of immigrant labor in the garment industry can be found in Joel Seidman, The Needle Trades (New York, 1942), pp. 30-37. An examination of regional rates of labor turnover in manufacturing for the years 1958 and 1959 discloses average total accession and separation rates for New York City of 4.9 and 5.3 per hundred workers, respectively, compared with 3.3 and 3.5 for the entire United States (U.S. Bureau of Labor Statistics Bulletin, No. 1116, "Occupational Wage Survey," pp. 129-35).

[6]In his comments on this paper, Easterlin investigates further the impact of Puerto Rican migration to the United States. In doing so, he draws attention to the detailed characteristics of in-migrants and out-migrants from New York and to factors other than migration which affect the size of the labor supply. In carrying out his analysis, Easterlin fills in some of the lacunae which must necessarily occur in a study such as the present one, which attempts merely to sketch the outlines of the problem so that the principal issues will stand out in bold relief.

In drawing attention to demographic factors which affect labor supply, such as the changing age distribution of the population, Easterlin is quite correct. In a detailed model of location, where labor force is a variable, demographic factors will undoubtedly appear.

However, in his analysis of the comparative labor force and demographic characteristics of in-migrants and out-migrants, I feel Easterlin is on much weaker ground. If we knew for certain how the influx of new migrants affected each sector of a labor market we could predict which individuals, having high propensities to migrate or jobs with greatest sensitivity to in-migration, would be most likely to be displaced. But it is not acceptable as a means of testing hypotheses regarding the impact of immigration to assume, as Easterlin does, that those individuals most directly competitive with the new immigrants are most likely to be forced to change localities. If such were strictly true, we would observe Puerto Ricans displacing Puerto Ricans, and there would be little or no cumulative population change due to immigration in any single locality; new immigrants would immediately be diffused throughout the economy! Such absurd conclusions can be avoided by remembering that substitution can take place throughout a continuous range of workers, ranked according to their labor force characteristics; those who will change localities, given the impact of of immigration, will be those with the highest propensities and greatest abilities to move. A priori one can make as good a case for predicting out-migration among those with high characteristics as among those with low characteristics.

[7]This explanation is consistent with Stouffer's theory of "intervening opportunities." Samuel A. Stouffer, "Intervening Opportunities and Competing Migrants," Journal of Regional Science, Spring 1960, pp. 1-26; and "Intervening Opportunities: A Theory Relating Mobility and Distance," American Sociological Review, December 1940, pp. 845-67.

[8]My estimates of foreign-born white immigration for 1950-60 indicate a net inflow of 242,000 to New York State and 774,000 to the other states. These are not insignificant amounts; but certainly the ethnic and occupational composition of the current European immigration must be radically changed from that of the pre-1920 period due both to the nature of the quota system and the difficulty of freely emigrating from the Iron Curtain countries. One would thus expect the labor force impact of European immigration to be different now than it was fifty years ago. For instance, the require-

ment that new immigrants must be guaranteed financial support before arriving in the United States probably induces a pre-selection of those potential immigrants who can easily find jobs and who have rather high skill levels. The elasticity of supply of such labor to any particular region is likely to be high, and the impact upon wage levels may thus be less than that of earlier immigrants.

The accompanying table gives some idea of the changing composition of European immigration over time.

Characteristics of European Immigrants to the United States

	Area of origin (proportion of total)				Occupations (proportion of total)		
Year	Northwestern Europe	Central Europe*	Eastern Europe	Southern Europe	Professional	Craftsmen, etc.	Laborers
1910	.20	.31	.23	.20	.01	.12	.22
1920	.34	.05	.03	.58	.02	.13	.19
1950	.17	.68	<.01	.08	.08	.17	.02

Source: U.S., Bureau of the Census, Historical Statistics of the United States, Colonial Times to 1957, Washington, D.C., 1960.

*Predominantly Poland in 1910, Germany in 1950.

[9]It is possible that part of the negative net migration of native whites to New York State is due only to a redistribution of population within the New York metropolitan region. However, Table 2, which shows comparative migration figures for the nine largest SMA's besides New York and for New York-Northeastern New Jersey separately, indicates that the negative net migration of native whites was largely away from the metropolitan region and not merely to areas within it which are outside New York State.

[10]In his comments on this article, Easterlin points out that data now available in the Current Population Reports of the U.S. Census Bureau (Series p-25, No. 247) suggest that my assertions regarding the apparent impact of Puerto Rican immigration on population flows internal to the United States may not be correct. Using the census data, one finds that between 1940-50 and 1950-60, total white immigration to New York State fell from -6000 to -72,000, while to the eight states used for comparison it rose from 1,792,000 to 1,990,000. This suggests an impact on white migration flows similar to that which I pointed out, but one not quite so large. However, the non-white behavior which I pointed out cannot be observed if one uses the census data as presented. They show non-white migration to New York increasing from 276,000 to 282,000, while it decreases to the other states from 1,046,000 to 987,000. My Figures 1 and 2 and the census data suggest, however, that the effect on white flows was greater than on non-white flows.

Unfortunately, the census data were not available until after my study was completed, but even if they had been, they could not have been used in exactly the way I have used the Lee estimates and my own. This is because the latter estimates break white migration into two component parts-- native white and foreign-born white. If one adds native white and foreign-born white migration together using my data sources, the apparent impact of Puerto Rican migration on white flows is again reduced. This suggests that some of the difference in the interpretation of the census data and those presented here may be due to the lumping together of native and foreign-born white information in the Census figures.

Whether it is appropriate to treat these two flows as one depends upon whether foreign-born white immigration is "exogenous" or "endogenous" to the system. That is, it depends upon the effects of immigration from one extra-continental source upon the immigration from another such source. The proper treatment of foreign-born white immigration is thus not immediately obvious. My predilection is to treat it as an exogenous factor; thus it should not be combined with the figure for native whites.

Furthermore, the data as reproduced in Figure 1 are not net of Puerto Rican immigrants. This is an error on my part which was not committed in Table 2 (see note 3 of table). If one subtracts Puerto Ricans from the data (I assumed on the basis of the 1950 Census of Population that non-whites constitute 8 per cent of Puerto Rican immigration), then the new census figures show a pattern more closely resembling the pattern which I asserted in the text could be inferred from an examination of Figures 1 and 2. The non-white migration flow, however, still appears to have been affected but little by Puerto Rican immigration if one relies upon the (corrected) Census estimates.

Finally, Easterlin points out that the data in my Table 2 contradict the evidence I find in Figures 1 and 2. Once again, it is my impression that the behavior of white migration in Table 2 might be due in part to the combination of native-born and foreign-born whites into one figure.

[11]Martin Segal, Wages in the Metropolis (Cambridge: Harvard University Press, 1960), p. 130.

[12]Women's and Children's Garments includes the following subcategories: Blouses, Dresses (unit priced and dozen priced); Suits, coats, skirts; Women's neckwear; Women's outerwear n.e.c.; Women's and Children's underwear; Children's dresses; Children's coats; Children's outerwear n.e.c.

[13]Ideally, data for some smaller area, such as the New York-Northeastern New Jersey Consolidated Area should have been used in place of data for the two states. Unfortunately, these are not

available prior to 1929. However, a comparison of the behavior of employment and earnings in all manufacturing for the New York-New Jersey industrial area with that for the two states for the years 1929-58 indicates that the state data probably are not misleading.

[14]U.S., Bureau of the Census, Census of Population: 1950, Vol. IV, Special Reports, Part III, Chapter D, "Puerto Ricans in the Continental United States."

[15]See, for instance, Roy B. Helfgott, "Women's and Children's Apparel," in Max Hall (ed.), Made in New York (Cambridge: Harvard University Press, 1959), pp. 48-49.

[16]Seidman, op. cit., Chapter 9.

[17]Leo Wolman, Ebb and Flow in Trade Unionism (New York, 1936), p. 181.

[18]Seldman, op. cit., p. 5.

[19]Wolman, op. cit., p. 124

[20]Leo Troy, Distribution of Union Membership Among the States, 1939 and 1953, National Bureau of Economic Research Occasional Paper No. 56 (New York, 1957), p. 24.

[21]Martin Segal, op. cit., p. 134, states: "How can we account for the lag in the rate of wage increases in the garment sector? The influx of Puerto Rican labor facilitated the lag but did not cause it. The chief cause, we believe, is the competitive pressure emanating from outside the New York Metropolitan Region." He also states [p. 102], "In several New York services and industries which Puerto Ricans have entered in fairly significant numbers--including hotels, restaurants, the cleaning of office buildings, and the manufacture of paperboard boxes, paints and varnishes, toilet preparations and radios--wages between 1947 and 1955 have risen about as much as, and sometimes more than, those of other industries in the Region. Such wages also have risen about as much as the wages of the same services and industries in other metropolitan areas. . . . On the whole, garment wages between 1947 and 1955 increased somewhat less in the Region than in other metropolitan areas, though not less than they increased in the country as a whole."
The data presented in Figures 3, 4, and 5 incidate that Segal's assertions regarding the behavior of relative wages may be wrong. It is doubtful whether wages in any of the industries we have studied have risen as rapidly in New York as elsewhere. Also, my calculations indicate that Women's and Children's Garments earnings rose more slowly than those in other manufacturing industries in the New York region during the period 1947-58. (See footnote 27.) It is essential to point out that the data used in this paper are taken from the Census of Manufactures and refer to all areas of the United States, whereas for the most part Segal's data are derived from the Bureau of Labor Statistics' Wage Structure Series and refer to major metropolitan areas. Also, Segal's data are for hourly wage rates, while those used here are for average per capita annual earnings. (Postwar data are available in the Census of Manufactures for hours worked. My calculations for Apparel and related Industries show a pattern for relative hourly wage rates similar to that for relative earnings.)

[22]Stricter regulation regarding hours of work now than those existing during the 1920's, however, may make a given earnings differential more favorable to New York employers now than previously.

[23]See, for instance, Helfgott, op. cit., chapter 3; and Charles S. Goodman, The Location of Fashion Industries (Ann Arbor, 1948), chapter 6. It is necessary to remember, of course, that a change in fashions may itself reflect changing factor endowments.

[24]Associated with this change in the demand for goods which require less skilled forms of labor has been a rise in "section" or mass production manufacturing of garments. Most of the areas producing Women's and Children's Garments outside New York use section system methods predominantly. (See, for instance, Helfgott, op. cit., chapter 3.)

[25]Helfgott, op. cit., p. 111.

[26]It is doubtful whether this similar pattern is due to changes in the industry mix of this group. Computing relative earnings using constant employment weights of 1899 leaves the series virtually unchanged.

[27]It was mentioned earlier that we should expect to notice a greater impact of Puerto Ricans upon those industries important in employing them than upon the remaining industries. That the data in Charts 3, 4, and 5 show fluctuations in relative earnings which are no wider for the Puerto Rican industries than for the non-Puerto Rican, even in the years following World War II, may at first sight seem to contradict this expectation. Actually a much more thorough analysis of the sequential impact of immigrants upon various industries is necessary before it becomes possible to make predictions regarding the differential behavior of the relative earnings series; the labor demand elasticities of the various industries will certainly be an important factor determining the magnitude of observed fluctuations. Furthermore, the reaction lag of the existing labor force to wage changes in industries which receive the initial impact of immigration is probably much shorter than the time between censuses of manufacturers.
In order to analyze the route by which new labor supplies affect wages among various industries, we should investigate the behavior of interindustry relative wages or earnings within markets receiving large immigration. In New York-New Jersey, the ratio of earnings in Women's and Children's Garments with respect to the non-Puerto Rican industries has been: 1947, .96; 1954, .72; 1958, .68.

Occupational Wage Differentials in Major Cities During the 1950's

by Martin Segal

Introduction - Female office worker salaries in sixteen cities - Manufacturing plant workers' salaries in sixteen cities - Fringe benefits - Ten-year comparisons - Comparisons of short-run variations in employment and prices - Are the differences to be explained by peculiarities of the female labor force? - Several interpretations presented and considered.

This paper is concerned with the course of occupational wage differentials in major cities during the decade of the Fifties. All the data presented below pertain to such cities. But there is no reason to believe that urban centers have displayed any differences in the behavior of occupational wage structures from other areas. Accordingly, the interpretation of the data is meant to pertain to the course of differentials in the economy as a whole.

My work on occupational differentials is not entirely completed. But judging from the examination of newly processed computations the results presented here will not be changed materially by further work. I believe, therefore, that it may be of interest to examine here the results of computations performed so far. These results are presented in the first two sections. The third and fourth sections will be concerned with the interpretation of the presented data.

TWO WORKER CATEGORIES

My computations of changes in occupational differentials are based on the data published by the Bureau of Labor Statistics in its occupational wage series of major labor markets. These data, though derived from very large samples, are not without their limitations. In particular, the industrial composition of each city's sample undergoes some changes

Author's note. The research connected with this paper was supported by the Amos Tuck School of Business Administration under a Sloan Research Grant; and by the Sloan Research Fund of the School of Industrial Management at M. I. T.

196

from year to year and this has an effect on the inter-temporal comparisons of differentials. Nevertheless, for the period that followed World War II and the immediate postwar inflation these are, except for the union rate series, the best available data relevant to our subject. This is especially true if one is to consider year-to-year changes in occupational differentials.

In this paper I am concentrating on two categories of labor: (a) female office workers in all industries; (b) male workers in manufacturing plants. The B.L.S. studies contain data for all numerically important office occupations in particular cities. As a result, I was able to use as my main measure of occupational differentials in office work the Pearsonian coefficient of variation. In addition, I have also used for office workers another measure--a ratio of average salaries of three skilled to average salaries of three least skilled occupations. The B.L.S. data on plant workers are limited to skilled maintenance men and to custodial and warehouse occupations. Since no data are provided for the wide range of semiskilled production workers, the only reasonable measure to be derived is a ratio between averages of selected skilled and unskilled occupations. I have then used this kind of a ratio to examine the course of occupational differentials among the plant workers.[1]

Turning to the substantive aspects of the computations I will consider first the over-all behavior of occupational differentials in the 1950's. The relevant data are presented in Tables 1 and 2.

Table 1 compares the dispersion of salaries of selected female occupation in office work in 1949 (or 1950) with that in 1961. As can be

Table 1. Occupational Differentials among Female Office Workers as Measured by Coefficients of Variation, Selected Cities, 1949-1961 (Fiscal Years)

City	1949	1961
New York (19)	0.114	0.108
Chicago (18)	.099	.104
*San Francisco (22)	.104	.110
Atlanta (17)	.091	.114
Boston (18)	.123	.098
Los Angeles (18)	.119	.127
Philadelphia (21)	.124	.129
Portland, Oregon (16)	.085	.122
St. Louis (17)	.108	.094
*Milwaukee (17)	.099	.100
*Denver (18)	.117	.129
*Memphis (16)	.101	.112
*Detroit (18)	.100	.145
New Orleans (16)	.108	.126
Minneapolis (18)	.116	.106
Dallas (10)	.093	.100

Source: B.L.S. Occupational Wage Survey, for relevant years and cities.

Note: The numbers in parentheses after each city indicate the number of occupations included in computations.

*The 1950 fiscal year data were used; the 1949 data not available.

seen, in twelve out of the sixteen urban centers for which data are available the dispersion--and thus occupational differentials in percentage terms--increased during the 1950's. The median change in the coefficients of variation was an increase of 6.2 per cent. These results are generally consistent with the computation of skill differentials change among female office workers by means of a skill ratio. Among the

Table 2. Ratios Between the Average of Six Skilled Occupations and the Average of Three Unskilled Occupations in Manufacturing Plants, Selected Cities, 1951-1961 (Fiscal)

City	1951	1961
New York	1.45	1.50
Chicago	1.47	1.54
San Francisco	1.37	1.35
Atlanta	1.66	1.74
Boston	1.37	1.42
Los Angeles	1.37	1.40
Philadelphia	1.48	1.44
Portland, Oregon	1.35	1.46
St. Louis	1.48	1.46
Milwaukee	1.41	1.41
Denver	1.46	1.47
Memphis	1.64	1.88
Detroit	1.27	1.38
New Orleans	1.58	1.70
Minneapolis	1.38	1.41
Dallas	1.63	1.55

Source: B.L.S. Occupational Wage Survey for relevant years and cities.

Note: In computing averages each occupation was assigned the same weight.

twelve cities for which such a ratio--using six identical occupations (three high paid and three low paid)--was constructed, narrowing of occupational differences took place only in four. In the rest differentials either widened or remained stable.[2]

Table 2 presents results of computations pertaining to manufacturing plant workers. As already indicated, the comparison here is between skill ratios. The year 1951 is the earliest for which data are available and is thus used for comparison with 1961. As in the case of female office workers, the data in Table 2 indicate that a majority of cities (eleven out of sixteen) experienced a widening of percentage skill differentials. In four cities differentials narrowed and in one (Milwaukee) they appear to have remained unchanged during the eleven-year period. The median change in the skill ratio was, however, only a small increase of 2.8 per cent.[3]

A closer examination of the two tables does not reveal any consistent regional pattern in the behavior of occupational differentials. The few cities that presumably experienced some narrowing of differentials are not concentrated in any broad region of the country. Equally interesting, only one of the sixteen cities, namely St. Louis, experienced narrowing of differentials both among the office workers and plant workers.[4]

Since the above computations pertain only to wages and salaries there is still the problem of a possible influence on differentials of the "fringe benefits." The available data on "fringe benefits" do not lend themselves to computations of skill differentials. Nevertheless it is reasonable to assume that taking into account the various supplementary remuneration payments would not change significantly the results presented above.

The effect of "fringe benefits" on occupational differences in compensation varies depending on the type of benefits. The programs of health, accident, and life insurance do in fact narrow occupational differences in compensation for work. But such programs are much less important in terms of outlays than payments for various types of paid

leave--primarily vacations--and also less important than retirement
plans.[5] The size of paid leave and retirement benefits is closely geared
to the salaries or wages of the employees. Accordingly, these types of
benefits are not likely to have any important effect in narrowing occupational difference in compensations. The very same applies to all kinds
of premium pay.[6]

    Neither the data presented above nor the consideration just cited
provide us with a complete picture of the developments in occupational
differentials during the 1950's. The totality of the evidence appears sufficient, however, to justify a conclusion that there has been no compression, in percentage terms, of the occupational wage structure either in
office work or in manufacturing plants of major cities. To be sure the
results are affected by the necessarily arbitrary choice of the terminal
years, the selection of occupations and the weights used.[7] But any significant narrowing of occupational differentials during the 1950's should
have been reflected in the computed indexes. Also it should be noted
that the results of the above computations are consistent with the available data on the behavior of income differences among broad occupational
groups.[8]

SHORT-RUN FLUCTUATIONS

    The next problem to be considered is that of a pattern of short-run
fluctuations in occupational wage or salary differentials. The possible
presence of such a pattern is not precluded, of course, by the fact that
over the entire period there has been probably little or no change in the
occupational wage structure.

    The attempts to discover some sort of a short-run pattern in the be-
havior of occupational differentials are handicapped by the nature of the
data. For one thing only four of the urban centers have been surveyed in
every fiscal year since 1949 or 1950. Secondly, the timing of these surveys has not necessarily corresponded to any cyclical behavior of the
economy. Nevertheless, it seems likely that, if the short-run behavior
of differentials would be critically determined by some economic variable,
even the present data pertaining to four important urban centers would
reflect such a relation. Thus it seemed fruitful to consider the problem.
Fortunately, the four urban centers--New York, Chicago, Atlanta, and
San Francisco--represent the four broad regions of the country.

    Recent discussions of short-run fluctuations in skill differentials
have emphasized the influence of two variables--strength of the demand
for labor and inflationary pressures. Accordingly, I tried to discover
whether there was in fact any distinct association between these two factors and the year-to-year changes in occupational differentials in the
four cities. As a measure of changes in demand for labor I used various
rates of unemployment (e.g., general, female, office work, etc.). Since
there are no reliable data pertaining to unemployment in individual areas,
I was compelled to use national rates. As a variable reflecting inflationary pressures I used the B.L.S. Consumer Price Index. This index is
available for individual cities.

    The results of my computations are summarized in Tables 3 and 4.
Table 3 shows the results of correlation between percentage change in
the coefficient of variation for office work salaries and unemployment as
measured by three different rates. This table also shows correlation
results with respect to the percentage change in the cost of living during
the coinciding period.[9]

Table 3. Occupational Differentials among Women Office Workers in Four Cities, 1949-1961. Simple Correlation between Percentage Change in the Coefficients of Variation and Selected Variables

City	General unemployment	Female unemployment	Office workers unemployment	Per cent change in consumer price index
Atlanta	.224	.227	.298	-.052
Chicago	.002	-.001	-.097	.086
New York	.081	.056	.001	.201
San Francisco	-.237	-.379	-.309	-.530

Source: B.L.S. Occupational Wage Survey, for relevant years and cities.
B.L.S., Unemployment, Employment and Labor Force, 1947-1961.
(Original and seasonally adjusted data), 1962. Data for office workers' unemployment and city consumer price index supplied by B.L.S. offices in Washington and Boston.

Note: The correlation is between percentage change in the coefficient of variation and average unemployment during the coinciding period; and between percentage changes in the coefficient of variation and percentage change in the Consumer Price Index during the coinciding period.

The results shown in Table 3 suggest that there was no obvious pattern of behavior of occupational differentials among female office workers either with respect to unemployment or with respect to changes in the cost of living. None of the correlation coefficients is statistically significant at even a 5 per cent level. Additional computations have also failed to reveal any pattern of short-run behavior in the coefficient of variation. Thus, on the assumption of non-linear relationship between unemployment and changes in occupational differentials, I computed correlation between the coefficients of variation and reciprocals of unemployment. But in this case, too, no statistically significant results were obtained.[10]

Table 4 shows correlation results between percentage changes in skill ratios for male manufacturing workers and selected variables. Here the results are somewhat more difficult to interpret than in the case of Table 3. The computations did not indicate any statistically significant association between changes in the skill ratio and changes in the cost of living. In the case of correlation with unemployment, however, a statis-

Table 4. Skill Ratios for Male Workers in Manufacturing Plants in Four Cities, 1951-1961. Simple Correlation between Percentage Change in Skill Ratios and Selected Variables

City	General unemployment	Manu-facturing unemployment	Male unemployment	Per cent change in consumer price index
Atlanta	-.117	-.017	-.134	-.009
Chicago	.696	.665	.689	.0465
New York	.034	.119	.048	-.604
San Francisco	-.150	-.144	-.129	.325

Source: Occupational Wage Survey, for relevant years and cities. B.L.S. Unemployment, Employment and Labor Force, 1947-1961, 1962. Data for manufacturing unemployment and city consumer price index supplied by B.L.S. offices in Washington and Boston.

Note: The correlation is between percentage change in the skill ratio and average unemployment during the coinciding period; and between percentage change in the skill ratio and percentage change in the Consumer Price Index during the coinciding period.

tically significant association is indicated by the behavior of skill ratio in Chicago. The correlation coefficients for that city are significant at a 5 per cent level. And they are higher when changes in skill ratio are correlated with the reciprocal of unemployment.[11] In contrast there is no significant relation between changes in skill ratios and unemployment in the three other cities. In addition I also computed the correlation coefficient between a change in the skill ratio and unemployment for Philadelphia--a city not included in the original study because the available data omitted one fiscal year. As in the case of Atlanta, New York, and San Francisco, there was no significant association between changes in the Philadelphia skill ratio and unemployment.[12]

I have no explanation for the difference of behavior between the skill ratio in Chicago and that in the other four cities. Certainly the Chicago case makes my conclusion pertaining to short-run changes in the skill ratio of manufacturing workers more tentative than in the case of occupational differentials among office workers. Nevertheless, in view of the results pertaining to the other cities the reasonable interpretation of the data must emphasize a lack of a clear pattern of short-run fluctuations in the skill ratio of manufacturing workers during the 1950's.[13]

EXPLAINING FEMALE SALARIES

Before turning to an interpretation of the above data, I want to consider the possibility that the results pertaining to salaries of women workers may represent an essentially different phenomenon from those pertaining to wages of male workers. Specifically the question is this: Does the apparent lack of narrowing of occupational differences among women office workers reflect conditions peculiar to the labor market for female work, or is it a result of the same forces that halted narrowing of occupational differences among male workers?

The reason that this question arises is that the pattern of the female labor force participation differs from that of male participation. I believe, however, that such differences that exist have not affected in a particular manner the course of occupational differentials among women workers. I suggest then that my results pertaining to occupational differences among women workers reflect basically the same forces that influenced the course of skill differentials among male workers.

The pattern of female labor force participation could have widened occupational differentials in the 1949-1961 period if the female labor force had been growing at an increasing rate. For then, on the assumption that the new entrants took least skilled jobs, the effect on differentials would have been analogous to immigration of unskilled labor. This did not happen, however. The female labor force has indeed been growing but the rates of growth, though varying from year to year, do not show any trend.[14]

Another possibility is that the cyclical pattern of female labor force participation has prevented occupational differentials from narrowing or that it even contributed to the tendency of widening such differentials. One may thus argue that, if short-run increases in demand for labor are accompanied by marked increases in the female labor force participation, the influx of new entrants--presumably taking low-skill jobs--could exert a relative downward pull on less skilled rates and either prevent any potential narrowing of occupational differentials or cause their widening.[15] Again, however, the available evidence does not appear to offer any support for this hypothesis. For judging from the work of Long and Hansen the behavior of the labor force, both male and female, during the 1950's shows about zero elasticity of labor supply with respect to changes in unemployment rate.[16]

The above considerations lead me to the conclusion that one may consider the course of occupational differences among women workers as reflecting essentially the same factors that influenced the behavior of the occupational wage structure among the male workers.[17]

INTERPRETATIONS

The interpretation of the data presented above can proceed most fruitfully, I believe, if one considers the behavior of the differentials in the 1950's against the background of the previous period of compression of the occupational wage structure. The reasons for this will become apparent in the following paragraphs.

As is well known, there are essentially two hypotheses that have been used to explain the narrowing of skill differentials during the 1940's. The first one of these, advanced in the most cogent form by Reder, views the compression of differentials as reflecting primarily the differences in the supply-demand conditions for different skills. Briefly, in a period of excess demand the supply of skilled can be augmented by lowering hiring requirements and substitution of less skilled for more. But this procedure cannot be used to increase the supply of the unskilled. Faced with relatively inelastic supply of unskilled the competition for labor during rapid expansion leads to proportionately larger wage increases for unskilled than for skilled.[18]

The second hypothesis has been suggested in a somewhat different form by different writers.[19] But the essential point is that inflation causes proportionately larger wage increases for lower skills because the lower paid workers and/or the unions insist on wage changes that would offset increases in living costs. Since lower paid workers spend larger proportions of their income on basic cost-of-living items, such wage policies (usually taking the form of flat cents-per-hour increases) lead to narrowing of occupational differentials in percentage terms.

I find some aspects of Reder's explanation--specifically, the emphasis on the expansion of skilled labor through promotion, etc.--useful in interpreting the narrowing of differentials during the 1940's. Nevertheless, his "excess demand" hypothesis seems unable to account for the complex of changes in the occupational wage structure of that period.

(a) Reder's hypothesis can at best explain only why the wages of the least skilled rose proportionately more than those of the rest of the workers. But as Douty and others pointed out, during the 1940's large increases in money wages were granted predominantly across the board in uniform cents-per-hour changes. This resulted in a compression of wage structure all along the line--among skilled, between skilled and semi-skilled, among semi-skilled, etc.[20] In order to use Reder's hypothesis to explain this general compression, one would have to assume that in general it is easier to substitute for more-skilled than for less-skilled workers.[21]

(b) Occupational differentials in percentage terms continued to decline during the post World War II inflation (1945-50). Yet this was hardly a period of significant excess demand for labor. Again, the Reder hypothesis fails to account for the narrowing of occupational differences in this period.[22]

(c) The narrowing of differentials among skills also took place in Britain and other countries. But as D. J. Robertson pointed out, the lowering of hiring requirements and the process of substitution visualized by Reder could not constitute the main factor responsible for such narrowing. Because of the prevalence of formal apprenticeship or formal training, the range of variation in hiring open to British employers is strictly limited. [23]

These points indicate a serious inadequacy in Reder's hypothesis as applied to the 1940's. In contrast, the factors stressed by Douty--the gearing of wage changes to make up for the cost-of-living increases, and the egalitarian tendencies in union and government policies reflected in flat cents-per-hour raises[24]--are able to account for the compression in the occupational wage structure. In particular, Douty's explanation gains in persuasiveness when one supplements it by taking into account some of the factors pertaining to the labor supply conditions that were stressed by Reder.

The fact that the wartime expansion began after a period of considerable unemployment and that the supply of skilled and semi-skilled was augmented to a degree by promotions and upgrading help to explain the acquiescence of workers on more skilled jobs to the egalitarian forms of wage changes. For if men find themselves employed after a period of lay-offs, if they are promoted to higher paying jobs, and if they experience, in addition, for the first time in years, a series of general wage increases they are not likely to be immediately concerned about the compression of percentage differentials.

Students of wage behavior are familiar with the fact that workers in a given productive set-up may be as much concerned with the relation of their wage changes to those of other workers as with the absolute amount of wages received. [25] But as Dunlop argued, it does not follow that equal percentage changes will always have a normative significance. [26] Whether the interest of workers in the relation of their wages to those of others will be reflected in concern about percentage or absolute differentials will depend on the economic circumstances of the time. The period of the 1940's characterized by a significant inflation and expansion of employment after a severe depression represented circumstances favorable to a workers' acceptance of wage changes that compressed occupational wage structures.

Analogous factors undoubtedly influenced also the demand side--i.e., employers' policies. On the one hand, the relatively large pool of unemployed and previously downgraded workers made possible staffing of skilled and semi-skilled jobs by upgrading, promotion, or accelerated training. [27] On the other, the concern with maintaining output, and thus keeping good relations with unions, and the desire to gear wage changes to the rising cost of living made the employers responsive to the egalitarian tendencies reflected in union demands and government policies. The result then was that the wage and salary changes took the form of flat cents-per-hour or dollars-per-week increases. Again one can reasonably argue that the employers' concern about preserving a "balanced" wage structure would not reassert itself until some drastic changes in such structure did in fact take place.

All told, the narrowing of differentials during the 1940's--consistent as it was with the long-run forces of a general rise of educational levels and curtailment of immigration--appears thus to have been a result of a combination of factors. The inflationary pressures influenced the form of wage changes that compressed the wage structure. And the timing of these pressures--i.e., after a period of prolonged unemployment-- facilitated the acceptance of a narrowing of differentials by more skilled workers and employers.

The period of the 1950's may be now interpreted as representing a change in those aspects of the economic environment that influence policies of employers--both organized and not organized--and of unions with respect to occupational wage structures.

(a) The inflationary pressures have lessened considerably. Moreover, the nature of increases in the cost-of-living index--with prices of food lagging behind those of other items--has probably had more adverse effect on the real income of higher paid workers than on that of the unskilled. In any case, wage increases were not negotiated or granted primarily to make up for cost-of-living changes.

(b) As annual wage increases became a customary and expected phenomenon--and as occupational differentials narrowed significantly-- the concern of employees about changes in relative standing centered more on the percentage comparisons rather than absolute size of raises. This led to the well-known internal difficulties in some industrial unions and eventually to changes in union negotiated contracts. In non-union establishments the changed attitudes of the skilled employees were undoubtedly reflected in informal grumbling, complaints, etc.

(c) American employers have been always concerned about the maintenance of skill differentials and "balanced" wage structure.[28] This concern has naturally increased after a period of pronounced compression of occupational wage structures during the 1940's.[29] In view of the changed attitudes on the part of the workers--whether unionized or not-- the employers were able during the 1950's to grant wage increases in a form that maintained (or even increased in some cases) the existing percentage differentials among skills.

(d) It should be emphasized that the developments in the supply-demand relationships during the 1950's supported the previously indicated reactions to the compression of occupational differentials. Because of the fluctuations in aggregate demand, employers (and also unions) have been less willing to augment the supply of skilled by training young workers than during the early 1940's. At the same time the broad changes in the job mix of the economy have favored the more- as against the less-skilled. Thus between 1950 and 1960 the number of craftsmen increased proportionately more than that of operatives and the number of non-farm laborers actually declined.[30] These factors strengthened the relative market position of the more skilled and provided thus further --and in some cases probably critical--reason for a change to wage increases formulated in percentage terms rather than equal absolute amounts.

The above reasoning--or perhaps rationalization--emphasizes thus several factors in the 1950's. First, there has been a reaction to the preceding period of narrowing--a reaction based on the well-known concern among the workers about relative wage standing in a work-place and the employers' traditional preference for the maintenance of existing wage differentials. Secondly, this reaction was facilitated by a lessening of inflationary pressures. And, finally, it was strengthened by developments in the labor market that favored skilled workers. If this interpretation is correct then it is not surprising that changes in occupational differentials showed no association with cost-of-living changes or fluctuations in unemployment. As a factor influencing differentials, cost-of-living changes have been pushed in the background by considerations of equity toward more skilled. And without the active influence of the inflationary pressures the relatively mild cyclical expansions of employment should not be expected to play a role as a factor influencing occupational wage structures.

The foregoing interpretations of the 1940's and 1950's rely heavily on the special characteristics of the institutional and economic setting of the two periods. As such these interpretations may well be classified in the ad hoc category. Indeed they are not meant to be construed as constituting a general explanation of short-run changes in occupational differentials. Thus, though I viewed inflation as a key factor contributing to the narrowing of differentials during the 1940's, I do not suggest that inflationary pressures are either a necessary or a sufficient condition for a compression of an occupational wage structure. One can reasonably assume that under circumstances different from those in the 1940's inflationary pressure need not lead to a narrowing of skill differentials.[31] And similarly factors other than inflation--government wage boards, social legislation, collective bargaining--may bring about narrowing of differentials that is consistent with long-run factors pertaining to the demand-supply relation for various types of skills.

It seems likely, in fact, that we may not be able to explain any particular short-run changes in the occupational wage structure in terms of a completely general theory of fluctuations of skill differentials.[32] The changes in the relative levels of pay for different skills are in the ultimate analysis a result of long-run changes in the complex of technological, educational, and other factors that influence the supply-demand relationships for these skills. But given the well-known characteristics of the labor market--downward rigidity of wage rates, the gearing of actual job training to expansions of output, institutional motivations in the administration of intra-firm wage structures--the impact of the long-run forces is brought into operation in short-run spurts--spurts that may be separated in time by many years and that many have as their immediate cause a host of factors peculiar to a particular period. Accordingly, a useful analysis of changes in occupational wage structures may well proceed on two different levels. On the one hand, it may seek to identify the long-run forces that eventually come into fruition by means of marked and bunched up in time changes in differentials. On the other, it may interpret the short-run changes in occupational wage differentials in terms of the socio-economic setting of the particular period. This latter type of an interpretation has been attempted in the preceding pages.

NOTES

[1]In computing the coefficient of variation for female office workers, each occupation was given the same weight. The same occupations were used for each year in each city; but the number of occupations varied from one city to another. In computing the ratios between skilled and unskilled wages in manufacturing, I used six identical skilled occupations and three identical unskilled occupations for each city. I computed the skill ratio in two ways: by assigning equal weight to each occupation and by weighting the occupations by employment weights derived from actual employment in 1956 in four major cities. Although some differences inevitably exist between these two ratios, the basic results are very similar.

[2]In computing this ratio, six identical occupations were used for each city. However, in view of the limited number of occupations, the results are probably less reliable than those indicated by the coefficient of variation.

[3]The results in Table 2 pertain to ratios in which each occupation has equal weight. The results using averages weighted by employment give basically similar results. Thus among the sixteen cities, twelve experienced a widening of skill ratios. In three cities the ratio narrowed and in one, remained the same. The main effect of weighting was to compress the range of variation between the terminal years.

[4]The 1949-1961 comparison of skill ratios for St. Louis office workers also revealed a narrowing of differentials. Similarly the skill ratio comparison for manufacturing workers computed from weighted averages also showed such narrowing.

[5]See, U.S., Bureau of Labor Statistics, Employer Expenditures for Selected Supplementary Remuneration Practices for Production Workers in Manufacturing Industries, 1959, Bull. 1308, January, 1962.

[6]In many cases, "fringe benefits" may, in fact, increase differentials in compensation. Thus Prof. D. Brown of M.I.T. pointed out to me that this may take place where pension rights are not vested and where the turnover among less skilled is larger than among the more skilled. For then the employer contributions on behalf of the less skilled will supplement the pensions of the senior, more skilled workers. A similar phenomenon may take place in situations where vacations' duration is geared to the length of service.

[7]After my own computations had been made, the B.L.S. published results pertaining to the movement of wages for skilled maintenance men and unskilled plant workers during the period of 1953-62. The data refer to twenty metropolitan areas. The B.L.S. computations show that in that period the wages of unskilled rose slightly more than those of skilled. For all industries the median difference over a nine-year period is less than one percentage point; for manufacturing, less than two percentage points. These results are not inconsistent with my own results. For one thing, the period covered is different since my own data for men pertain to 1951-61 rather than 1953-62. Secondly, the B.L.S. data use somewhat different occupational mix in the indexes. The B.L.S. has not computed any measures of occupational differences for women office workers. (See, Wage Trends for Occupational Groups in Metropolitan Areas, Bureau of Labor Statistics, Summary Release, October, 1962.)

[8]U.S., Bureau of the Census, Current Population Reports -- Consumer Income, Series P-60, No. 33, p. 6.

[9]This is the period between the two dates for which salary or wage data were available for a particular city.

[10]I also computed correlation between occupational differentials changes and changes in cost of living lagged by one period. Again no significant pattern of association was evident.

[11]The coefficient of correlation between percentage change in the skill ratio for Chicago and the reciprocal of general unemployment was -0.733. However, for the other cities there was no significant association between these variables.

[12]The correlation coefficients for Philadelphia were -0.461 (general unemployment); -0.461 (male unemployment); and -0.526 (mfg. unemployment). None of them are statistically significant and the signs are opposite of the coefficients for Chicago.

[13]Professor J. Mincer of Columbia suggested that, in view of the mild cyclical fluctuations, it would be useful to consider the change in differentials up in 1956 and after that year, the earlier period being presumably characterized by tighter labor markets. An examination of the work-sheets suggests, however, that stability of differentials or even widening was characteristic of both periods.

[14]The statement is based on computation of annual percentage increase in the female labor force from 1948 to 1961. The basic data are in B.L.S., Unemployment, Employment and Labor Force, 1946-61, 1962.

[15]This type of labor force behavior is viewed by Perlman as one of the critical factors that, in the absence of inflation, contribute to a widening of occupational differentials. Perlman posits such behavior with respect to both men and women but the argument would obviously apply with much greater force to female labor force participation. (R. Perlman, "Forces Widening Occupational Differentials," Review of Economics and Statistics, Vol. XL, pp. 112-13.)

[16]See W. L. Hansen, "The Cyclical Sensitivity of Labor Supply," American Economic Review, June, 1961, and the literature cited therein.

[17]Occupational differentials narrowed among office workers during the 1940's. Using the data collected by the National Industrial Conference Board, I computed coefficients of variation pertaining to dispersion of office work salaries for eleven cities and for the country as a whole in 1943 and 1949. These computations showed a narrowing of differentials in the country and in ten out of the eleven cities for which data (using the same occupations in both years) were available. The N.I.C.B. does not have separate data for men and women, but judging from the occupations included in the survey, an overwhelming majority of the collected salaries pertain to women office workers. Thus it appears that occupational differences among women have behaved, with respect to time periods, similarly to those of male workers. (The data are in National Industrial Conference Board, Clerical Salary Survey, for various years.)

[18]M. W. Reder, "The Theory of Occupational Wage Differentials," American Economic Review, December, 1955.

[19]Thus Perlman (op. cit.) appears to consider inflation as a possible long-run influence narrowing differentials. He also emphasizes egalitarian pressures reflected in the decisions of government wage boards. Douty emphasizes, in addition, the "broad underlying forces affecting the supply of workers at different skill levels" (H. M. Douty, "Union Impact on Wage Structures," Proceedings of Sixth Annual Melting of Industrial Relations Research Association, 1953). See also D. J. Robertson, Factory Wage Structure and National Agreements, Cambridge, 1960, p. 173.

[20]See, for instance, J. Stieber, "Occupational Wage Differentials in the Basic Steel Industry," Industrial and Labor Relations Review, (January, 1959), p. 173; R. M. Macdonald, "Pulp and Paper," in L. G. Reynolds and C. H. Taft, The Evolution of Wage Structure, New Haven, 1956, pp. 108-28. In an unpublished paper, D. Dacy, a graduate student at Harvard, shows that during the war years there was also a percentage compression of union rates among the skilled journeymen in the building trades.

[21]Reder does recognize that his hypothesis cannot explain the over-all changes in the occupational wage structure. But the problem appears to be largely ignored in his article. (Reder, op. cit., p. 838.)

[22]In a note to be published shortly by the Review of Economics and Statistics, Robert Evans of M.I.T. has tested whether "excess demand" or the inflation hypothesis provide a better explanation for year-to-year decreases in the building trades skill ratio during the 1940's. His conclusion was that "changes in the cost of living provide, by far, the best explanation for the observed year-to-year changes in the skill ratio" (Robert Evans, "Wage Differentials, Excess Demand for Labor, and Inflation: A Note").

[23]D. J. Robertson, op. cit., p. 175. Lowering of hiring requirements, upgrading, informal skill training are undoubtedly more common in the United States than in Britain. But in many strongly unionized areas the formal apprenticeship requirements in, say, construction are well enforced also in this country. Accordingly, Reder's assumption concerning substitution for skilled labor may well be too strong even with respect to the United States. Of course, in such areas as college teaching, Reder's hypothesis has an obvious validity. As all of us know, a new Ph.D. or even graduate student may be a good substitute for (or an improvement upon) a full professor.

[24]Douty, op. cit., p. 73.

[25]M. Segal, "Interrelationship of Wages Under Joint Demand," Quarterly Journal of Economics (August, 1956), and literature cited therein.

[26]J. T. Dunlop, "Discussion" on Douty's paper, Proceedings of Sixth Annual Meetings of Industrial Relations Research Association, 1953.

[27]That these methods could, in fact, augment the supply of skills is a reflection of the long-run trends, such as increasing educational levels, that had operated toward the relative increase in the supply of skills. But because workers acquire their skills primarily on the job, through factory training, or apprenticeship, the effect of such long-run forces cannot come to a fruition without a marked expansion of employment. (For analysis of the long-run forces, see P. G. Keat, "Long Run Changes in Occupational Wage Structure, 1900-1956," Journal of Political Economy, Dec., 1956.)

[28]For a discussion of some important factors explaining the attitudes and policies of the employers, see Lloyd G. Reynolds, The Structure of Labor Markets, New York, 1951, pp. 239-40; also Macdonald, op. cit., pp. 120-21.

[29]See Macdonald, op. cit., p. 122, for description of employers concern in one industry.

[30]U. S. Department of Commerce, U. S. Census of Population, 1960, U. S. Summary, General Social and Economic Characteristics, Table 89, pp. 1-219. One should note, however, that the changes in the occupational composition in the 1940-1950 period also showed a relative increase in the number of skilled and semi-skilled as compared with unskilled.

[31]Consider, for instance, a situation in which strong inflationary pressures begin after a prolonged period of full employment. If most of the skilled workers have been employed on their jobs for a lengthy period, they need not be as willing to tolerate a narrowing of skill differentials as they were in the 1940's. Or consider that strong inflationary pressures last for several years and that their first impact was, in fact, reflected in a narrowing of differentials. Then, if such narrowing is inconsistent with the actual educational attainments of the workforce, and in the absence of such employer motivations as predominate during a wartime period, the employers would tend, after some lag, to substitute more skilled for less skilled workers. Accordingly, by bidding up skilled wages, the employers would, in fact, tend to reverse the compression of the wage structure. Neither one of the above posited conditions applies to the period of the 1940's. Nevertheless, they imply to me that inflation need not always cause a narrowing of occupational differentials.

[32]This seems to be the position taken by D. J. Robertson (Robertson, op. cit., p. 176). Notice also that Reder who views his hypothesis as a theory of short-run changes in the skill ratio has to resort to the consideration of special institutional factors during particular short-run periods in order to "explain" the behavior of skill differentials. (Reder, op. cit., pp. 842-44.)

Labor Mobility and Wage Improvement

by Robert L. Bunting

The classical question - The BOASI continuous work history sample; wealth of data and their limitations - Test of the mobility hypothesis - Review of literature - Findings.

For approximately two centuries economists have been asserting that the labor market puts workers more or less where they belong -- with respect to their productive ability -- after allowances have been made for their tastes. The argument has been that prices simultaneously provide the signals as to where resources are needed and the stimulus required to get them there. The greater the need, the more clear the signal and the greater the incentive for movement. Equilibrium occurs when the optimum allocation has been achieved in the collective eyes of the ultimate prime movers, the sovereign consumers.[1]

During the late 1930's, the 1940's, and early 1950's a substantial amount of empirical work was done which led many students of labor markets to feel that these markets were not performing their allocative function in accordance with the expectations engendered by traditional theory. Questions of central importance were raised: questions about the breadth of knowledge of job opportunities, about the strength of the incentive provided by market prices, even about the basic notion of rational conduct on the part of workers in the use of their personal resources.[2]

It is my judgment that there is rather less concern about these matters currently -- that interested scholars seem to have moved toward the view that the market mechanism is more effective than these critics had suggested. This revised attitude has several bases. As suggested in the preceding paragraph, the revolt grew primarily out of the belief of some labor economists that their research findings tended to discredit traditional theory as a device for analyzing modern labor market problems. This line of criticism has been received unsympathetically; it seems safe to say that it has been rejected as a statement of fundamental error in the body of theory.[3] Moreover, some of the empirical work which led to the revolt has been subjected to methodological criticism; questions have been raised about its intrinsic merit and the significance of some of the conclusions drawn from it.[4] Finally, there have been several more recent empirical studies indicating that postwar mobility flows have been fairly heavy and generally in accordance with predictions of conventional theory.[5] Thus there seems to be the beginning of a new consensus which puts us in a healthier position than before the controversy: with a more solid basis for believing that the core of our theory is sound, with a better awareness of short-run processes and problems, and with many unanswered questions about market efficiency.[6]

The above statement implies the belief that most economists are no longer seriously bothered by the question of "worker rationality." It incorporates the judgment that most economists feel workers do not change jobs capriciously or randomly, but that their moves tend toward positions of higher net advantage. (Most of us would hasten to add that these moves take place within a framework of incomplete knowledge of present alternatives and future events, both of which contribute to error and less-than-perfect choices.) Within this context, the questions which have been raised about the relative significance of "the wage rate" remain incompletely answered. We have evidence that the wage rate is an important consideration in the eyes of mobile workers, but little work has been done in the direction of trying to say how important it is in comparison with the other components of the job package.[7]

This paper bears on the above issue by presenting evidence on the extent to which a large number of workers who moved actually did or did not experience increases in wages. Existing evidence on this matter is of two general types. There are small sample studies which get directly at the question but which provide rather inconclusive answers in the form of numbers or percentages of mobile workers showing wage improvement.[8] There are others which approach the question indirectly, by investigating whether movement is from low-income areas or occupations toward high-income areas or occupations.[9] This study represents an effort to combine the advantage of the more direct approach of the former group with the broader scope of the latter.

DESCRIPTION OF THE DATA

The Bureau of Old-Age and Survivors Insurance has proved itself a valuable source of data for purposes of mobility research; the continuous work history samples, used by the Bureau for purposes of constantly checking the impact of the old-age and survivors insurance program, have been explored enough to give firm promise of future fruitfulness in this regard.[10] There are several characteristics of these data which underlie this promise: the scope of the data is extensive geographically, occupationally, and industrially; large numbers of employees can be investigated at relatively low cost; the data are based upon periodic reports of employers to the agency and hence certain difficulties are avoided which make for error in other types of data; finally, these data open the possibility of unique opportunities to "follow" samples of workers through time -- an opportunity, it should be added, which has not been thoroughly exploited.

There are certain serious shortcomings to these OASI data as sources of information about labor mobility. Chief among these are the lack of occupational information (firms do not report what workers do; they only report who they are and how much they earn) and the gaps in program coverage.[11] These limitations are obviously more serious for certain types of investigations than for others: e.g., little can be done by way of tracing worker movement through successive occupations, but industry information is good and the geographic location of employers is indicated by state and county.

The sample of data used in this study was drawn from the Bureau's continuous work history sample; it consists of those workers in the larger sample who were employed at some time during 1953 in North Carolina, South Carolina, and Georgia.[12] The number of workers in this sample was 30,666. For each of these the Bureau provided employer-employee (IBM) cards; these cards contained wages paid, by quarter employed, for each employer-employee relationship. Thus if a certain worker was employed by a single employer throughout the year, there was only one employer-employee card for him; it provided wages earned in each of the four calendar quarters of the year. But for a worker who was employed by four separate employers -- say, one in each quarter of the year -- there would be four such employer-employee cards; this worker could be "followed" from one employer to another (assuming all four employers were in covered employment) because, as indicated above, the quarters in which the wages were earned were identified.

The sample described above had to be modified in two ways before it could be used for purposes of the wage improvement experiment. First, an important restriction upon its usefulness lay in the wage credit limitation which, in 1953, was $3,600: employers paid taxes only on the first $3,600 of each employee's total wages. If an employee who worked all year for the same employer earned no more than this throughout the year, all of his earnings were reported and he appeared on all four quarterly reports of his employer. If he earned $7,200 during the year, he would appear only on the first two quarterly reports of his employer. If he earned $7,200 during the year, half from one employer for the months January through June and half from another employer for the months July through December, all $7,200 would be reported -- $3,600 by each employer.

Basically, the experiment to be described below consists of a comparison of the wages of mobile workers before and after their job changes in the light of what happens to the wages of workers who do not change jobs. It was necessary to drop the people earning more than $3,600 from the sample because of the uncertainties surrounding the actual wages they received. To illustrate, consider the hypothetical case of a worker whose records show that he received $3,600 from one employer in the first calendar quarter of 1953 and $3,600 from another in the third quarter. This makes it possible to say only that he moved from one job in which he was earning income at a rate not less than $14,400 per year to another in which he was earning at a rate not less than $14,400 per year. The data make it impossible to answer the crucial question: was he earning income at a more rapid rate in the second job?

Similar problems arise in connection with interpreting the earnings of non-mobile workers in the more-than-$3,600 class. For example, suppose the data reveal that a certain worker received $2,500 in the first calendar quarter and $1,100 in the second; the presumption is that this worker was earning income at the rate of $10,000 per year and that he continued in the same employment for the remainder of the year.[13] But, as will be made more clear below, the purpose of bringing the non-mobile workers into the investigation is to provide a control group against which the earnings of the mobile workers may be viewed. Thus the question to be raised in the experiment is <u>not</u> "Did mobile workers improve their earnings by changing jobs?" but rather "Did mobile workers improve their earnings by changing jobs <u>more than those workers who did not change jobs?</u>" (Or perhaps: "Did the earnings of mobile workers decrease less than those of non-mobile workers?") Thus, assuming that the hypothetical non-mobile worker continued to earn at the rate of $10,000 throughout the year establishes by assumption the information being sought. It was these considerations which led to the decision to delete from the sample all workers whose total earnings in covered employment exceeded $3,600. This decision caused 4,101 workers to be dropped from the sample, leaving 26,565 workers.

The discussion above considers only those workers who worked, or appeared to work, throughout the year. There are, however, many workers in the sample for whom wages were reported for only one, two, or three calendar quarters -- seasonal workers; workers who came into the labor force after the year began, such as college graduates; workers who were working at the beginning of the year, dropped out for a quarter or two, then went back to work for the same or another employer; and so on.

The form of the data makes it impossible to use one-quarter workers in the wage improvement experiment. The reason is that wages are allocated by quarter in which earned. Thus the investigator can see, for example, that there are workers who earned wages from two employers in the third calendar quarter of 1953, but he cannot tell which of these employers was first and which was last. The data seem to show that the worker was hired by one firm for a portion of the three-month interval and moved to the other employer; but which employer was "to" and which was "from" cannot be distinguished. So, all one-quarter workers were deleted from the sample. There were 3,565 such workers; removing them left a total sample of 23,000 workers who (a) earned less than $3,600 in covered employment in 1953 and (b) who earned wages in covered employment in not less than two separate quarters in that year.[14]

THE WAGE IMPROVEMENT EXPERIMENT

As indicated above, the sample was used to try to test the hypothesis that workers improve the wage portion of their job packages through mobility. The method of doing this involved comparing the changes in wages of mobile workers with those of non-mobile workers. For these purposes, mobile workers were defined as those who appeared on the quarterly reports of two or more firms; non-mobile workers were those who appeared on the report of only one employer.[15] Specifically, for each worker in this sample, total covered wages earned in the first and last calendar quarters of employment in 1953 were determined:[16] the arithmetic mean of wages earned by non-mobile and mobile workers in the last quarter of employment was compared with that earned in the first.[17] The significant matter is, of course, how the "wage improvement" of the two groups compare.

The results of the experiment are shown in accompanying Table 1. Approximately 15,000 of the workers in the sample did not change employers during the year; their last quarter wages were 6 per cent higher than their first quarter wages. There were more than 8,000 workers who did change jobs, and their last quarter wages were almost 11 per cent higher than their first-quarter wages. As they stand, therefore, the data provide strong support for the hypothesis that wage improvement is an important consideration in the eyes of mobile workers.

Testing the Data on Mobility

Of course, the important questions are these: Do the data really say what they seem to say? And, do they say what they seem to say for the reason suggested by the hypothesis? Most of the remainder of this paper

Table 1. Mean Covered Wages, First and Last Quarters,[1] and Wage Improvement Percentages for Non-Mobile and Mobile Workers

Mobility status[2]	Number of workers[3]	Mean wages[4]		Wage improvement percentages[5]
		First quarter	Last quarter	
Non-mobile	14,856	$443.01	$469.43	5.96
Mobile	8,113	351.25	389.43	10.87

[1] These are the first and last calendar quarters of 1953 in which the workers concerned earned wages in covered employment. They necessarily correspond with the first and fourth calendar quarters of that year only for those workers who earned covered wages in all four quarters.

[2] Non-mobile workers are those who were reported by only one covered employer in 1953; mobile workers appeared on the quarterly reports of two or more firms during the year.

[3] The data in this Table refer to those workers in the BOASI continuous work history sample who were employed at some time during 1953 by a covered firm located in North Carolina, South Carolina, or Georgia. Four groups were deleted: self-employed workers, workers earning $3,600 or more, one-quarter workers, and thirty-one workers the sex or race of whom was unknown.

[4] First quarter mean wages were computed by summing covered wages earned in the first quarter of employment by all employees in the class and dividing by the number of employees in the class; last quarter mean wages were figured similarly.

[5] These were computed by expressing the difference between last and first quarter mean wages as a percentage of first quarter mean wages.

will be concerned with trying to find negative answers to the above questions; failure in this endeavor will leave the data as support for the proposition that workers generally do improve the wage component of their job packages through mobility. First, however, it is important to see that the wage improvement percentage of mobile workers presented in Table 1 is too small.

Voluntary or Involuntary? Traditional mobility theory states that a worker tends to change to another job when its net advantages exceed those of his present job. (Cost of movement may be viewed as a negative aspect of the new job.) The emphasis is upon voluntary action: he moves from job "A" to job "B" because the aggregate cost-gains relationship makes it to his advantage to do so. Layoffs can be incorporated into this scheme only by making the job change a two-step affair: an involuntary change (from employment at job "C" to unemployment) followed by a voluntary one (from unemployment to job "D"). The theory is not designed to explain the first action -- the worker took it against his will; it is supposed to explain the second, the move from a job package containing no income and much leisure to one containing less leisure but more income.

The problem which arises in this connection is that it was impossible to tell from the OASI data which moves were voluntary. Thus the data in Table 1 include all moves. Even the most extreme critics of conventional theory do not argue that wages are a negative component of the job package and they generally agree that voluntary movers are more apt to improve their job packages than involuntary movers. Consequently, there will be no disagreement with the conclusion that the presence of involuntary mobile workers in the data biases the wage improvement results: the improvement percentages of mobile workers would surely be larger than they are if it had been possible to exclude those who changed jobs involuntarily.

It is possible to be a little more explicit about the magnitude of this problem. The average of monthly separation rates for the nation as a whole in 1953 was 4.3 per cent; those of "quits" and "layoffs" were 2.3 per cent and 1.3 per cent, respectively.[18] "Quits" are generally thought of as the voluntary component of total separations.[19] Most of the workers in the "layoff" class may be presumed to have gone back into the market for another job after separation, since their separations were imposed upon them. Moreover, some of the workers who are not accounted for by these two major sub-classes of total separations -- people who were discharged or retired, for example -- might also return to the market. Since not all of the "quits" necessarily go to other jobs, these numbers suggest that a sizable portion of the mobile workers in the sample investigated here might be involuntarily mobile -- probably between 20 and 50 per cent. If it were possible to compensate for this aspect of the data, the average improvement percentage of voluntary movers in Table 1 would surely rise.

Of course, mobile and non-mobile workers are different groups. The question which arises is this: Is the wage improvement which has been observed really that, or is it something which grows out of some difference between the two groups other than mobility? In what follows, an effort will be made to take some of these other differences into ac-

Table 2. Mean Covered Wages, First and Last Quarters,[1] and Wage Improvement Percentages for Non-Mobile and Mobile Workers, by Number of Quarters in which Wages were earned in Covered Employment

Mobility status[2]	All workers[3]	Wage-earning quarters		
		Four	Three	Two
Number of workers				
Non-mobile	14,856	10,205	2,168	2,483
Mobile	8,113	4,840	1,982	1,291

Mean wages[4]								
	First Qtr.	Last Qtr.	First Qtr.	Last Qtr.	First Qtr.	Last Qtr.	First Qtr.	Last Qtr.
Non-mobile	443.01	469.43	507.18	542.39	357.57	355.26	253.89	269.28
Mobile	351.25	389.43	419.96	466.09	278.48	312.21	205.41	220.60

Wage improvement percentages[5]				
Non-mobile	5.96	6.94	-0.65	6.06
Mobile	10.87	10.98	12.11	7.39

Note: The footnotes to this table correspond identically with those of Table 1.

214

count to see if the wage improvement percentages still obtain -- or perhaps to see if these percentages change in any way such as to give indirect evidence about the validity of the hypothesis.

Part-Year Workers. In his work with OASI data Bogue found that mobility rates for two- and three-quarter workers were higher than for those who had employment during all four quarters of the year.[20] If the mobile group contains disproportionately large numbers of part-year workers, the possibility exists that what appears to be a wage improvement phenomenon is really something associated with part-year employment. Thus the sample was divided into three groups according to the numbers of quarters worked and the experiment was repeated. The results, as presented in Table 2, show variation in the wage improvement percentages, but in all cases the percentage of the mobile group is greater than that of the non-mobile group.

Sex, Race, and Age. Similarly, one of the most pervasive findings in mobility studies has been a difference of mobility patterns between classes of workers based on sex and race. The four-quarter workers were used to investigate the possibility that these characteristics might, in some fashion, be related to what has been referred to as wage improvement. Table 3 provides the same sort of information as Table 2: the data show variation in the wage improvement percentages among sex-race classes, but the percentages of the mobile classes are distinctly higher in all cases.

Table 3. Wage Improvement Percentages for Four-Quarter Workers[1] within Race and Sex Classes

| Race-sex class | Non-mobile[2] | | Mobile[2] | |
	Number	Wage improvement percentage[3]	Number	Wage improvement percentage[3]
White males	3,782	6.37	2,278	9.33
Negro males	1,611	9.60	1,273	12.06
White fem.	4,030	6.62	1,088	14.09
Negro fem.	782	6.51	201	14.57

[1]Four-Quarter Workers are those for whom wages were reported in all calendar quarters of 1953. See Note 3, Table 1.

[2]See Note 2, Table 1.

[3]See Note 5, Table 1.

Again: many mobility studies have shown that there is a strong inverse relationship between age and mobility. Thus it is conceivable that the "wage improvement" phenomenon observed in the data is associated with the disproportionate representation of younger workers among the mobile group. The data of Table 4, which show wage improvement percentages within age classes, leave little doubt that the better showing of mobile workers is related to age; but they indicate with equal clarity that taking age into account does not diminish the tendency for the data to show what they consistently have shown in the previous experiments.

Hypermobile Workers. Finally, several studies have indicated that there is a hypermobile fringe -- a small group of workers who change jobs very often.[21] Elsewhere it has been shown that the sample under investigation here contains such an element. Specifically, 3.6 per cent of the

Table 4. Wage Improvement and Age[1]

| Age (1) | Non-mobile workers | | Mobile workers | |
	Number (2)	Wage improvement percentages (3)	Number (4)	Wage improvement percentages (5)
Less than 20	436	17.63	557	44.85
20 - 29	2,534	10.69	1,761	12.35
30 - 39	2,937	6.55	1,242	8.64
40 - 49	2,378	4.08	805	3.48
50 - 59	1,323	5.68	368	6.00
60 and over	597	2.14	107	3.85
All ages	10,205	6.94	4,840	10.98

[1] This table contains data on Four-Quarter Workers only.

workers were seen to have had more than three employers during the year; these workers accounted for almost 35 per cent of all job changes.[22] One need not speculate as to the causes of such extra mobility in order to suggest the desirability of separating out its influence; in a discussion of the gains to be made by reasoned change on the part of employees among employers, most investigators do not have in mind the sort of worker who moves 6, 10, or 15 times in a single year. Thus Table 5 has much the same format as Table 4 except that mobile workers are broken into two groups: those who had two or three employers during the year and those who had four or more. The resulting figures leave no room for concern that the strong wage improvement position of mobile workers is based upon the presence of a few especially active job changers.

Table 5. Wage Improvement Percentages, by Age and Number of Employers[1]

| Age (1) | Number of employers | | | | | |
| | One[2] | | Two or three[3] | | Four or more[3] | |
	Number (2)	Percentage (3)	Number (4)	Percentage (5)	Number (6)	Percentage (7)
Less than 20	436	17.63	430	51.50	127	28.20
20 - 29	2,534	10.69	1,263	14.34	498	6.32
30 - 39	2,937	6.55	893	9.18	349	7.10
40 - 49	2,378	4.08	588	5.33	217	-1.95
50 - 59	1,323	5.68	275	3.94	93	11.91
60 and over	597	2.14	88	-2.03	19	34.66
All ages	10,205	6.94	3,537	12.31	1,303	7.04

[1] This table contains data on Four-Quarter Workers only.

[2] These are Non-Mobile Workers.

[3] These are Mobile Workers.

Multiple Jobholders. In footnote 15 a data difficulty was mentioned which must now be given explicit attention. For purposes of this study, workers who had more than one employer during the year are defined as mobile; thus multiple jobholders -- workers who held two or more jobs concurrently -- are included within this classification. No doubt many multiple jobholders actually did change their primary or secondary jobs during the year and so actually were mobile in the sense appropriate to this paper. The point being made, however, is that all multiple jobholders are automatically classified as mobile by the definition employed here, and this fact undoubtedly leads to the misclassification of many of them. This becomes a problem comparable with those discussed above: the class "mobile workers" is disproportionately weighted with multiple job-holders, so it could be that this underlies the relatively strong wage improvement shown by that class. Especially troublesome is the possibility that "moonlighting" -- as multiple jobholding is popularly called -- is strongly seasonal: for example, if the only kind of multiple jobholding were among retail clerks during the Christmas season, the classification of all such multiple employees as mobile would, by itself, produce an appearance of strong wage improvement among mobile, four-quarter workers relative to non-mobile, four-quarter workers.

Unfortunately our knowledge of multiple jobholding is not great; most of what has been learned about this class of workers is contained in a series of reports on surveys made for the Bureau of Labor Statistics by the Census Bureau. The last four of these reports relate to surveys conducted in July, 1957, July, 1958, December, 1959, and December, 1960.[23] They show multiple jobholders to have comprised from 4.5 to 5.3 per cent of total employment. However, they present other information which makes it obvious that these percentages are too large for the body of data being investigated here. Thus the percentages of multiple jobholders who had at least one job in agriculture were, for the four indicated years, 40, 36, 28, and 29, respectively. Since agricultural workers were very sparsely represented in the sample of data being investigated here,[24] these workers would be included in the data, if at all, primarily as non-mobile workers. These same four surveys showed that non-agricultural, self-employed workers accounted for these percentages of multiple jobholders: 16, 18, 19, and 20. Since self-employed workers were deleted from the sample, such workers would appear in the computations, if at all, as non-mobile workers. Thus it seems that about 50 per cent of the multiple jobholders who might have been included in such a sample as the one being used here were (a) not part of the sampled population, (b) deliberately deleted from the sample, or (c) included in the sample as non-mobile workers.

Thus it may be that as many as 2.5 per cent of the approximately 23,000 workers being investigated actually are multiple jobholders -- perhaps between 550 and 600 workers. What can be said about the possibility that these workers could account for what has been called "wage improvement"?

Consider the following very crude computations: Suppose that the number of workers who were multiple jobholders was 1,000 rather than 600, so that these workers accounted for approximately one out of eight of those classified as mobile.[25] In December of 1960, the median total number of hours worked by non-agricultural, wage and salary workers was 50; the median number of hours spent on the secondary jobs of these workers was 10; thus, their secondary jobs could have boosted their incomes by 25 to 33 per cent.[26] Such an increase for 1,000 workers could be thought of as an increase of 3 or 4 per cent for all 8,000 mobile workers combined. So it seems that the presence of multiple jobholders could account for the observed "wage improvement." Note, however, that this would require that all multiple jobholders work only at their primary jobs in their first quarters of covered employment and at both primary and secondary jobs in their last quarters of covered employment. In fact, the authors of the two most recent B.L.S. reports have both commented upon what appears to be a high degree of stability in the number of non-farm jobs held by multiple jobholders. (They were referring primarily to the stability of this number in the face of cyclical changes in the level of unemployment, but both acknowledged the constancy of the figure despite the change from July to December as the survey month.) In a word, such rough evidence as is available does not seem to permit a satisfactorily complete explanation of the observed wage improvement by reference to multiple jobholders. It is equally clear that the evidence is too rough to permit a denial of a multiple jobholder effect as a partial explanation.

It is possible to obtain more direct evidence on the most likely multiple jobholding effect, referred to earlier: the rise in incomes of those dual jobholders especially associated with the Christmas season. The two- and three-quarter data presented in Table 2 are aggregates of various part-year patterns. Workers employed in the first and second calendar quarters as well as workers employed in the first and third calendar quarters are two-quarter workers. Breaking these aggregates into all possible patterns and recomputing wage improvement percentages makes it possible to see whether dual jobholding during the pre-Christmas weeks was the source of the extra percentage points of wage increase shown by the "mobile" group. These part-year employment patterns and their wage improvement percentages are shown in Table 6. There seems to be little consistency about the percentages of the small, non-continuous patterns: 1-3, 1-4, and 2-4. The larger, non-continuous three-quarter patterns (1-2-4 and 1-3-4) show "wage improvement" as predicted by the hypothesis.[27] The still-larger, continuous two- and three-quarter patterns show mixed results; the pertinent observation is that less relative wage improvement is shown for the mobile workers than for the non-mobile workers of the 3-4 and 2-3-4 patterns. It is clear that the most obvious seasonal pattern which could explain the wage improvement findings -- dual employment of Christmas workers -- does not do so.

Table 6. Wage Improvement Percentages for all possible Employment Patterns of Two- and Three-Quarter Workers

Calendar quarters (1)	Non-mobile workers		Mobile workers	
	Number (2)	Percentage (3)	Number (4)	Percentage (5)
Two-Quarter Workers[1]				
1 - 2	901	- 34.53	324	- 30.85
1 - 3	105	32.44	89	- 21.19
1 - 4	85	- 7.49	184	9.27
2 - 3	432	37.30	181	40.30
2 - 4	41	123.06	110	57.43
3 - 4	919	62.88	403	41.96
All	2,483	6.06	1,291	7.39
Three-Quarter Workers[2]				
1-2-3	1,153	- 19.69	700	- 14.00
1-2-4	166	- 26.88	313	- 24.54
1-3-4	149	42.57	332	63.21
2-3-4	700	61.11	637	38.36
All	2,168	- 0.65	1,982	12.11

[1]Workers employed in two calendar quarters of 1953.

[2]Workers employed in three calendar quarters of 1953.

218

Affirmative Indications

In what has gone before, an effort has been made to take account of
several mobility-related variables and one "built-in" definitional problem; the purpose was to search for other explanations, or suggestions of
explanations, of the observed wage improvement shown by the OASI data.
These explorations have produced little evidence pressing toward a dis-
belief in the obvious conclusions. It seems appropriate then to re-
examine them with the opposite question in mind: Do they contain evidence suggesting that it is indeed relative wage improvement of mobile
workers which underlies the differences in the percentages? There
seem to be several justifications for a cautious affirmative.

(1) Columns 2 and 3 of Table 4 show that the wage improvement
percentages of non-mobile and mobile workers decline with increases in
age. Most observers would surely predict such a result -- a fact that
supports the idea that the percentages reflect real-world wage improvement.

(2) Hypermobile workers were put into a separate class in Table 5
because of the speculation that at least some of them -- casual workers
and workers with personal characteristics which place them close to the
"unemployable" line, for example -- were workers for whom the wage
improvement hypothesis was not strictly appropriate. The much more
systematic relationship between the percentages of columns 3 and 5 as
compared with that of columns 3 and 7 -- as well as the "neater" and
less erratic descent of the percentages in column 5 as compared with
those of column 7 -- is consistent with the view that the data actually are
wage improvement percentages.

(3) The year of the data sample was 1953. This was the third of
three "good" years which followed the 1949 recession. It was also a
transitional year -- the year in which strong indications of 1954's reces-

Table 7. Monthly Labor Turnover Rates, July 1952 - June 1954

Year and Class	Monthly rate											
	J	F	M	A	M	J	J	A	S	O	N	D
1952												
New Hires							3.3	3.9	4.4	4.1	3.3	2.6
Quits							2.2	3.0	3.5	2.8	2.1	1.7
Layoffs							2.2	1.0	0.7	0.7	0.7	1.0
1953												
New Hires	3.4	3.3	3.5	3.5	3.3	4.2	3.3	3.3	3.0	2.4	1.7	1.1
Quits	2.1	2.2	2.5	2.7	2.7	2.6	2.5	2.9	3.1	2.1	1.5	1.1
Layoffs	0.9	0.8	0.8	0.9	1.0	0.9	1.1	1.3	1.5	1.8	2.3	2.5
1954												
New Hires	1.4	1.3	1.4	1.2	1.4	1.9						
Quits	1.1	1.0	1.0	1.1	1.0	1.1						
Layoffs	2.8	2.2	2.3	2.4	1.9	1.7						

Source: U.S., Bureau of Labor Statistics, Employment and Earnings, Vol. 6,
No. 11 (May, 1960), p. 43.

sion made their appearance. The labor market effects of these changes are pointed to by the 1953 monthly turnover data reproduced in Table 7. They show a steady rise in the layoff rate beginning in July and a fall in "new hires" and "quits" in the last quarter. In other words, voluntary moves as a percentage of total moves declined sharply in the last few months of the year. This would lead to the prediction that the percentages associated with the five continuous patterns of employment shown in Table 6 behave more or less as they do. That is, it would be expected that a rise in the proportion of involuntary mobility would be associated with a decline in the wage improvement percentage of mobile workers relative to that of non-mobile workers.

CONCLUSION

This paper has described the results of an experiment designed to further our understanding of the extent to which increases in wages are associated with labor mobility. The experiment consisted of a comparison of the percentage changes in wages of mobile and non-mobile workers. In general, the comparisons seem to indicate that workers who voluntarily changed jobs during 1953 experienced wage increases several percentage points greater than those who did not move. It is obvious, however, that the percentages shown in the preceding section of this paper could reflect something other than the tendency for workers to make relative wage gains through mobility. Thus an effort was made to take into consideration several variables known to be associated with mobility. This effort did not have the effect of weakening the apparent results of the original experiment; rather, the attempts to shake the initial findings probably worked in the direction of strengthening them. Nevertheless, none of the tests satisfactorily accounted for occupational or industrial seasonality, and so the findings warrant only tentative acceptance.

NOTES

[1]For a modern general statement, with appropriate qualifications, see Kenneth E. Boulding, Economic Analysis (3rd ed.; New York: Harper and Brothers, 1955), Chapters 9 and 10.

[2]A summary of much of this work is provided in Chapter 5 ("Mobility and the Process of Labor Allocation") of: Herbert S. Parnes, Research on Labor Mobility (New York: Social Science Research Council, 1954).

[3]Perhaps it would be more accurate to put the matter more or less as Reder does: Much of the criticism of the competitive hypotheses is directed primarily at its inability to deal adequately with short-run problems; this criticism misses the mark in the sense that the theory's supporters do not present it as a short-run predictive or descriptive device. See: Melvin Reder, "Wage Determination in Theory and Practice," A Decade of Industrial Relations Research, 1946-56, ed. Neil W. Chamberlain, Frank C. Pierson, Theresa Wolfson (New York: Harper and Brothers, 1958), pp. 84-86.

[4]The questionnaire techniques used have been subjected to a good deal of criticism; see, for example, "The Methodology of Positive Economics" in Milton Friedman, Essays on Positive Economics (Chicago: University of Chicago Press, 1953), pp. 3-43. Simon Rottenberg picks up and expands the Friedman attack in his article, "On Choice in Labor Markets," Industrial and Labor Relations Review, Vol. 9, No. 2 (January, 1956), pp. 183-99.

[5]See: Donald J. Bogue, A Methodological Study of Migration and Labor Mobility in Michigan and Ohio in 1947, Scripps Foundation Studies in Population Distribution, No. 4 (Oxford, Ohio: 1952); Dale E. Hathaway, "Migration from Agriculture: The Historical Record and Its Meaning," American Economic Review, Vol. 50, No. 2 (May, 1960), pp. 379-91; Donald E. Cullen, "Labor-Market Aspects of the St. Lawrence Seaway Project," The Journal of Political Economy, Vol. LXVIII, No. 3 (June, 1960), pp. 232-51; Robert L. Bunting, "A Test of the Theory of Geographic Mobility," Industrial and Labor Relations Review, Vol. 15, No. 1 (October, 1961), pp. 76-82.

[6]See: Allan M. Cartter, Theory of Wages and Employment (Homewood, Ill.: Richard D. Irwin, 1959), pp. 178-80.

[7]Parnes comments upon the need for such research as follows: "Manpower policy based on the assumption that wage differentials are the major factor inducing workers to change jobs will be quite different from policy based on the assumption that wages are of relatively minor importance. The question is also important for the personal policies of individual employers." Parnes, op. cit., p. 9.

[8]Charles A. Myers and George P. Shultz, The Dynamics of a Labor Market (New York: Prentice-Hall, 1951), pp. 33-34; Charles A. Myers and W. Rupert Maclaurin, The Movement of Factory Workers (New York: John Wiley and Sons, 1943), pp. 64-66; Herbert G. Heneman, Jr., Harland Fox, and Dale Yoder, Minnesota Manpower Mobilities, Industrial Relations Center, Bulletin 10 (Minneapolis: University of Minnesota Press, 1950), pp. 18-20.

[9]See Bunting, loc. cit., Larry Sjaastad, "Occupational Structure and Migration Patterns" in Labor Mobility and Population in Agriculture, a volume assembled and published under the sponsorship of the Iowa State University Center for Agricultural and Economic Adjustment (Ames, Iowa: Iowa State University Press, 1961), pp. 8-27.

[10]See: Bogue op. cit., Franklin M. Aaronson and Ruth A. Keller, Mobility of Workers in Employment Covered by Old-Age and Survivors Insurance, U.S., Social Security Administration, Bureau of Research and Statistics, Report No. 14 (July 1946, mimeo.); Paul Eldridge and Irwin Wolkstein, "Incidence of Employer Change," Industrial and Labor Relations Review, Vol. 10, No. 1 (October 1956), pp. 101-107; Robert L. Bunting, "Labor Mobility: Sex, Race, and Age," The Review of Economics and Statistics, Vol. XLII, No. 2 (May, 1960), pp. 229-231; Robert L. Bunting, Lowell D. Ashby, and Peter A. Prosper, Jr., "Labor Mobility in Three Southern States," Industrial and Labor Relations Review, Vol. 14, No. 3 (April, 1961), pp. 432-45; Bunting, Industrial and Labor Relations Review, Vol. 15, No. 1, pp. 75-82; Isadore Blumen, Marvin Kogan, and Philip J. McCarthy, The Industrial Mobility of Labor as a Probability Process (Ithaca, N.Y.: Cornell University, 1955).

[11]These and other limitations of the data are discussed more fully in the three articles by the present author cited in the preceding footnote.

[12]Self-employed workers were deleted from the sample for purposes of the work described here.

[13]Of course, the "and that he continued in the same employment for the remainder of the year" may be in error. He may have moved into non-covered employment, he may have retired, etc.

[14]The sample used in the following computations actually included 22,969 workers; thirty-one workers were deleted because sex or race was unknown.

[15]Thus workers classed as non-mobile here could actually have been mobile: e.g., they could have changed jobs by moving to other firms which were not covered by OASI. Similarly, some of the employees could have been classified here as mobile without actually having gone from one employer to another; they could have been working for two or more employers concurrently. This second possibility is a potentially serious source of difficulty; it will be given more attention later.

[16]The qualifying phrase "of employment" is important. To illustrate, some workers earned covered wages only in the second and third calendar quarters of 1953; for these workers, the second calendar quarter was the first quarter of employment and the third calendar quarter was the last quarter of employment.

[17]Thus for workers who earned wages in three calendar quarters, the wages earned in the "middle" quarter were dropped; for four-quarter workers, the second and third quarter wages were dropped.

[18]U.S., Bureau of Labor Statistics, Employment and Earnings, Vol. 6, No. 11 (May, 1960), p. 43.

[19]U.S., Bureau of Labor Statistics, "Technical Note on Measurement of Labor Turnover" (March, 1959) (mimeographed), pp. 2-3.

[20]See Bogue, op. cit., Chap. 8.

[21]See W. S. Woytinsky, Three Aspects of Labor Dynamics (Washington: Social Science Research Council Committee on Social Security, 1942), pp. 23-28; Lloyd G. Reynolds, The Structure of Labor Markets (New York: Harper and Brothers, 1951), pp. 27-28; U.S. Department of Labor, Bureau of Labor Statistics, The Mobility of Tool and Die Makers, 1940-1951, Bull. No. 1120 (Washington: U.S. Government Printing Office, 1952).

[22]Bunting, The Review of Economics and Statistics, op. cit., Table 5.

[23]There have been six multiple jobholding reports; another is in progress at the time of this writing. The first four were Nos. 30, 74, 80, and 88 of the Bureau of the Census's Series P-50 Current Population Reports. The last two, written by Gertrude Bancroft and Jacob Schiffman, respectively, appeared in the October, 1960, and October, 1961, issues of the Monthly Labor Review; these were issued as Special Labor Force Reports Nos. 9 and 18. The information in this paragraph on agricultural and self-employed multiple jobholders was drawn from the Bancroft and Schiffman articles.

[24]See footnote 20 in Bunting, Industrial and Labor Relations Review, op. cit., pp. 81-82. In another context it was estimated that the original sample of approximately 30,000 workers included between 100 and 200 agricultural workers.

[25]Another estimate (based on a small random sample) of the number of multiple jobholders in these data indicates that 1,000 may not be too large. See footnote 3 in Bunting, The Review of Economics and Statistics, op. cit., p. 229.

[26]The medians were obtained from Tables F and G in the (Schiffman) Special Labor Force Report #18. The percentage distribution of total hours worked, contained in Table F, does not permit one to determine the extent of skew in the total distribution; it probably is skewed (slightly) in the direction of smaller numbers, so the arithmetic mean of total hours worked is probably a little below the median. Table G, presenting hours worked on the secondary jobs of multiple jobholders, definitely is skewed toward large numbers so there is no question but that the arithmetic mean in this case is above the median. Hence the arithmetic mean of hours worked on the second job expressed as a ratio to those worked on the primary job is probably between 10/40 and 15/40. The range indicated in the text above, 1/4 to 1/3, was made by assuming that hourly earnings on the secondary job are less than on the primary job.

[27]Combining the five non-continuous patterns, 1-3, 1-4, 2-4, 1-2-4, and 1-3-4, gives a class of 546 non-mobile workers with a wage improvement percentage of 5.14 and a class of 1,028 mobile workers with a percentage of 10.30. These percentages are very close to those for the whole sample, as shown in Table 1.

Level and Structure in Wage Rates
of the Metropolitanized Work Force

by William Goldner

Use of large cross section of metropolitan labor market surveys to study occupational wage structures - Comparisons with firm's wage structure - Principal determinants of metropolitan wage structure are unionization, size of labor market, rural-urban income ratio, and employment in durable goods manufacturing - Advantages of factor analysis as a method include use of artificially created variables: (1) Measurement of deviations from composite average of all areas; (2) Measurement of skilled-unskilled wage differential - Method described - BLS data assessed - Reasons for equal weighting - Behavior of aggregated occupational structure range over time - Unique performance of teamsters' wage levels - Findings and their significance - Stability of ranking and spacing of occupational wage rate throughout the country.

The traditional approach to the analysis of occupational wage structures has usually involved a careful piecing together of aggregate data and the use of this single framework as the basis for developing hypothesis, testing and inference, and policy evaluation. The process characteristically involves a large case study or a very limited number of observations; the sample is not representative, and the method is primarily descriptive. With the availability of a large cross section of metropolitan labor market surveys accumulating over a time span that exceeds a decade, it is time to attempt a more comprehensive analysis of wage structures based on a larger number of observations.

Metropolitan labor markets are somewhat independent economic units within which the factors influencing supply and demand work themselves out. Consequently, with the development of standarized and meaningful measurements and the availability of comprehensive data on a uniform basis through this cross section of labor markets, it is possible to place

Author's note. Substantial assistance was granted to this study from several sources: The Institute of Business and Economic Research at the University of California, Berkeley, provided funds for clerical and statistical assistance; the Computer Center made computer time available; the Institute of Industrial Relations provided an environment in the past which generated some of the incentive to attempt the new methodology used here; Mrs. Carol Daniels, Mrs. Pat DeVito, and Robert Hesse were co-operative and patient in fulfilling a wide range of services without which this paper would never have been completed.

these labor markets in a spectrum along which quantitative data can be varied and their interrelated effects analyzed.

The models that underlie this discussion are not complex. The occupational wage structure in a metropolitan labor market essentially reflects the corresponding structure within the firm. Occupational wage rates within a firm are determined primarily by long-term, and secondarily by short-term, influences on the availability of different skills; i.e., by labor supply. This ordering or structure is frequently rationalized in terms of wage and salary standards or in provisions of collective bargaining agreements and is administered through a complex system intermingling time spent on the job, preparation and education, experience, and job characteristics or skill application. These systems are validated periodically against the market and therefore reflect limits imposed by economic influences.

The general level of this occupational structure is set by the demand for the firm's output. This product demand translates itself into the generalized demand for workers which the firm projects on the demand side of the labor market.

The transformation of the intrafirm structure and wage level to composites reflecting conditions in the metropolitan labor market is an aggregation problem. With reasonable interfirm mobility of workers, wages for each occupation tend toward some norm. Consequently, we still can look to the supply side of the metropolitan labor market for the main influences on the occupational wage structure. Similarly, aggregate demand for labor reflecting the general level of economic and productive activity in the labor market over a long-term or medium-term horizon determines the level of the market. Because nationally pervasive influences have such strong influence on local economic activity, wage levels for most metropolitan labor markets tend to move concordantly, with resulting rigidity in intermetropolitan differentials.

In this author's 1954 study,[1] the variables identified as having the most substantial influence on the metropolitan wage level were: unionization, the relative size of the metropolitan labor market, the rural-urban income ratio, and the proportion of employment in durable goods manufacturing. The study not only specified this list of associated variables explaining the metropolitan wage level; regression analysis provided estimates of their relative importance. The four variables which together explained 80 per cent of the total variance of the dependent variable, i.e., a coefficient of multiple determination of .80, individually accounted for the following proportions of the total variation:[2]

Unionization	.46
Log SMA population	.15
Rural-urban income ratio	.11
Proportion of employment in durable goods manufacturing	.08

In general, the explanatory factors which have been identified above exert their strongest influence on the supply side of the labor market. If the logic of our model is approximately correct, then the first three variables will influence the market wage level through their effects on occupations and the relative supplies of those skills.

Notice that the above model implies market-wide influences on labor supply rather than impacts on each occupation. The complexity of dealing with a long list of jobs in many labor markets makes necessary some device for abstracting from this detail. This has normally been accomplished by assuming a structure of job rates weighted by standard employment weights. This approach tends to insure that intermetropolitan labor market comparisons are composed of the same elements except for the actual occupational wage rates.

The relatively simple models which have been described above involve the occupational wage structure and by implication describe the intermetropolitan structure of wage rates. The elaboration of these models has been carefully avoided, not to shirk responsibility for theorizing, but rather because this points directly toward a new methodology, factor analysis, which is compatible with computer operations.[3]

The methodology of factor analysis may take several alternative forms. In this study, the principal factor solution was adopted. In the principal factor approach, a complex structure is reduced to a lesser number of variables. This compression may afford substantial economy of description and also has side benefits in the form of the scoring techniques which allow characteristics to be quantitatively expressed.

In this study, twenty-five occupational wage rates in each metropolitan labor market constitute the variables. These have been reduced to two quantitative measurements which embody 80 per cent of the variation reflected in the original structure of rates. Analysis of these two artificially created variables suggests that they measure meaningful concepts used in labor market analysis. The first factor measures deviations of the metropolitan wage level from the composite average of wage levels of all metropolitan areas. This factor is a direct representation of the intermetropolitan wage structure and therefore reflects geographical differences in area wage levels. The second factor converts the skilled-unskilled wage differential in individual labor markets into a quantitative measurement and provides the basis for analyzing geographical differences in wage differentials. In addition to its geographical significance, its internal structure is such that it provides a much more accurate and multi-variable measurement of the skilled-unskilled differential than is obtained using traditional methods.

This study therefore attempts to attain several objectives. One is to display the properties of a new and economical methodology for the analysis of structures and particularly of the occupational and geographical structure of wage rates. In addition to the methodological innovations, it attempts to enlarge the area of analytical findings with regard to these structures. Finally, it suggests other directions in which the methodology can be applied meaningfully in labor market analysis.

METHODOLOGY

Factor analysis[4] is concerned with two basic problems. The first deals with methods for transforming a complex set of variables which are structurally interrelated into one or several artificially composed variables. The second problem concerns the description of these hypothetical variables in terms of the original data.

The basic data from which factor analysis starts is a correlation matrix in which each variable in the study is correlated with every other variable. In this study, individual metropolitan labor markets constitute the units of observation. The individual occupational wage rates are the twenty-five variables used in the analysis. The correlation matrix was made up of the correlations across a large number of metropolitan labor markets for each pair of occupations.

It turns out that the method of principal axes is a statistically opti-mal solution which in this study meets criteria of statistical simplicity and economic meaningfulness. As will be observed in the following pages, the analysis attains an extremely parsimonious description of the ob-served data because the complexity of twenty-five variables is reduced to two quantitative measurements. The meaningfulness of these syntheti-cally constructed factors turns out to be central to the analysis of wage structures, for the first factor measures deviations from the composite wage level, and the second constitutes a standard index of skill differ-entials.

To suggest the descriptive summarization that this process provides, the following tabulation presents the proportion of variance accounted for by each successive factor.

Factor	1951-52	1959-60	1960-61
1	.730	.699	.691
2	.093	.096	.104
3	.033	.045	.043
4	.025	.032	.029
5	.023	.024	.024
6-25	.096	.104	.109

The high degree of interrelationship in the structure of occupational wage rates is reflected by the extremely high proportion of variance in the first factor.

An important question is the number of factors that may be signifi-cant to the study. Rules of thumb, more sophisticated tests of signifi-cance of the number of factors, and meaningfulness of the factor content all point to two factors as the limit beyond which no significant gain will be obtained.[5] Therefore, this study confines its attention to the first two factors.

A second purpose served by the factor analysis methodology is the analysis and description of the synthesized factors. Several descriptive characteristics of each factor in terms of the original variables are pro-vided. First, there is the proportion of total variance accounted for by each factor. Second, there are several alternative methods of assessing the contribution that each original variable imparts to the respective fac-tors. This assessment frequently leads to the identification of meaning-ful concepts with the particular factors. Third, each factor can be scored and quantitatively expressed for each unit of observation.

DATA AND DATA PROCESSING

The data used in this study embody wage rates collected by the Labor Market Wage Survey program of the U. S. Bureau of Labor Statistics. This program has generated wage data in varying numbers of metropoli-tan labor markets each year since 1949. Presently, the eighty labor markets in the 1960-61 series of surveys constitute a sample designed to be representative of all standard metropolitan statistical areas (SMSA's). This sample design represents a most progressive step forward in BLS statistical activities and must be highly commended.

In the interim since 1951-52, when forty metropolitan areas were surveyed, the Labor Market Survey program had been substantially reduced. Between seventeen and twenty metropolitan areas were covered until the expansion of the sample to sixty areas in 1959-60, and the further augmentation of that sample to a total of eighty in 1960-61. Because of the small samples and the necessity to delete certain metropolitan areas from them, individual year-by-year studies were not made. Instead, only three fiscal year programs were incorporated into the analysis, 1951-52, 1959-60, and 1960-61. It was felt that this almost decade-long span might be sufficient to show secular changes in the structural relationships. However, no formal attempt was made to analyze influences that might be associated with a short cycle.

In the very long list of occupations surveyed by BLS, wage rate averages were absent in certain occupations in a substantial number of the area surveys. Consequently, it was necessary to delete from the total list of occupations several that had substantially incomplete data. Machine tool operators, millwrights, plumbers, and sheetmetal workers were deleted for this reason. Similarly, among occupations at lower skill levels, elevator operators, janitors (women), packers (women), truck drivers, heavy (other than trailer type), truck drivers, power (other than forklift), crane operators, and packers A and B were either insufficiently reported in particular annual cross sections or were not consistently defined through all of the surveys. With these job deletions, a structure of twenty-five occupations remained which constitute the twenty-five variables in this study.

It is also appropriate to mention the significant exclusions to complete coverage:
1. Government establishments and the construction and extractive industries.
2. Small establishments; surveys covered establishments employing fifty or more workers except in twelve of the largest areas, where the minimum size was 100 in manufacturing, public utilities, and retail trade.
3. Occupations characteristic of particular industries; wage data are provided only for occupations common to manufacturing and non-manufacturing industries.

These data were treated in several ways to prepare them for the requirements of the factor analysis procedure. These treatment processes are important in understanding differences between the results of this study and somewhat comparable data published by BLS in several of its bulletins,[6] and in two articles in the Monthly Labor Review.[7]

First of all, the factor analysis procedure requires that there be data in every cell. Consequently, for some SMSA's where reporting standards did not allow the publication of data, it was necessary to create a synthetic number. The procedure for creating these synthetic numbers was as follows: First, for the SMSA involved, the mean value in a particular occupation within the region containing the SMSA was estimated. Similarly, the standard deviation within this region was estimated. Then, a random number was selected from a table of randomly ordered deviations from the mean of a normal distribution and the product of the random number and the standard deviation was added to the mean estimate. This enabled all cells in the table to be filled with randomized numbers, by an estimating process which maintained the mean wage level within regions and also preserved the dispersion of SMSA averages within regions. Other estimating procedures were considered but were too costly of time for the value they offered.

Synthetic numbers appeared to be concentrated in certain SMSA's. Therefore, an arbitrary criterion was established. Any SMSA was deleted from the study data which had more than five synthetic numbers out of the total of twenty-five occupational categories. Thus, of the forty SMSA's sampled in 1951-52, only thirty-one met this criterion and are included in our study. Similarly, in 1960-61, of the eighty SMSA's which the BLS surveyed, only seventy-two met the criterion.

It is worthwhile making explicit the direction of bias that this procedure incorporates into the data processing for factor analysis. Since the synthetic figures are randomized for regions rather than reflective of the influence of indigenous locational chacteristics of the individual SMSA, the correlations between job rates are reduced where these synthetic figures have been used. The study therefore understates the structural interrelationships which would be determined had real data been available instead of the artificial substitutes.

Descriptive Characteristics of the Data

In order to orient the reader to the scope and content of the material used in this study, several summaries of the data are presented in this section. Table 1 shows the unweighted average wage rates for the twenty-five occupations in the three matched cross-section samples that were

Table 1. Unweighted Average Wage Rates in Specified Occupations for a Matched Selection of 31 SMSA's

Occupation	Rank in 1951-52	1951-52 ($)	1959-60 ($)	1960-61 ($)
Tool and die makers	1	2.06	3.00	3.10
Pipefitters, maint.	2	1.98	2.86	2.97
Machinists, maint.	3	1.96	2.86	2.97
Electricians, maint.	4	1.95	2.87	2.97
Carpenters, maint.	5	1.88	2.68	2.78
Engineers, stny.	6	1.85	2.69	2.79
Mechanics, maint.	7	1.84	2.70	2.80
Painters, maint.	8	1.81	2.60	2.71
Mechanics, auto., maint.	9	1.73	2.58	2.68
Truck drivers, heavy	10	1.60	2.46	2.56
Oilers	11	1.53	2.26	2.34
Firemen, stny., boiler	12	1.51	2.26	2.39
Shipping clerks	13	1.50	2.21	2.30
Trucker, power (forklift)	14	1.50	2.20	2.28
Helpers, trades, maint.	15	1.49	2.22	2.28
Shipping and receiving clerks	16	1.48	2.22	2.31
Truck drivers, medium	17	1.47	2.28	2.38
Receiving clerks	18	1.45	2.06	2.15
Guards	19	1.43	2.07	2.15
Truck drivers, light	20	1.41	1.98	2.05
Order fillers	21	1.34	1.99	2.07
Packers (new)	22	1.33	1.88	1.96
Laborers, mtl. handling	23	1.32	1.98	2.06
Watchmen	24	1.16	1.60	1.64
Janitors (men)	25	1.13	1.67	1.72
Range		.93	1.40	1.46
Range/Tool and die makers		.45	.47	.47
Range/Helpers		.62	.63	.64
Range/Laborers		.71	.71	.71

used. This particular table involves only data for thirty-one SMSA's which were identical in the three specified cross-sections. Therefore, there is no effect from the change in sample size or sample coverage. In the table, occupations are ranked by the 1951-52 average wage rates. Except for the shift in the level of all of the wage rates, the remarkable stability of the occupational structure is apparent. The position of jobs in 1951-52 is virtually unchanged in the structure of 1960-61. In fact, for the several years shown here, correlations between years are of the order of .97 or .98. The most noticeable shifts in rank include an upward movement of truckdrivers, medium, from the seventeenth rank to the twelfth; a less marked upward movement of laborers; and a seemingly significant downward shift in the ranking of truckers, power (forklift).

It is interesting to compare the results of the 1960-61 averages with those reported in the BLS studies.[8] Because BLS used a smaller number of occupations in its study, there are only thirteen comparisons possible. The differences, all of the same sign, vary from zero to plus six cents. It is significant and explainable that the BLS results are consistently higher than the unweighted averages which are presented here. The explanation lies in the fact that larger metropolitan labor markets have higher rates and the inclusion of these higher rates weighted by the substantially larger employment in those areas results in some positive differentials above the levels of the data in this study. The fact that they are not remarkably large is noteworthy.

The argument for not weighting each labor market by its relative employment should be mentioned. Each metropolitan labor market is a more or less independent entity in which factors influencing supply and demand are working out their internal adjustments. The influences on and effects of this continuously equilibrating process within each labor market are most significant for the understanding of the indigenous occupational structure, and for observing intermetropolitan wage differences. To weight each occupation in an SMSA by the relative importance of SMSA employment allows the averaging to be swamped by the largest SMSA's. That the size of the SMSA is a relevant factor in explaining the SMSA average wage level is well established and thoroughly documented. However, the methodological requirements for analysis of structures is different from the procedures for estimating national aggregates. These arguments have been persuasive in prescribing equal weights for each labor market in this study.

The absolute range of rates in the twenty-five occupations included in the occupational structure has increased from 93 cents in 1951-52 to $1.40 in 1959-60, and $1.46 in 1960-61. The stability which has already been remarked upon also is reinforced by the ratios of the range to particular key jobs in the structure. These ratios, shown at the bottom of Table 1, are remarkably constant in relative terms through the time period. In other words, the range of the occupational structure has increased in absolute terms but remains constant in relative terms through time.

Quite a different relationship emerges when we study the dispersion of individual SMSA job averages among the sample of thirty-one SMSA's. Table 2 shows the standard deviations of the job averages around the composite average (i.e., the unweighted average of the thirty-one SMSA job averages).

Although these measurements fluctuate randomly to a certain degree as we move down the table, there is a clearly observable tendency for

Table 2. Standard Deviations of Distribution of SMSA Occupational Wage Rate
Averages for a Matched Selection of 31 SMSA's

Occupation	1951-52	1959-60	1960-61
Tool and die makers	17.65	19.38	19.27
Pipefitters, maint.	16.48	16.45	18.39
Machinists, maint.	12.31	19.78	20.41
Electricians, maint.	12.46	19.20	19.11
Carpenters, maint.	16.02	32.81	23.73
Engineers, stny.	16.39	26.11	25.57
Mechanics, maint.	12.56	22.05	22.34
Painters, maint.	18.06	25.83	28.22
Mechanics, auto., maint.	18.79	20.93	22.23
Truck drivers, heavy	27.35	32.56	32.84
Oilers	15.91	23.46	24.00
Firemen, stny., boiler	23.07	32.50	34.85
Shipping clerks	13.43	25.37	26.02
Trucker, power (forklift)	17.66	26.21	26.80
Helpers, trades, maint.	16.73	23.04	23.36
Shipping and receiving clerks	16.00	20.82	20.83
Truck drivers, medium	28.32	33.90	33.90
Receiving clerks	15.22	23.25	24.82
Guards	16.68	27.38	29.90
Truck drivers, light	29.24	45.16	47.78
Order fillers	17.28	31.04	32.50
Packers (new)	18.91	30.00	31.02
Laborers, mtl. handling	20.62	29.83	32.02
Watchmen	16.00	24.95	25.72
Janitors (men)	17.92	26.89	27.35

the standard deviation to increase. Because the occupations listed are
ranked by the wage rate level in 1951-52, this means that there is some
relationship between the absolute geographical spread of wage rates and
the average level of occupational wage rates.

In general, the dispersion is narrow for a group of skilled occupations, signifying some greater tendency toward uniformity among metropolitan labor markets for these jobs. On the other hand, the Truckdriver
and Trucker categories are extremely wide, which suggests that their
place in the occupational structure within each SMSA differs considerably
among metropolitan areas.

Except for the selected skills and the Truckdriver classifications,
the dispersion does not vary too substantially in different parts of the
structure. In fact, although the increases in dispersion to match the increasing wage levels through time are proportional to those increases,
there is no comparable tendency for dispersion to increase as we move
up the occupational scale.

Further insights into the complexity of the occupational wage structure can be gleaned by studying the distribution of correlation coefficients
which make up the structural data embodied in the factor analysis. Data
summarizing these distributions are presented in Table 3. First of all,
since there are 25 occupations, each correlated with every other occupation, there are 300 correlation coefficients in the matrix. The distribution of values of these 300 coefficients is shown in the table along with
the median and mean of the coefficients.[9] Because the distribution of
correlation coefficients has an absolute upper limit of one, it is characterized by skewness toward lower values. This is illustrated by the
relative positions of the median and mean which show the median ex-
ceeding the mean consistently in all distributions. In addition, there

Table 3. Distributions of Correlation Coefficients in Specified Time Periods

Class interval	1951-52			1959-60	1960-61
	40 SMSA's All Ind.	29 SMSA's Mfg. Ind.	31 SMSA's All Ind.	31 SMSA's All Ind.	31 SMSA's All Ind.
Total	300	300	300	300	300
.90 - .99	4	2	16	5	3
.80 - .89	32	57	92	60	28
.70 - .79	97	90	82	100	84
.60 - .69	60	70	46	57	64
.50 - .59	39	49	28	32	50
.40 - .49	29	19	16	22	19
.30 - .39	22	13	5	17	18
.20 - .29	7	-	4	6	16
.10 - .19	7	-	7	1	12
.00 - .09	3	-	3	-	5
-.00 - -.09	-	-	1	-	1
Median r	.67	.70	.75	.72	.64
Mean r	.63	.68	.70	.68	.60

seems to be a characteristic decline in the average intensity of the structural relationship through time for the thirty-one matched SMSA's. It is interesting to observe that despite the consistency of the structure that has been previously mentioned, the distribution of correlation coefficients fluctuates quite significantly. The sensitivity of these correlation coefficients to slight changes in randomized as well as persistent effects are apparently reflected in these data.

Two other comparisons are worthy of comment: if wage rates for manufacturing industries alone are used as a basis of analysis, there is a decrease in the level of structural consistency and intensity. Contrary to general impressions, there are no gains in homogeneity of the occupational wage structure by confining the analysis to manufacturing industry wage rates.

A second comparison illustrates one of the analytical steps taken in the early stages of the study when randomized data were used in all forty SMSA's surveyed for 1951-52. In this cross section, no SMSA's were eliminated for lack of published data. The effect of including nine SMSA's which had many randomized synthetic numbers is clearly apparent in the lower average correlation coefficient. This supports the inference that the inclusion of additional synthetic data using the methods followed in this study serve to lower the level of structural interrelatedness.

ANALYTICAL FINDINGS

Against the background of descriptive material and comparisons with other studies, it is now appropriate to turn our attention to the analytical findings which flow from this particular study.

Attention has already been called to the importance of the synthesized components which embody all of the multilateral relationships between the variables in this study. From that discussion, it is clear that two factors, or components, serve to explain approximately 80 per cent of the variation in the twenty-five occupational variables. Expressing these components in quantitative form would comprise a substantial economy in description. In addition, if these two synthesized components have some recognizable characteristics which are analytically meaningful, then the artificiality associated with them can be disregarded and the meaningful concepts substituted.

The First Factor: SMSA Wage Level

In order to determine if there is any particular significance to these synthetic components, it is necessary to study the structure of factor coefficients. These coefficients represent the correlations between the twenty-five original variables and the newly expressed quantifiable variables for the first or second component. It is important to remember that all of the variables are expressed in standard-score form. Table 4 presents the values of these factor coefficients. They can be thought of as the individual coefficients in a twenty-five-term regression equation, in which the variables are z-scores of the original data for each occupation.

Looking at Table 4 in which the occupations have been ranked by the size of the factor coefficient, we may ask what would be measured if

Table 4. Occupations Ranked by Factor Coefficient--First Factor

	Occupation	1951-52	1959-60	1960-61
1.	Truck drivers, medium	.956	.896	.878
2.	Receiving clerks	.941	.939	.944
3.	Laborers	.940	.886	.900
4.	Mechanics, auto	.940	.912	.922
5.	Order fillers	.939	.860	.879
6.	Shipping clerks	.922	.896	.892
7.	Janitors (men)	.915	.892	.894
8.	Firemen	.907	.867	.792
9.	Oilers	.905	.855	.848
10.	Packers (men)	.904	.838	.849
11.	Carpenters	.903	.908	.905
12.	Truck drivers, heavy	.885	.792	.788
13.	Truck drivers, light	.880	.889	.902
14.	Shipping and receiving clerks	.877	.860	.834
15.	Tool and die makers	.873	.754	.806
16.	Mechanics	.872	.835	.863
17.	Truck drivers, power (forklift)	.860	.923	.920
18.	Engineers	.841	.865	.856
19.	Watchmen	.834	.726	.739
20.	Helpers	.828	.818	.828
21.	Electricians	.781	.782	.690
22.	Painters	.749	.866	.837
23.	Machinists	.742	.846	.763
24.	Guards	.587	.491	.580
25.	Pipefitters	.304	.537	.504
σ_α		.079	.086	.110

individual scores were weighted by the factor coefficients. First, all of the coefficients are significant. Second, all except a few, which are in the bottom ranks of the table, are extremely high and are fairly close to unity. In effect, then, we are weighting the individual z-scores by coefficients or weights which are somewhat equivalent for all occupations. The individual z-scores, it should be remembered, represent the average occupational wage rate of a particular SMSA expressed as a deviation from the composite mean of all SMSA rates for that occupation. The first factor then must measure the wage level for that SMSA. For example, high-wage SMSA will have high positive z-scores, all virtually equally weighted by the factor coefficients. The summation of these cross products would result in an aggregate which would be high and positive. A low-wage area would have relatively large negative z-scores as deviations from the composite average wage rate in a particular occupation. These negative scores multiplied by the factor coefficients would result in a relatively high negative aggregate. It is important to understand that these measurements are transformations of the original SMSA wage data into scored form rather than an explicit substitute for the SMSA average wage rate. Nevertheless, they have many of the characteristics which are similar to area wage rate averages and do reflect the structure of geographical wage relationships. We therefore identify the first factor quite clearly as a measure of the wage level in specific metropolitan labor markets.

It is interesting to observe the fairly consistent pattern of these first factor coefficients through time, although there are differences which may reflect the impact of the randomized numbers that were placed into the data matrix. In general, these individual coefficients do not change significantly when measured against the size of the standard errors which are shown at the bottom of the table. This is further confirmation of the rigidity and stability of the occupational wage structure. For the jobs included here, their relative importance in contributing to the measurement of the SMSA wage levels has not changed substantially in the decade covered by the data.

An additional point of practical significance is the catalog of occupations with high first factor coefficients. These coefficients measure the contribution that each job rate makes to the aggregate; high coefficients therefore represent jobs whose places in the occupational wage structure are firmly notched and consistently ranked from one area to another. Those jobs with low coefficients at the bottom of the table are less meaningful in marking the level of the occupational wage structure. The practical considerations stemming from the lineup of occupations involve wage and salary standarization procedures. Bench-mark job rates are used to orient an internal wage structure to the appropriate market level. The top twenty occupations in Table 4 are appropriate bench-mark jobs; the bottom group is to be avoided for this purpose.

Table 5 presents the scores derived from the procedure described above for the thirty-one matched SMSA's which represent the core of the data used in this study. The table embodies several significant analytical findings. First, the scores for 1951-52 display an extremely high correlation with weighted wage rate averages which were calculated in traditional fashion by the author in another study using the same data.[10] For the thirty-one comparable SMSA's, the correlation between these two alternative methods of expressing the area wage rate averages is .983. The second significant analytical comparison is the relative position of particular metropolitan areas on this scale in each of the specified years and the shifts in position which have occurred. The table confirms the well-established relationships with regard to regional incidence of high and low wage rates. In the Northeast, there is a mixture of wage levels exhibited in particular labor markets. Most of the large labor markets are above the mean; that is, they have positive factor scores. Most of the smaller metropolitan areas fall below the general average in the region. Southern metropolitan labor markets are all characterized by negative scores, although some of them are shifting their positions substantially upward toward the national wage level. The North Central Region is characterized by a full configuration of positive scores and even among these some of them are changing significantly upward. Western metropolitan labor markets include the highest scores in the table

Table 5. First Factor Scores in Region and Code Order

SMSA's	1951-52	1959-60	1960-61	Regression study wage rates 1951-52
Northeast				
01 Albany	-0.01	-0.29	-0.35	$1.39
02 Allentown	-0.47	-0.21	-0.26	1.33
03 Boston	-0.18	-0.60	-0.49	1.37
04 Buffalo	0.52	0.72	0.78	1.47
05 Newark	0.93	0.66	0.64	1.53
06 New York	0.67	0.27	0.37	1.49
07 Philadelphia	-0.03	0.04	-0.07	1.37
08 Pittsburgh	0.69	1.03	1.11	1.51
09 Providence	-1.05	-1.75	-1.67	1.27
10 Worcester	-0.61	-0.64	-0.65	1.33
South				
01 Atlanta	-1.59	-1.09	-1.20	$1.12
02 Birmingham	-1.10	-0.10	-0.27	1.16
03 Houston	-0.17	-0.48	-0.43	1.25
04 Jacksonville	-1.64	-2.06	-2.10	1.04
05 Memphis	-1.57	-1.82	-1.80	1.07
06 New Orleans	-1.95	-1.81	-1.58	1.03
07 Richmond	-1.26	-1.38	-1.44	1.13
North Central				
01 Chicago	1.24	1.15	1.10	$1.57
02 Cincinnati	0.05	0.08	0.19	1.39
03 Cleveland	0.59	0.85	0.72	1.47
04 Detroit	1.77	1.43	1.53	1.65
05 Indianapolis	0.01	0.11	0.08	1.39
06 Kansas City	0.13	0.17	0.10	1.40
07 Milwaukee	0.77	1.01	0.99	1.51
08 Minneapolis	0.28	0.67	0.64	1.43
09 St. Louis	0.53	0.70	0.65	1.43
West				
01 Denver	-0.61	-0.16	-0.08	$1.28
02 Los Angeles	1.14	1.07	1.10	1.57
03 Phoenix	-0.16	0.15	0.03	1.31
04 San Francisco	1.85	1.66	1.82	1.68
05 Seattle	1.23	0.60	0.54	1.58

but also include some scores which are negative; i.e., below the composite average.

The changing positions of particular labor markets are also noteworthy. Inspection of the first factor scores shows that a few labor markets characteristically monopolize the extremes of distribution. In the central ranges of scores, however, there is an intense clustering of labor markets. This suggests that small shifts in the scores are not to be considered of great significance.

As would be expected, there are many more significant shifts in the eight-year span from 1951 to 1959 than there are in the one-year period from 1959 to 1960. Table 6 contains arrows connecting rank shifts of three or more places in the scale. Upward and downward shifts characterize labor markets in both halves of the distribution. High wage areas which move downward toward the center include Seattle, Newark, and New York. Low wage areas that moved down and away from the center of the distribution are Albany, Houston, Boston, and Providence. On the other hand, upward shifts were reflected by Pittsburgh, Cleveland, Buffalo, and Minneapolis among those above the average wage level.

Table 6. First Factor Scores in Rank Order and Shifts in Rank, for 31 Matched SMSA's

1951-52		Shifts[1] over 8-year period	1959-60		Shifts[1] over 1-year period	1960-61	
SANFR	1.85		SANFR	1.66		SANFR	1.82
DETRO	1.77		DETRO	1.43		DETRO	1.53
CHICA	1.24		CHICA	1.15		PITTS	1.11
SEATT	1.23		LOSAN	1.07		CHICA	1.10
LOSAN	1.14		PITTS	1.03		LOSAN	1.10
NEWAR	0.93		MILWA	1.01		MILWA	0.99
MILWA	0.77		CLEVE	0.85		BUFFA	0.78
PITTS	0.69		BUFFA	0.72		CLEVE	0.72
NYORK	0.67		STLOU	0.70		STLOU	0.65
CLEVE	0.59		MINNE	0.67		NEWAR	0.64
STLOU	0.53		NEWAR	0.66		MINNE	0.64
BUFFA	0.52		SEATT	0.60		SEATT	0.54
MINNE	0.28		NYORK	0.27		NYORK	0.37
KANSA	0.13		KANSA	0.17		CINCI	0.19
CINCI	0.05		PHOEN	0.15		KANSA	0.10
INDIA	0.01		INDIA	0.11		INDIA	0.08
ALBAN	-0.01		CINCI	0.08		PHOEN	0.03
PHILA	-0.03		PHILA	0.04		PHILA	-0.07
PHOEN	-0.16		BIRMI	-0.10		DENVE	-0.08
HOUST	-0.17		DENVE	-0.16		ALLEN	-0.26
BOSTO	-0.18		ALLEN	-0.21		BIRMI	-0.27
ALLEN	-0.47		ALBAN	-0.29		ALBAN	-0.35
DENVE	-0.61		HOUST	-0.48		HOUST	-0.43
WORCE	-0.61		BOSTO	-0.60		BOSTO	-0.49
PROVI	-1.05		WORCE	-0.64		WORCE	-0.65
BIRMI	-1.10		ATLAN	-1.09		ATLAN	-1.20
RICHM	-1.26		RICHM	-1.38		RICHM	-1.44
MEMPH	-1.57		PROVI	-1.75		NORLE	-1.58
ATLAN	-1.59		NORLE	-1.81		PROVI	-1.67
JACFL	-1.64		MEMPH	-1.82		MEMPH	-1.80
NORLE	-1.95		JACFL	-2.06		JACFL	-2.10

[1]Shifts are shown only where 3 or more ranks have been bridged.

Those lower in the scale of wage levels that moved upward included Phoenix, Denver, Birmingham, and Atlanta. In the one-year span from 1959 to 1960, the only change in position involving three or more scores was the Cincinnati labor market. Because this area was rather centrally situated in the distribution both before and after the move, it is not certain that this represents a substantial change in its position.

The upward movement of high wage cities may spotlight the significance of union strength in these metropolitan centers. On the other hand, the comparable and even steeper upward shifts of low wage cities are not universally associated with strong unionization, although Birmingham, the center of the steel industry in the South, may have had its extremely sharp rise due to the reinforcing influence of unionization on top of the basic economic health of Southern industrial centers.

Perhaps the most significant element in the presentation of the first factor scores is the evident relationship between these scores and area wage levels.

The Second Factor: The Multilateral Skill Differential

A more interesting and perhaps more significant product of the factor analysis methodology emerges from interpretation of the second factor which generates the following analytical findings. Table 7 presents

Table 7. Occupations Ranked by Factor Coefficient--Second Factor

	Occupation	1951-52	1959-60	1960-61
1.	Pipefitters	.839	.635	.584
2.	Electricians	.532	.568	.633
3.	Painters	.518	.291	.329
4.	Machinists	.442	.396	.560
5.	Mechanics	.350	.344	.309
6.	Tool and die makers	.275	.162*	.163*
7.	Carpenters	.233	.280	.290
8.	Helpers	.075*	.082*	.053*
9.	Shipping & rec. clerks	.073*	.154*	.022*
10.	Mechanics, auto	.057*	-.103*	-.101*
11.	Oilers	-.035*	.230	.170*
12.	Shipping clerks	-.059*	.071*	.046*
13.	Engineers	-.070*	-.014*	-.113*
14.	Firemen	-.075*	-.138*	.234
15.	Packers (men)	-.087*	-.131*	-.124*
16.	Guards	-.088*	.320	.308
17.	Receiving clerks	-.159	-.066*	-.089*
18.	Order fillers	-.175	-.228	-.221
19.	Truck drivers, medium	-.186	-.299	-.322
20.	Janitors (men)	-.238	-.348	-.350
21.	Truck drivers, heavy	-.245	-.464	-.501
22.	Laborers	-.280	-.396	-.382
23.	Watchmen	-.283	-.474	-.474
24.	Truck drivers, light	-.291	-.291	-.279
25.	Truck drivers, power	-.346	-.174	-.210*

*Coefficients not significant at .05 level.

the second factor coefficients for the particular occupations in the study where these occupations have been ranked by the size of the factor co-efficient in 1951-52. First, it should be noticed that instead of a consistent set of positive factor coefficients, the second factor has both positive and negative coefficients; that is, it is bipolar in its content. In addition, ten of the factor coefficients are so small that they cannot be deemed to be significant when measured against the standard error for that particular year. Occupations which had non-significant factor coefficients in 1952-53 have been starred in the table in order to assist in the analytical process. What remains, then, is a set of factor coefficients for seven relatively skilled occupations, all of which have positive factors and a remaining group of semi-skilled and unskilled occupations which have negative factors.

Let us direct the same kinds of questions toward this configuration of factors that were put to the first factor coefficients. That is, what would the pattern of coefficients represent if individual labor market wage rate scores were multiplied by the particular regression coefficients in Table 7? If all of the wage rates in a particular labor market were high, then the individual z-scores would all be positive and high; some of these would be multiplied by the high coefficients in the skilled category and others would be multiplied by the negative coefficients in the unskilled categories. When these linear combinations were added together, the net effect would be something close to zero. In other words, an undistorted or "normal" structure of occupational wage rates would not be reflected in any significant aggregation of scores using the second factor coefficients.

What would be true, however, if the skilled wage rates were low and the unskilled wage rates were high? Here, the low deviations for the skilled occupations would generate negative z-scores; the high wage rates for the unskilled occupations would generate positive z-scores; then the linear combination of factor coefficients and z-scores would result in a substantial negative aggregate. This means that a negative aggregate score reflects a compression of differentials between the skilled and the unskilled wage rates. It has the special characteristic, however, that this is measured in a multilateral fashion rather than in a summary fashion where one typical job at the bottom and another at the top are used to establish skilled-unskilled wage differentials.

We might follow with another example. Suppose the skilled occupations were very high at the same time that the unskilled occupations were very low in a particular labor market. The high occupational wage rates would generate positive z-scores multiplied in turn by the positive factor coefficients. The low unskilled wage rates, generating negative z-scores, would be multiplied by negative factor coefficients resulting in positive linear combinations throughout the structure. The multilateral aggregation of this type of labor market structure would yield a high positive measurement. The second factor can clearly be identified as a measure of the skilled-unskilled wage differential. The innovational aspect of this concept is the multilateral nature of the measuring device, incorporating all of the jobs into the area aggregate score.

Table 8 presents data of second factor scores derived from the factor analysis procedure. Inspection of the scores reveals an even more

Table 8. Second Factor Scores in Rank Order and Shifts in Rank, for 31 Matched SMSA's

1951-52		Shifts[1] over 8-year period	1959-60		Shifts[1] over 1-year period	1960-61	
HOUST	2.45		HOUST	2.38		HOUST	2.50
JACFL	2.39		BIRMI	2.09		BIRMI	2.02
RICHM	1.33		RICHM	1.53		PHOEN	1.35
CHICA	0.52		PHOEN	1.23		RICHM	1.12
DETRO	0.46		DETRO	0.79		NORLE	0.90
STLOU	0.45		ATLAN	0.53		DETRO	0.90
ATLAN	0.40		NORLE	0.52		CHICA	0.53
NEWAR	0.38		CHICA	0.48		MEMPH	0.42
MEMPH	0.37		MEMPH	0.43		STLOU	0.31
PHOEN	0.36		STLOU	0.41		INDIA	0.28
SANFR	0.26		INDIA	0.33		MILWA	0.22
MINNE	0.24		KANSA	0.22		LOSAN	0.19
MILWA	0.18		MILWA	0.05		KANSA	0.01
DENVE	0.11		LOSAN	-0.04		JACFL	0.00
KANSA	0.07		JACFL	-0.06		ATLAN	-0.05
INDIA	0.03		CINCI	-0.08		PITTS	-0.09
BIRMI	0.02		CLEVE	-0.15		CINCI	-0.10
LOSAN	-0.01		BUFFA	-0.18		BUFFA	-0.14
PHILA	-0.05		MINNE	-0.23		MINNE	-0.27
ALBAN	-0.14		PITTS	-0.26		CLEVE	-0.33
BUFFA	-0.21		ALBAN	-0.32		DENVE	-0.33
CLEVE	-0.22		DENVE	-0.40		SANFR	-0.52
SEATT	-0.35		SANFR	-0.45		ALBAN	-0.54
NYORK	-0.56		SEATT	-0.64		NEWAR	-0.75
CINCI	-0.77		WORCE	-0.72		PHILA	-0.75
PITTS	-0.86		PHILA	-0.80		BOSTO	-0.80
BOSTO	-0.94		NEWAR	-0.90		WORCE	-0.82
WORCE	-1.25		BOSTO	-0.91		NYORK	-0.90
NORLE	-1.27		ALLEN	-1.02		SEATT	-0.98
ALLEN	-1.54		NYORK	-1.14		ALLEN	-1.05
PROVI	-1.86		PROVI	-2.70		PROVI	-2.03

[1]Shifts are shown only where 3 or more ranks have been bridged.

significant clustering of large numbers of labor markets in the central
range of the data than was true for the first factor. Variations from
"normal" dispersion of skill differentials therefore are very slight for a
substantial number of labor markets. Contrarily, only a few metropolitan areas at the extreme ends of the distribution have substantially ab-
normal or unbalanced wage structures. Twenty of the thirty-one labor
markets fell within .50 of the center of the distribution in 1951-52. However, in 1959-60 and in 1960-61, more cities moved away from the central position toward the extremes. Table 8 also has arrows connecting
shifts in position of three or more ranks. It will be observed that many
more shifts of this size took place among the second factor scores than
was true among the first factor scores. This applies both during the
eight-year period, 1951 to 1959, and in the one-year period, 1959 to
1960. It is also true that upward and downward shifts took place from all
sectors of the distribution. In the eight-year period, from 1951 to 1959.
substantial narrowing of skill differentials represented by decreases in
second factor scores was exhibited by Jacksonville (Florida), Chicago,
St. Louis, Newark, San Francisco, Minneapolis, Denver, Philadelphia,
and New York. Widening of differentials, which involves an increase in
the second factor score, was displayed by Phoenix, Kansas City, Indianapolis, Birmingham, Los Angeles, Buffalo, Cleveland, Cincinnati, Pittsburgh, and Worcester. In the one year period, 1959 to 1960, downward
movements were reflected by Atlanta, Cleveland, and Seattle. This
represented a significant narrowing of skill differentials in these three
labor markets. The single widening movement that appears on the table
is Pittsburgh. Although changes did take place in the first factor scores,
there was a certain stability to the rankings and a much fewer number of
significant shifts than among the second factor scores. Skill differentials
and shifts in the dispersion of occupational wage rates appear to be much
more sensitive to labor market changes than the measures of wage level
among the same cities.

In order to test the relationship between the two factor scores it is
appropriate to cross-classify them. This cross-classification enables
us to associate high and low wage levels with narrow and wide differentials and provides the extremely interesting configuration of data which
is shown on Table 9. This table has been assembled to show the persist-
ence of both a particular wage level and a particular type of skill differential structure through time. Labor markets remaining in the same cell
of the table through the three time periods have been marked off. This
suggests that, although there is some shifting both in level and in differential structure, particular labor markets maintain a consistent combination of these two labor market characteristics through time. Nineteen of
the thirty-one labor markets in the study maintained their level and skill
structure through the full time period. Eleven others did not shift in the
one year comparison, 1959 to 1960. In addition, two labor markets remained in the same cell from 1951 to 1959. The point emphasized in this
particular exhibit is the strong tendency toward stability both with regard
to level and with regard to skill differential structure for a large proportion of the labor markets. A smaller proportion were those which shifted
positions either with regard to wage level or with regard to skill differential structure.

Implicit in this analysis is another significant element in the analysis
of wage structures. A standard inference from viewing aggregate data
over long time periods is the phenomenon of narrowing relative differ-

Table 9. Metropolitan Labor Markets Classified by Skill Differential Structure and Wage Level in 1951-52, 1959-60, and 1960-61

Metropolitan wage level	Skill Differentials								
	Narrow (-0.6 to -1.9)			Normal (-0.5 to +0.5)			Wide (+0.6 to +2.5)		
	1951-52	1959-60	1960-61	1951-52	1959-60	1960-61	1951-52	1959-60	1960-61
High (+1.0) (to) (+1.8)	NYORK BOSTO WORCE ALLEN CINCI PITTS	NYORK BOSTO WORCE ALLEN SEATT PHILA NEWAR	NYORK BOSTO WORCE ALLEN SEATT PHILA NEWAR	SANFR LOSAN SEATT NEWAR	SANFR LOSAN PITTS MILWA	SANFR LOSAN PITTS MILWA	CHICA DETRO	CHICA DETRO	CHICA DETRO
Medium (-1.0) (to) (+1.0)				CLEVE DENVE MINNE KANSA INDIA ALBAN BUFFA PHILA PHOEN MILWA	CLEVE DENVE MINNE KANSA INDIA ALBAN BUFFA CINCI STLOU	CLEVE DENVE MINNE KANSA INDIA ALBAN BUFFA CINCI STLOU	HOUST STLOU	HOUST BIRMI PHOEN	HOUST BIRMI PHOEN
Low (-1.0) (to) (-2.1)	PROVI NORLE	PROVI	PROVI	MEMPH ATLAN BIRMI JACFL	MEMPH JACFL	MEMPH ATLAN JACFL	RICHM JACFL	RICHM ATLAN NORLE	RICHM NORLE

entials as the general wage level increases. However, when analysis is focused on individual labor markets, it is quite clear that a wide variety of combinations characterizes the movement of skill differentials in relationship to wage level.

For example, Seattle moved from high to moderate wage levels and narrowed skill differentials in the process. Pittsburgh moved in the opposite direction, increasing its level and shifting its structure from narrow to normal. Most shifts, however, involved changes in skill structure with no change in area wage level.

It is interesting that the one unoccupied cell in Table 9 is the high wage-narrow differential combination. However, there are high wage markets with normal structures and with wide differentials. The same is also true of the other end of the spectrum: low wage levels are matched by wide differentials as would be expected, but there are also some labor markets with narrow differentials and low wage levels.

Finally, there are some clues to the labor markets that might be considered typical or normal in the sense that they are characterized by medium levels of wage rates and with relatively balanced wage differential structures. These labor markets are not the ones that are characteristically studied and from which policy inferences are made. In fact, they are noteworthy for their having been overlooked in most small sample, case-study-type analyses.

Geographical Distributions of Wage Levels and Differentials

Inspection of the data has already revealed that there is a tendency for individual labor markets in regions of the United States to have somewhat related wage levels and skill differentials. The generalized expression of this areal distribution is possible to map now that a large sample of metropolitan areas is included in the annual BLS Wage Survey program. Most of the analysis up to this point in the study has been in terms of thirty-one matched SMSA's in order to reinforce the comparability of the data. For mapping, however, seventy-two SMSA's were used, reflecting data collected in the year 1960-61. It is important to recognize that the values plotted in the charts are scores and not actual wage rates or skill differentials.

Figure 1 shows the spatial configuration of wage levels in the United States. The high wage ridge through the Midwestern industrial centers of Cleveland, Detroit, and Chicago falls off rather sharply as we move into the southern areas of the Central states. The Great Plains area, west of the Mississippi, is a very shallow trough covering a hugh expanse of space. It is significant that along each of the coastal areas there are tendencies toward higher wage levels, with a ridge on the West Coast and a lower ridge along the Atlantic shoreline from New York south. The coastal influence along the Gulf is discernible as a rather flat hill looking into the much lower elongated valley that circumscribes the area of low southern wage rates. The deepest part of this trough is toward the Old South in the Piedmont areas away from the coast line.

The distribution of wage differentials differs considerably from the spread of wage levels. This already has been suggested by the less than simple relationship between levels and differentials which has been discussed. As a generalization, in Figure 2 a wide area through the central sector of the nation is characterized by rather regular wage structures. Only in the northern areas of the Central industrial belt and in New Eng-

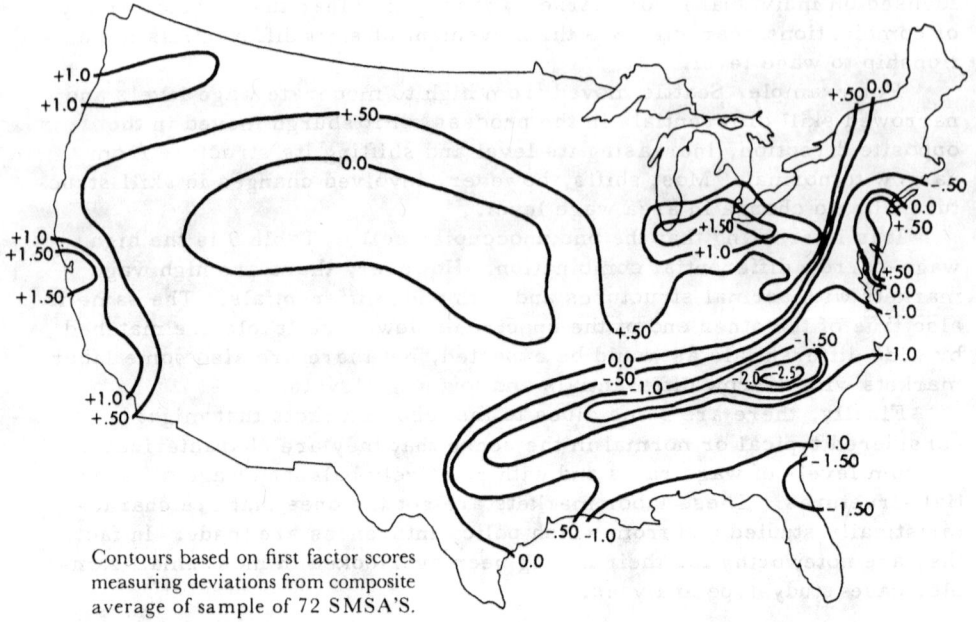

Contours based on first factor scores measuring deviations from composite average of sample of 72 SMSA'S.

Figure 1. Geographical distribution of metropolitan wage levels: 1960-61.

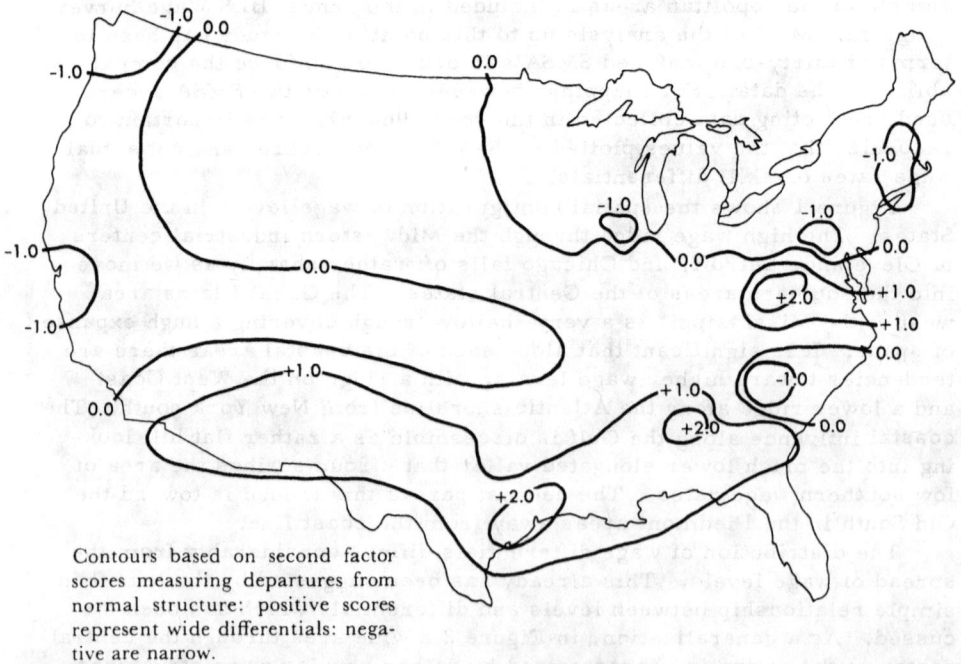

Contours based on second factor scores measuring departures from normal structure: positive scores represent wide differentials: negative are narrow.

Figure 2. Geographical distribution of metropolitan skilled-unskilled wage differentials: 1960-61.

land do differentials characteristically narrow. The strip along the southern edge of the nation is characterized by slightly wider differentials and includes two mounds where rather significantly wider differentials cluster. A similar area of wider differentials appears in the northern Appalachian area, in West Virginia, and in southern Pennsylvania.

SIGNIFICANCE OF FINDINGS

In this final section, inventory is taken of the particular contributions and innovations which are demonstrated by the material and analysis in this study.

Methodological Innovations

Use of the techniques of factor analysis has not been widespread in economic analysis and more especially in labor market studies. Factor analysis is an economical and easily available tool which will come into more frequent use as researchers become aware of its existence and the possibilities which stem from its characteristics.

The application of factor analysis to the data in this study resulted in the development of two significant measurements; one, a measure of the deviations from the composite wage level, and two, a measure of skill differentials in which multilateral relationships are included as part of the scoring procedure. The measure of wage levels correlates closely with other types of estimation of the same concept. The quantification of skill differentials, by incorporating all of the relationships in the skill structure, may be presumed to be more accurate than previous measures of this concept.

In the domain of measurement, many studies are concerned with the wage variable or labor cost variable in econometric equation systems, in metropolitan economic and transportation models, and in many other situations. The scores for the wage concepts developed here may properly be used to summarize a much more complex set of data. On the other hand, the study reinforces the notion that the wage level should be complemented by some measure of skill differentials in attempting summarization of the labor market elements in these models.

The Occupational Structure

This study of the occupational structure emphasizes the substantial stability which characterizes the wage rates associated with occupations in their ranking and spacing through time and in diverse metropolitan labor markets. Although this stability is characteristic of most of the occupations in the structure, there are certain occupations which do not sit consistently in the same place. Consequently, their presence does not contribute much to the description of the structure.

The consistently located occupations in the structure can be extremely useful in the administration of wage and salary standardization systems. In these systems, bench-mark jobs are identified and their level adjusted in relationship to local labor market surveys. The objective identification of these bench-mark jobs is a significant contribution of this study.

Structural changes appear to be more sensitive than shifts in the level of individual labor market averages. Here, too, there is a strong tendency for labor markets to maintain both their position in the distribution of area wage rates and their configuration of skilled-unskilled wage differentials. However, the differentials change more frequently than the levels.

Locational and Geographical Distributions of Metropolitan Wage Rates

Because the enlarged sample of BLS community wage surveys provides the basis of more accurate mapping, it is interesting to look at the geographical pattern of wage levels and wage differentials. Wage levels characteristically rise at important metropolitan shoreline centers. Inland areas away from the shore tend to have lower wage rates. This characteristic is true in the South as well as on the East and West Coasts.

In viewing the national contours describing metropolitan wage levels, caution should be used in projecting ceteris paribus onto the map. Situations are not identical in metropolitan economies with regard to several variables, some of which are also not measurable. The productivity-influencing factors in the metropolitan economy are often overlooked. These include the quantity and age of the community's social capital, and the values flowing from the skills of the human resources that are contributing to metropolitan product. Wage levels cannot be translated into labor costs until these items are quantified.

There does not appear to be an exact relationship between wage level and the skilled-unskilled wage differential. The geographic pattern is different enough to suggest that these two concepts do not respond to identical influences.

NOTES

[1]William Goldner, Factors Affecting the Wage Levels of Metropolitan Labor Markets, unpub-lished doctoral dissertation (Berkeley: University of California, 1955).

[2]Ibid, p. 131.

[3]The economy of time resulting from the use of computers can best be expressed by reporting that the materials studied, only one-third of which have been integrated into this analysis, took less than forty minutes of computer time. Individual jobs were usually of three to five minutes' duration.

[4]The authoritative treatment of factor analysis is found in Harry H. Harman, Modern Factor Analysis (Chicago: University of Chicago Press, 1960). In addition, the following treatments of portions of this methodology will be found extremely helpful: M. G. Kendall, A Course in Multi-variate Analysis (New York: Hafner Publishing Company, 1957), see particularly Chapters 2 and 3; Philip H. Dubois, Multivariate Correlation Analysis (New York: Harper and Brothers, 1957), see particularly Chapters 11 and 12; Robert Ferber and P. J. Verdoorn, Research Methods in Eco-nomics and Business (New York: The Macmillan Company, 1962), see particularly Section 3.6.

[5]Harman, op. cit., pp. 362 ff.

[6]U. S., Bureau of Labor Statistics, Wages and Related Benefits, 82 Labor Markets -- 1960-61, Bull. No. 1285-83 (Washington: U.S. Government Printing Office, 1962).

[7]Alexander N. Jarrell, "Job Pay Levels and Trends in All Metropolitan Areas," Monthly Labor Review, Vol. 85, No. 5 (May, 1962), pp. 510-16; and Toivo P. Kanninen, "Wage Differences Among Labor Markets," Monthly Labor Review, Vol. 85, No. 6 (June, 1962), pp. 614-20.

[8]Jarrell, op. cit., p. 512.

[9]The average values of the correlation coefficients in the data matrices are necessary in eval-uating the significance of the factor coefficients (factor loadings). See Harman, op. cit., p. 441.

[10]Goldner, op. cit.

244

Comments on Part IV

COMMENT by Richard A. Easterlin*

On Fleisher

Dr. Fleisher has done a worthwhile job of assembling historical data on migration and industrial wage trends so that the recent Puerto Rican migration to New York can be placed in perspective. This approach has the obvious advantage of enabling one to draw on earlier experience with seemingly similar movements in order to assess the effects of current migration. So far as I know, however, there is no accepted analysis of the causes or effects of the earlier movements, so that Dr. Fleisher's task is correspondingly magnified. Given the limited space available he has made a quite suggestive start on this subject. I feel, however, that much more needs to be done to establish the facts relating to this large problem and that the conceptual framework needs broadening before any generalizations of the type sought by Dr. Fleisher can be established. To indicate what I have in mind, I have organized some information below bearing on his proposition that "Puerto Rican immigration curtailed the movement of both native whites and Negroes to New York, but the impact on the movement of native whites seems to have been greater." I wish to stress that my primary interest is in illustrating additional basic research needs regarding both the current and earlier migrations, rather than a definitive evaluation of the proposition. The comments focus on New York, because I believe interregional comparisons are of uncertain value until the New York situation has been fully investigated. The basic points are enumerated below:

1. More work is needed to establish reliably for the several population groups the magnitude of the net migration movements to New York in the post-1940 period (see Table 1). Dr. Fleisher's state estimates for Negroes show a noticeable decline between the 1940's and 1950's, while a recent Bureau of the Census estimate for non-whites shows a slight rise. The latter appears more consistent with the figures for non-white migration to the New York-Northeastern New Jersey SMA shown in Dr. Fleisher's Table 2. With regard to the white population other than Puerto Rican, Dr. Fleisher's state estimates, as I read them, indicate a noticeable rise in net out-migration, a result which seems supported by the Census estimates, though the noticeably lower level of the latter in 1940-50 is disturbing. However, the SMA estimates in Fleisher's Table 2 show a slight decline in net out-migration, and it is not certain which geographical unit the author considers appropriate for his purpose. Finally, estimates of the net migration of Puerto Ricans are needed in the paper for the census decades, 1940-50 and 1950-60, to permit comparison of magnitudes between the Puerto Rican and non-Puerto Rican movements for identical time periods.

2. More information is needed on the characteristics of the several population groups and the migrants therein.

* Note. The author is grateful for helpful discussions to Richard T. Geruson, of the University of Pennsylvania, who is currently engaged in an historical analysis of the causes of Puerto Rican migration as a doctoral dissertation in economic history, and for statistical assistance to Chantal de M. Dubrin of the National Bureau of Economic Research. With regard to the comment on Professor Segal's paper, I wish to acknowledge also an earlier conversation with Professor W. Lee Hansen, of the University of California, Los Angeles, out of which these thoughts grew.

Table 1. Alternative Estimates of Net Migration to New York, by Color,
1940-50 and 1950-60

(thousands)

Line	Population class	Geographic unit	1940-50	1950-60
1	Negro	New York State	+240	+130
2	Non-white	New York State	+276	+282
3	Non-white	N.Y.-No'eastern N.J. SMA	+273	+291
4	White, except Puerto Rican	New York State	-285	-406
5	White, except Puerto Rican	New York State	-130	-360
6	White, except Puerto Rican	N.Y.-No'eastern N.J. SMA	-262	-234

Sources:

Line 1   Read from Fleisher, Figure 1.

Line 2   U.S., Bureau of the Census, Current Population Reports, Series
P-25, No. 247 (April, 1962), p. 7.

Lines 3, 6   Fleisher, Table 2.

Line 4   From figures for total white, obtained by summing native and
foreign-born white in Fleisher's Chart 1, decade estimates of
Puerto Rican net migration to New York State were deducted.
Since no decade figures for Puerto Rican migration are given in
the Fleisher paper, they were estimated as equal to the inter-
censal change in the native population (all classes) residing in
New York State and born in U.S. outlying areas. See U.S.,
Census of Population, 1960, New York General Social and Economic Characteristics, PC(1) 34C, p. 34-210.

Line 5   From estimates for the total white population, given in the source
for line 2, estimates of Puerto Rican net migration, derived as in
line 4, were subtracted.

    The author's procedure of evaluating the consistency of relative price
and quantity movements with alternative explanatory hypotheses has a
reasonably sound basis in theory, though it fails to take account of the
possibility of adjustments taking place through channels other than wage
rates, e. g. , via movements in unemployment rates. However, here, as
is so often the case, relevant price data are not easily obtained, and the
results are at best mixed. One of the most striking features of the relative earnings data in his Figures 3, 4, and 5 is the substantial similarity
in the amplitude and timing of the major swings shown by industry groups
characterized by significant differences in the employment of immigrants.
(This is touched on by the author in footnote 27 of his paper.) Moreover,
examining the relation in the 1900-1940 period between these swings and
fluctuations in immigration (as shown by the foreign-born white curve in
Chart 1), one finds that in the decade of highest immigration, 1900-10,
relative earnings fell in all three industries; in the decade of next highest
immigration, 1920-30, relative earnings rose in all three; and in the
decade of lowest immigration, 1930-40, relative earnings fell in all
three. There appears little evidence in these data of a systematic effect
of immigration on relative earnings prior to 1940, in terms of which the
current impact of Puerto Rican migration can be assessed.
    On the other hand, it is possible to obtain data on the characteristics
of the various groups which can be used not only to clarify the facts relating to their migration but also to test the author's view regarding
their causal connections, since his argument implies that the several
groups are similar and compete in substantially the same labor markets.
Table 2 presents 1950 data for those aged 25-44 (the age span in which
migration is concentrated) for four population groups in the New York-

246

Table 2. Selected Characteristics of Population Aged 25-44 of New York-Northeastern New Jersey Standard Metropolitan Area, by Sex, Color, and Parentage, and of Puerto Rican Population Aged 25-44 of Continental United States, 1950

Selected characteristics	All classes (1)	Native white, native parentage (2)	Native white, foreign or mixed parentage (MALES) (3)	Foreign-born white (4)	Non-white (5)	Puerto Rican (6)
Median school years completed	11.3	12.2	11.9	9.9	9.1	8.1
EMPLOYMENT STATUS						
Population, thousands	2,003	655	914	253	181	50.8
Per cent in labor force	93	91	96	95	86	86
Unemployment rate of labor force	5	5	5	5	10	14
OCCUPATIONAL DISTRIBUTION						
Total civilian employed, thousands	1,766	571	827	227	141	36.7
Total civilian employed, per cent	100	100	100	100	100	100
Professional and technical	12	16	12	10	4	4
Farmers and farm managers	...	...	...	...	...	...
Managers, off., prop., ex. farm	14	12	16	18	4	5
Clerical and sales	19	22	20	13	11	9
Craftsmen and foremen	19	18	19	22	11	12
Operatives	20	18	21	20	29	32
Private household	...	...	...	...	1	1
Service, except priv. household	8	8	6	10	21	26
Farm laborers and foremen	...	...	...	...	1	2
Laborers, except farm and mine	6	5	5	5	16	8
Occupation not reported	1	2	...	1	1	1
INCOME (BOTH SEXES)						
Median income in 1949 of persons with income, dollars						
Aged 14 and over	2,459	2,540	2,636	2,468	1,858	1,654
Aged 25-44	2,758	n.a.	2,920	2,760	n.a.	1,786

(continued)

Table 2 (concluded)

Selected characteristics	All classes	Native white, native parentage	Native white, foreign or mixed parentage (FEMALES)	Foreign-born white	Non-white	Puerto Rican
	(1)	(2)	(3)	(4)	(5)	(6)
EDUCATION						
Median school years completed	11.1	12.2	11.7	9.1	9.3	7.8
EMPLOYMENT STATUS						
Population, thousands	2,241	731	992	294	224	54.3
Per cent in labor force	36	34	34	36	56	44
Unemployment rate of labor force	5	4	4	4	7	9
OCCUPATIONAL DISTRIBUTION						
Total civilian employed, thousands	783	243	322	101	117	21.6
Total civilian employed, per cent	100	100	100	100	100	100
Professional and technical	12	17	11	10	5	4
Farmers and farm managers	...	...	...	...	...	...
Managers, off., prop., ex. farm	4	4	5	5	1	1
Clerical and sales	38	47	46	27	9	9
Craftsmen and foremen	2	2	2	2	2	2
Operatives	27	19	28	35	33	74
Private household	7	1	1	8	35	2
Service, except priv. household	8	7	6	12	13	7
Farm laborers and foremen	...	...	...	...	...	...
Laborers, except farm and mine	...	1	...	...	...	1
Occupation not reported	1	2	1	1	1	1

Sources: Cols. 1 and 5: U.S. Census of Population: 1950. Vol. II, Characteristics of the Population, Part 32, New York: Median school years completed, p. 235. Participation rate and unemployment, pp. 241-42. Major occupation group, pp. 305-308. Median income, p. 235.

Col. 2: For each characteristic, the underlying absolute data used in deriving the percentages or medians for the population groups in columns 3 through 5 were subtracted from those underlying column 1.

Cols. 3 and 4: U.S. Census of Population: 1950. Special Reports. Nativity and Parentage; P-E, No. 3A, p. 3A-67.

Col. 6: U.S. Census of Population: 1950. Special Reports. Puerto Ricans in Continental U.S., P-E, No. 3D, p. 3D-13.

Northeastern New Jersey SMA and for Puerto Ricans on the U. S. main-land, most of whom were located in the same SMA. Compared to Puerto Ricans, the three white groups are markedly better educated, differ noticeably with respect to labor force participation and unemployment rate, are typically concentrated in higher occupational classes, and have noticeably higher incomes. On the other hand, there is a noticeable similarity between the characteristics of non-whites and Puerto Ricans in the New York area.

In Tables 3 and 4 an attempt is made to single out some of the indus-tries which are currently providing substantial employment to Puerto Ricans, and to determine the historical importance of these industries in providing employment to the native white population, using 1930 data. Although there are problems arising from comparability in industry and occupation classes over time and in detail on color and nativity, this limited set of data suggests that Puerto Ricans are currently finding employment in industries which employ a disproportionately high per-centage of the non-white, not the white population, and that historically the employment in these industries of native whites has been dispropor-tionately low.

Column 1 of Table 5 provides a sample of evidence on the characteristics of out-migrants (chiefly white) from the New York area between 1949 and 1950. To judge from their education and occupational characteristics, this is hardly a group which was pushed out because of close competition with incoming Puerto Ricans.

Table 6 indicates that native white in-migrants to New York State from the South have always been relatively small in number in this century, and that even in 1920-40 when net native white migration to New

Table 3. Per Cent Distribution of Employment by Color for Selected Industries, New York-Northeastern New Jersey Standard Metropolitan Area, 1950

Industries	All classes	White	Non-white
All industries	100	92	8
Apparel and other fabricated textile products	100	90	10
Laundering, cleaning, and dyeing services	100	69	31
Eating and drinking places	100	89	11
Hotels	100	84	16
Medical and other health services	100	88	12

Source: U.S. Census of Population; 1950, Vol. II, Characteristics of the Population. Part 32. New York, p. 395.

Table 4. Per Cent Distribution of Gainfully Occupied by Color and Nativity for Selected Occupations, New York City, 1930

Occupations	All classes	Native white	Foreign-born white	Non-white
All occupations	100	53	41	6
Operatives (n. e. s.), clothing industries	100	30	65	5
Laundry operatives	100	33	30	37
Waiters	100	24	61	'15

Source: Fifteenth Census of the U.S.: 1930. Vol. IV, Occupations, by States, pp. 1130-1134.

Table 5. Out-migrants and In-migrants from 1949 to 1950, New York State
Economic Area G, by Selected Characteristics, 1950

Selected characteristics	Out-migrants (1)	In-migrants (2)
Total, 1 year old and over	131,805	115,835
Per cent white	94	89
Per cent male	51	49
Per cent 1 to 13 years old	18	16
Per cent 14 to 24 years old	24	26
Per cent 25 to 44 years old	38	41
Per cent 45 years old and over	19	17
Total, 25 years old and over	75,845	67,720
Median school years completed	11.4	11.4
Males, 14 years old and over	55,200	48,090
Per cent in labor force (incl. armed forces)	72	80
Per cent in armed forces	11	7
Per cent of labor force unemployed	8	12
Females, 14 years old and over	52,870	49,650
Per cent in labor force	29	44
Per cent of labor force unemployed	12	9
Per cent distribution by major occupation group of employed civilian males, 14 and over	100	100
Professional, technical and kindred workers	24	24
Farmers and farm managers	2	...
Managers, officials, and proprietors except farm	17	14
Clerical and kindred workers	7	10
Sales workers	11	11
Craftsmen, foremen, and kindred workers	13	12
Operatives and kindred workers	11	11
Service workers	8	11
Farm laborers and foremen	2	...
Laborers except farm and mine	4	5
Occupation not reported	1	1
Median family income in 1949	$3,759	$3,427

Note: New York State economic area G includes the following counties:
Bronx, Kings, Nassau, New York, Queens, Richmond, Rockland, Suffolk, West-
chester. The data for out-migrants include approximately 21,000 intrastate
migrants and those for in-migrants about 25,000 intrastate migrants.

Source: U.S. Census of Population: 1950. Special Reports. Population
Mobility-States and State Economic Areas, P-E, No. 4B, p. 114.

York was positive, the principal sources have been New England, other
Middle Atlantic states, and the East North Central division. If one may
judge from the 1949-50 data (Table 5, column 2), these in-migrants too
have been relatively well-educated and high up the occupational ladder
compared to Puerto Ricans. These data hardly suggest that the imposi-
tion of immigration restrictions resulted in a major flow of native whites
from the South to New York in the interwar period which was curtailed or
diverted after World War II because of an influx of competing migrants
from Puerto Rico.

All in all, the figures cast considerable doubt on the view that the
recent rise in Puerto Rican migration has been an important factor
influencing the net migration of native whites to New York. They are
more consistent with the view that Puerto Ricans compete with non-whites,
but in this connection it should be noted that the Census estimates

Table 6. Change in Number of Native White Persons Living in New York and Born Outside the State, by Place of Birth, 1900-10 to 1940-50

(thousands)

Birthplace	1900-10	1910-20	1920-30	1930-40	1940-50
U.S. except New York	162	150	379	170	128
New England	29	16	102	19	10
Mid-Atlantic except New York	72	66	146	86	46
East North Central	24	30	46	20	11
South	23	17	54	29	41
West North Central and West	15	21	31	17	21

Source: Everett S. Lee, et al., Population Redistribution and Economic Growth, United States, 1870-1950, I (Philadelphia: American Philosophical Society, 1957), p. 279.

indicate that the share of New York State in the total gain through migration of non-white population for all in-migration states hardly changed between 1940-50 and 1950-60 (see table below). My basic point, however, is that data such as these need to be mobilized for effective study of both recent and earlier migrations.

	1940-50	1950-60
Net non-white migration (thousands)		
a. New York	276	282
b. All states with net in-migration		
of non-whites	1,572	1,703
Ratio, a to b, per cent	18	17

Source: U.S. Bureau of the Census, Current Population Reports, Series P-25, No. 247 (April, 1962), p. 7.

3. The conceptual framework should be broadened to take account of the possible relation between migration and other sources of labor force change. It is well known that labor force participation rates have changed over time, and particularly in the period since 1940. Also the influence on labor force change of natural growth of the working age population has varied over time, as is illustrated in Table 7. Note that in 1940-60 normal demographic processes would have resulted for the first time, in the period covered by these figures, in an actual reduction in the working age population, aged 20-44, in New York. There is at least a suggestion here that the impact of Puerto Rican migration needs to be considered within a framework broad enough to take account of the recent deficit in the supply of younger working age persons from the local population.

On Segal

My only comment on Professor Segal's paper is of a related nature. In his thoughtful analysis of the factors responsible for the disparate movement of occupational wage differentials in the 'forties versus the 'fifties, Professor Segal draws a useful distinction between longer run forces and changes in the socio-economic setting of a particular period. He explicitly chooses the latter approach though favoring both lines of inquiry. I should like to suggest one set of longer term forces that might bear examination.

Young persons typically earn less than old, and one would expect that changes in the relative earnings of young and old would give rise

Table 7. Change in Population 20-44 Years Old Due to Aging and in Labor Force 20-44 Due to All Causes, New York State, 1890-1900 to 1960-70

(thousands)

	Change in population 20-44 years old due to aging (1)	Change in labor force 20-44 due to all causes (2)
1890-1900	+374	+438
1900-10	+261	+814
1910-20	+316	+244
1920-30	+181	+715
1930-40	+152	+252
1940-50	-118	-125
1950-60	-579	+ 21
1960-70	+138	

Source and Method:

Column 1. For each decade, beginning of decade figure for population aged 35-44 was subtracted from that for population aged 10-19. No deduction was made for deaths during the decade among the population aged 10-34 at the beginning. Data are from U.S. Census of Population: 1960, New York, General Population Characteristics, PC(1) 34B, pp. 34-57. The 1940-50 and 1950-60 figures include an allowance for net migration of armed forces of -106 thousand and -53 thousand, respectively, from U.S., Bureau of the Census, Current Population Reports, Series P-25, No. 227, p. 6.

Column 2. Through 1940, data are from E. S. Lee, et al, Population Redistribution and Economic Growth, pp. 539-40. (Figures for those aged 20-24 from 1920 through 1950 were obtained from the basic census reports.) Prior to 1920, figures shown are for those aged 16-44. Data for 1950 are from U.S., Census of Population: 1950, Vol. II, Characteristics of the Population, Part 32, New York, p. 32-250; for 1960, from U.S. Census of Population: 1960, New York, General Social and Economic Characteristics PC(1) 34C, pp. 34-219, with the change for the 20-44 in 1950-60 group assumed the same as that for 18-44.

Table 8. Comparative Number and Education of Young and Middle-Aged Adults, United States, 1890-1960

	Ratio of population aged 15-24 to that aged 45-54, per cent (1)	Years of school completed by male population aged 25-29 in excess of that by male population 45-54 (2)	Change from preceding date Col. 1 (3)	Col. 2 (4)
1890	252	n.a.	-	-
1900	233	n.a.	-19	-
1910	215	n.a.	-18	-
1920	177	0.3	-38	-
1930	173	0.5	-4	+0.2
1940	154	1.7	-19	+1.2
1950	128	3.3	-26	+1.6
1960	117	2.3	-11	-1.0
1970	147	0.5	+30	-1.8

Sources: Column (1): U.S. Census of Population: 1960, United States Summary. General Population Characteristics, PC(1) 1B, pp. 1-153. Figure for 1970 estimated as ratio of those aged 5-14 to 35-44 in 1960.

Column (2): The derivation of these figures is explained in my paper in the American Economic Review, LI (December 1961), No. 5, p. 898, n. 31.

252

to changes over time in cross-section wage or income distributions covering persons of all ages. Other things being equal, a decline in the relative number of young persons and/or an increase in their educational advantage over older persons would be expected to raise their relative earnings and reduce wage and income differences in aggregate distributions.

Table 8 brings together some hastily assembled data designed to throw light on historical movements in the relative number and education of young versus middle-aged persons in the U.S. population. The following observations are suggested by the figures in columns 1 and 3:

   (a) the decades in which the greatest decline in relative number of young persons occurred were 1910-20 and 1940-50, periods in which significant declines in inequality occurred;
   (b) the decade in which the smallest decline in the relative number of young persons occurred was 1920-30, a decade in which inequality rose;
   (c) the decade in which the next smallest decline occurred was 1950-60, a period in which the previous downward trend was arrested and possibly reversed.

Columns 2 and 4 show in addition that:

   (d) major advances in the educational advantage of young over middle-aged persons occurred abruptly in the 'thirties and 'forties, the period hailed as bringing a "great social revolution" in our income distribution;
   (e) the 1950's, the period in which doubts began to rise about the permanence of this revolution, saw the beginning of an important reversal in the educational advantage of young over middle-aged persons.

Together, the illustrative figures in Table 8 add up to a picture consistent enough with our knowledge of movements in wage and income differentials in this century to suggest the need for further inquiry along this line.

COMMENT by W. Lee Hansen

On Bunting

Professor Bunting is to be commended for making use of the tremendous stock of BOASI data; in this case, to test hypotheses about labor mobility and earnings improvement. His careful examination of the data leads him to conclude that job changers improve their relative earnings position by such moves and, further, that the improvement patterns observed are not dominated by seasonal, age, sex, and color factors. The obvious implication is that job changing provides a method of enhancing the individual's earning position. On the basis of Bunting's careful exploration of the data, one is initially inclined to be persuaded. Upon reflection, however, there appear to be several other factors that still require investigation before great confidence can be placed in his findings.

Let us consider the act of a job change. A job change will ordinarily involve certain costs. First, if some time elapses between the time a person leaves one job and begins another, a loss of earnings is incurred. Second, to the extent that job changes require a change of residence, the prior acquisition of new skills, etc., there will be additional direct costs that must be offset by the higher earnings stream that results from such a job change.

Consider only the first cost, loss of earnings. If a job change is involuntary, there is almost certain to be some period of time in which earnings are cut off, simply because a search for a new job must be undertaken. The same will be true in the case of voluntary job changes, though the earnings loss will probably be considerably smaller on the assumption that voluntary job changers ordinarily arrange to take up a new job with little or no time lost. Hence, when a job change is made, whether it be voluntary or involuntary, the move will very likely reduce the level of earnings in the quarter in which the move is made. If we then make a two-quarter comparison and if a job change and earnings loss takes place in the initial quarter, we would expect an increase in actual earnings from the first to the second quarter even if the rate of earnings in both jobs were similar; however, if the job change takes place in the final quarter, then we would expect a decline in earnings from one quarter to another. Thus, to the extent that job changers do in fact move to higher paying jobs, Bunting's earnings improvement estimate is probably overstated if changes occur in the initial quarter and understated if changes occur in the final quarter. To permit an accurate before and after comparison Bunting requires data which show the levels of full-quarter earnings in the two different jobs. If these are not available, then information on the quarter of the change and an estimate of the earnings loss while changing jobs is needed. It is to be hoped that Bunting can obtain these kinds of information from his sample data so as to permit him to indicate more precisely the magnitude of the earnings improvement associated with job changes.

If we are to conclude that job changers do in fact improve their earnings position, two other considerations become important. First, if substantial amounts of earnings are lost in finding a new job, it may be some time before these losses are recouped even at the new, higher earnings level. And second, the other costs of job changing also become important when we try to indicate that job changers are in some sense "better off" after a change. While these costs will presumably be offset by the new, higher level of earnings being received, this may not be the case when such costs are large. In other words, is it conceivable that the present value of the higher earnings stream, after due allowance is made for all costs, could be less than that of the previous earnings stream. Consequently, even though job changers may improve their earnings position, it is still not evident that their net position is improved; this will depend in addition upon the magnitude of the costs that must be incurred in accomplishing a job change.

On another aspect of the paper, it would be of interest to know to what extent Bunting's conclusions are a function of the particular geographical area, the industrial structure of the area, and the time period under consideration. More elaborate tests, involving comparisons with other regions, would be required to throw additional light on these matters.

While these comments may have sounded overly critical, they are not meant to suggest that Bunting has been unsuccessful in his efforts. On the contrary, he has embarked upon a fascinating investigation that should do much to produce new knowledge of and insights into labor mobility and earnings patterns.

On Segal

Professor Segal shows that in the 1950's there has been an apparent widening of occupational wage differentials in major urban labor markets, at least as measured by skill ratios for female clerical workers in all industries and for males in manufacturing. In addition, he shows that short-run fluctuations in these differentials have been negligible over the same period. Since no pattern of regional variation appears in the changing differentials, Segal concludes that his findings are indicative of trends in occupational differentials for the economy as a whole. To the extent that such a conclusion is warranted, this marks a reversal in the widely noted narrowing of occupational differentials which persisted through the 1940's.

First, several minor points. Since no mention is made of the quality of the data employed, we must assume that the changes observed do not represent a statistical mirage. As to the comparisons themselves, it is difficult to interpret them because the occupational groups used in calculating the skill ratios have not been indicated, even in footnotes. Second, it should be possible to provide some further checks on the conclusion that a general widening of occupational differentials occurred, since this is what Segal really discusses in his paper. For example, the annual census income survey data would be useful in this respect, permitting the measurement of not only differentials among occupations but those within occupations as well. Third, it is difficult to know whether Segal is really concerned with occupational differentials or with skill differentials of wage workers in industries where unions enter into wage negotiations. Obviously, the latter category is an important part of the whole, but his explanations seem to run largely in terms of differentials among groups where unions play a key role. Fourth, as to why cyclical fluctuations fail to exert any effect on occupational differentials, one can only suggest that most adjustments to the mild recessions we have experienced took the form of layoffs or reductions in hours rather than changes in earnings differentials. In view of the short duration of these cycles (an average of thirty-six months), considerable variation in earnings would have been the order of the day had not these other adjustment mechanisms been available.

Let us now shift our attention to Segal's interpretation of his findings. He places great weight upon various institutional forces, as is indicated by some of the explanations included as evidence. For example, Segal cites the recent "concern of employees" with percentage differentials and changes in earnings, "considerations of equity toward the skilled" voiced by employers, the maintenance by employers of a "balanced" wage structure, and so on. These institutional responses to a long period of narrowing differentials combined with the advent of favorable demand conditions for skilled workers can be considered, Segal feels, as providing at least a tentative explanation of the change in differentials observed.

While the importance of demand factors can hardly be doubted, the other "institutional" evidence would seem to indicate a response to these demand conditions rather than an independent force as suggested by Segal. It certainly is true that demand forces have been strong in the period of the 1950's and that they have been selective in favor of the more skilled groups. Although this in itself is or may be an "explanation" of the widening occupational differentials, the really interesting question, and the one to which Segal does not address himself, concerns the mechanism of adjustment by which occupational differentials change.

One suggestion of the mechanism emerges from some of my own work which takes account of the way in which earnings differentials appear to change. Given the strength of demand forces, especially at the higher income-skill occupations, and the smaller cohorts of new workers entering the labor force in the 1950's, new workers are attracted into growing occupations only by bidding up the starting earnings levels. When this occurs, average incomes in those occupations rise relative to other occupations by virtue of the relative increase in the earnings of younger workers. As time passes these changes may be transmitted through to other age groups so as to further widen the differentials. In addition, the impact of changing levels of education among different occupational groups is completely ignored by Segal. A casual examination of some of the educational attainment data by occupational level indicates that this quality dimension may be extremely important.

These comments clearly indicate that we must press much harder in attempting to measure changing occupational differentials, in trying to account for them, and in suggesting the mechanism that gives rise to them.

COMMENT by Robert L. Aronson

On their own terms, it is difficult to be critical of the papers presented in this session of our conference. Methodologies are not only appropriate to the purposes described but in several cases also represent improvement in analysis of wage and mobility relationships. I am especially impressed by Goldner's attack on the problem of relating wage levels to skill differentials; and Bunting's effort to test the wage-mobility relationship is rather successful in avoiding a major weakness of some earlier studies of this problem. Along the explanatory dimension, the gains are somewhat more modest, though such a judgment may be premature. Despite these riches, I am also uneasy that we may miss our target--a fuller understanding of the interior structure and behavior of metropolitan labor markets. Although the bulk of my remarks will be directed to this aspect of the papers, I also wish to comment on a few minor points of substance and technique.

Partly because they do focus on problems of long-standing interest in the field of labor market economics, the papers before us throw into bolder relief the question of the appropriate frame of reference for study. A substantial portion of the questions studied by labor economists are studied in an urban or metropolitan setting, of course, because this is the principal locus of the behavior of interest to us. Except on a broad regional basis, however, locality and the various socio-economic attributes that it may represent have only rarely been studied as explicit variables. Yet, it seems to me, that our objective is, or in this context ought to be, to learn how labor markets internal to the large metropolitan areas are structured and how they function in matching job opportunities and labor supplies. More particularly, we may wish to know the extent to which such markets are stratified or "balkanized," to use Clark Kerr's apt phrase, the degrees of monopsony or competition within such markets, how much mobility there is within them and along what dimensions, and, of course, the extent to which several economic and institutional variables that can be identified as characteristically urban influence wage movements and the behavior of occupational wage differentials.

With the exception of Goldner's study, the papers presented, though meritorious on other grounds, do not take us far in the direction described. Segal is explicitly concerned with the national wage structure and, especially, with updating our knowledge of the behavior of skill differentials during the past decade, though he does turn up some interesting intercity variations in the differential. Fleisher's focus on New York City is largely a matter of historical and locational accident, though migration would clearly be relevant to the analysis of changes in the industrial composition of metropolitan areas, or of intercity differences in the movement of wages. Bunting's data on wage-mobility experience in a three-state area may represent a situation which is substantially metropolitan—— I have no knowledge of the distribution of BOASI coverage in that area—— but I am uneasy that only about a fifth of the total population of the area was located in SMAs in 1950. Goldner's framework is more explicitly metropolitan in focus, but I regret that, in this paper at least, he has been so determinedly sparing in his investigation of explanatory variables.

Of course, the difficulties of dealing with urban labor market phenomena, per se, should be understood. Much of the data we might desire for analysis, especially on the critically important supply side, are not available to us. Segal, for example, had to use national unemployment rates as a measure of short-run changes in labor demand, even though acknowledging wide differences between cities with respect to the levels and rate of change in unemployment.[1] To take another example, it is doubtful, at least in the region studied by Bunting, that the Continuous Work History sample data can be broken down within the limits set by the rules of statistical inference to show wage-mobility patterns within metropolitan areas.[2] In the domain of labor market institutions and institutional behavior, of course, we are terribly deficient in reliable and continuing sources of information.

Because of the unavailability of much of the data that we might desire for metropolitan labor market analysis, I think we must be alert against excessive concern for methodological purity. I grant Goldner and Bunting their criticisms of earlier studies as deficient in method and unsuccessful as attacks on conventional labor market theory.[3] Yet, I am apprehensive that the gains achieved by these earlier studies may be overlooked as we become further interested in metropolitan labor markets. Some of the studies provided rich insights into some of the variables which may explain wage relationships and mobility patterns. This conclusion, I think, could be demonstrated by evidence from the papers by Professors Segal and Bunting. Second, labor market behavior involves worker and employer decision processes which, probably, can be better understood through studies in depth; moreover, for the purpose of implementing labor market policy, the path by which Bunting's mobile workers moved to higher-paying jobs, for example, may be as important as the knowledge that mobility is induced by wage differentials. Finally, we may be compelled by the nature of the problems we wish to study to use a case-study approach; studies of intra-firm mobility and internal wage structures, or the influence of pension plans, etc., on mobility in local labor markets are illustrative possibilities.

Professor Goldner's paper, in my judgment, strongly invites comparative studies of individual metropolitan labor markets. To my knowledge, for the first time, a satisfactory method of relating skill differentials to the community wage level has been shown. This alone is an important development, since the relation between wage structures and general wage levels has heretofore been quite conjectural and widely acknowledged as one of the least satisfactory of the central issues in wage-employment theory. But his preliminary findings also reveal a lack of consistent relationship between the level and spread of occupational rates, which suggests that one cannot assume homogeneity of metropolitan labor markets. Given the means of obtaining wage data on an establishment basis, moreover, it should be possible to extend Goldner's method to disclose the degree and pattern of interdependence between the general wage level, external wage structure, and occupational wage changes within the firm. My only uncertainty concerns Goldner's concluding comment on the need for further study of intermetropolitan labor mobility. For obvious reasons, we know little about such mobility, but I am dubious that it has much association with particular wage configurations. Thompson, in his Ohio study, showed that intermetropolitan movement effected relatively little change in the size of urban populations. [4]

On substantive matters in the other papers under discussion, I have only a few comments. Professor Bunting's conclusion about the association between voluntary mobility and differences in wage improvement among workers may need to be tempered, since his method does not permit a statement about the relative weight of wage differences in directing labor flows. His analysis of age, sex, and race differences in wage improvement experience only tells us the extent to which the propensity to move may be damped or facilitated. Robert Raimon observes that the wage difference thesis works well in explaining long-distance migration, but may fail for the local labor market where much of the mobility is initially involuntary. [5] The data used by Professor Bunting probably would not permit distinguishing local labor market from interstate or intraregional mobility.

With respect to the short-run behavior of the female office worker differential discussed by Professor Segal, I have a few additional suggestions to make about the factors that may account for its recent stability. Splitting the data between manufacturing and non-manufacturing industries may reveal either a weighting problem or offsetting movements in these two components of the differential, since the strength of labor demand may be quite different in the two sectors. Another possibility is that the over-all female labor force participation rate may be too crude as a measure of cyclical changes in female labor supply to office work employment, since there may have been above-average increases in participation rates in those age and marital classes in which female office workers are typically concentrated. Finally, it may be worthwhile to correlate changes in office worker wage differentials with male unemployment rates, on the supposition that an "additional worker" effect is operative in the labor markets investigated by Professor Segal.

[1] The need and desirability of reliable labor force and employment data on a local basis and at intervals more frequent than the decennial census has long been recognized, but I am doubtful that we shall have such data, or even much improvement in the data available, in the near future. For a recent discussion of this problem, see President's Committee to Appraise Employment and Unemployment Statistics, Measuring Employment and Unemployment (Washington, 1962), pp. 190-98.

[2] This is apparently difficult even in industrially more populous states. See Donald J. Bogue, A Methodological Study of Migration and Labor Mobility in Michigan and Ohio in 1947 (Scripps Foundation, 1952), esp. pp. 18-21 and Chap. XI, pp. 81-85.

[3] It may be noted, however, that the authors of the post-war labor market studies were usually modest in their claims, and that at least one of them has publicly recanted. See Charles A. Myers, "Labour Market Theory and Empirical Research," in John T. Dunlop, ed., The Theory of Wage Determination (New York: The Macmillan Co., 1957), pp. 317-26.

[4] Warren S. Thompson, Migration Within Ohio, 1935-40 (Scripps Foundation, 1951).

[5] Robert L. Raimon, "Interstate Migration and Wage Theory," Review of Economics and Statistics (forthcoming).

COMMENT by J. B. Lansing

On Bunting

I should like to raise a question about the interpretation of the difference in wage improvement percentages between mobile and non-mobile workers which is reported in the paper by Robert L. Bunting. Is it possible that some part of this difference is due to the statistical phenomenon known as regression toward the mean? The question arises since all workers earning over $3,600 have been deleted from his study.

We can imagine a situation in which regression toward the mean would be important, and then raise the question whether the observed facts fit that model. Suppose we were dealing with a homogeneous group of workers who were similar in training, experience, and all other variables known to have a systematic effect on wages. Within this group, however, we would not expect wages to be identical. There would be some random fluctuation in earnings from one worker to the next associated with differences in the jobs in which they found themselves. The actual distribution of wages at a given point of time about the mean for the group would very likely be normal.

Suppose then we arbitrarily required each worker to leave his job and find a new job. We may assume that all workers have new jobs at a second point of time. The wage rate earned by each worker in his new job would be independent of his wage rate in the old since we speak of a homogeneous population of workers. We would then expect that of those below the group mean at time (1), half would be below and half above at time (2). Of those above the mean at time (1), half would be below and half above at time (2). There would be no change at all in the mean wage or the distribution of wages. But if we followed the sequence of events for individual workers, those below the mean to begin with would show an increase on the average, while those above the mean would show a decrease on the average.

Suppose we now relax the assumption that all workers must change jobs. We may at random designate certain workers who will change jobs and others who will not change. It will still remain true that those who do change jobs, and hence earn different wages, will regress toward the mean.

The question which I wish to raise, then, is whether the facts of the labor market may not correspond in some degree to such a model as the above.

COMMENT by Martin Segal

On Fleisher

The following comments are limited to only one aspect of Professor Fleisher's paper -- his analysis of the impact of Puerto Rican migration on the wage behavior of the New York Region in the post World War II period. I cannot claim any special knowledge of the most recent developments either in the movement of the New York wages or in the pattern of Puerto Rican migration. Accordingly, my discussion is based almost entirely on the material contained in the study of New York Region's wages completed over three years ago, and on the statistics used by Fleisher.[1]

Fleisher's paper contains two conclusions pertaining to the impact of Puerto Rican migration. First, the author concludes that the migration had a significant effect in that it caused a decline in the relative earnings

(relative with respect to the nation as a whole) of those New York industries that employ large numbers of Puerto Ricans. Secondly, he concludes that the Puerto Rican migration has been a more important factor
in determining the course of wages in the Women's and Children's
Garment industries than the changes in technology and demand that have
characterized these industries, and that have exposed the New York
Region's firms to strong competitive pressure from producers located in
other -- and generally lower wage -- areas.

Fleisher's conclusions differ from the interpretation of the Puerto
Rican impact suggested in my study. [2] In a brief and simplified form this
interpretation is as follows: Technological changes, developments in
transportation, and changes in tastes have exposed some industries that
previously found important locational advantages in the New York Region
to competition from manufacturers locating and expanding in relatively
low wage areas. The bulk of the New York industries exposed to such
competition is found in the sector of garment and related industries. The
impact of increasing inter-regional and inter-area competition has been
reflected in the wage policies of employers and unions of the New York
industries. As a consequence of these policies, wages of the garment
industries have lagged markedly behind those of the other sectors of the
Region's economy. However the maintenance of such a lag -- and this
means, of course, the maintenance of relatively low wage levels in the
New York garment sector -- would have been hardly possible without
the migration of Puerto Ricans. The new migrants provided in effect
labor supply for the lower paid jobs in the garment sector -- jobs that
would not be filled at the prevailing wages by natives of the New York
region. Thus the Puerto Rican workers facilitated the lag in the rate of
wage increases of the New York garment and related industries. To use
the familiar terminology, Puerto Rican migration is viewed, in this interpretation, as a necessary but not a sufficient condition for the fact that
the New York garment wages have risen only about as much as those of
garment industries in the new rural centers of production, and that they
have been able to lag significantly behind the wages of the rest of the
Region's economy. In this sense the Puerto Rican migration may be said
to have influenced the course of wages in the New York Region.

Fleisher's interpretation of the influence of Puerto Rican migration on
New York's wages is supported, in his view, by the behavior of New York-
New Jersey's relative earnings during the 1947-58 period. During the
1947-54 period the average annual migration of Puerto Ricans was about
8,000 less than during the following four years. Fleisher's data for what
he calls "Women's and Children's Garments" show that New York-New
Jersey relative earnings declined in 1947-54 but not in 1954-58. The other
so-called "Puerto Rican industries," when considered one by one, apparently do not exhibit such a pattern. Fleisher therefore aggregates them
but even then, as evidenced in his Figure 4, the aggregate shows a decline
in relative earnings both in 1947-54 and 1954-58. [3] Finally the non-Puerto
Rican industries -- presumably the rest of manufacturing grouped together -- show relative earnings behavior very much like that of Women's
Garments, i.e. declining in 1947-54 and remaining stable in 1954-58.

Even if Fleisher's statistics reflected correctly the actual course of
relative wages, they would still provide, I believe, slim support for his
interpretation. A difference of 8,000 in the influx of Puerto Ricans implies a change in the annual increase of the Puerto Rican labor force of
about 4,500. Of these, probably 50 per cent would find jobs in a variety

of local services (hotels, restaurants, laundries) that are not considered by Fleisher. It does not seem likely that a reduction of 2,250 in the annual increase of Puerto Rican workers would produce immediately a reversal in the course of relative earnings of strongly unionized industries employing a few hundred thousand workers. [4] Indeed, insofar as a reversal, of the course of relative earnings in garments did take place in the late fifties, it can be explained more satisfactorily, as I shall indicate below, in terms of factors that are ignored by Fleisher, and that are perfectly compatible with the emphasis on the critical role of inter-regional competition. But before considering the more basic problems of interpretation, I wish to discuss some troublesome aspects of the statistical data presented by Fleisher.

The wage data used by Fleisher are average annual earnings of broad industrial categories composed of several four-digit industries. For example, his "Women's and Children's Garments" category consists essentially of ten four-digit industries. The individual Puerto-Rican industries, other than "garments," are three-digit classifications, each consisting of several industries. And the non-Puerto Rican group consists of an unspecified but presumably large number of manufacturing industries. All these data pertain not to New York City or to the New York Metropolitan Region but to the two states of New York and New Jersey.

As noted, Fleisher's interpretation relies heavily on the evidence indicated by the short-run behavior of his earnings series. But, in my view, Fleisher's data, regardless of whatever they may show, are not capable of providing support for a hypothesis concerning the impact of Puerto Rican migration on the earnings of the New York Metropolitan Region.

The reasons for this view are illustrated by the example of one of Fleisher's Puerto Rican industries -- "Canning, Preserving and Freezing." The industrial classification of "Canning" is composed of several different industries (e.g. "Pickles and Sauces," "Frozen Fish", etc.). In 1947 -- and undoubtedly also today -- a majority of the employees in the New York Region's "Canning" classification worked in one particular industry, namely "Pickles and Sauces" (S.I.C. 2035). While this industry employs mostly Negroes, it does have some Puerto Rican workers. Accordingly, when indicating in my study the industries employing both Puerto Ricans and Negroes, I included in the list what I called "canning plants." [5] This in turn led Fleisher to classify "Canning, etc." as a Puerto Rican industry.

Now a possible testing of a hypothesis pertaining to the impact of Puerto Ricans may involve an analysis of the course of earnings of the "Pickles and Sauces" in the New York Region as compared with that of the earnings for the country. A comparison of the earnings of the entire "Canning, etc." classification would already be highly misleading. The reason is that the industrial mix of this classification is very different in the New York Region from that in the entire nation. For instance, the particular industry that employs some Puerto Ricans and constitutes the bulk of "canning" in the New York Region employs only 10 per cent of the workers of "canning" in the country as a whole. It is incorrect to assume that in a period of a few years the other component industries -- e.g., "Canned sea foods" or "Frozen fruit juices" -- would experience the same percentage changes in earnings as "Pickles and Sauces." Differences in the rate expansion and demand, in technological changes, or in

timing of contracts may cause significant differences in percentage
changes in earnings among the different component industries of one
broad classification. As a result, a comparison of earnings of broad
classifications is likely to be affected in a significant way by the industrial
mix characterizing individual areas. [6] In the case posited above the earnings of the New York Region's "Canning" would reflect primarily the
course of "Pickles and Sauces" earnings. But for the country as a whole,
the aggregate earnings of "Canning" would reflect mainly the course of
wages of the other component industries.

Another problem arises because Fleisher's data pertains to the states
of New York and New Jersey rather than to the New York Metropolitan
Region. For one thing, the "Canning" employment of the New York Region
constitutes less than 17 per cent of the equivalent employment for the two
states. Thus Fleisher's earnings figure is dominated by wages of workers
outside of the area of Puerto Rican migration. Secondly, the industrial
mix of "Canning" for the two states is again different both from that of
the New York Region and from that of the entire country. Accordingly,
a comparison of the New York-New Jersey earning course in "Canning"
with the national data does not throw any light on the problem of the
Puerto Rican impact on the wages of "Canning" in the New York Region,
let alone on the wages of the one particular component industry that
actually employs some Puerto Ricans in New York City.

What was said above about "Canning, Preserving, Freezing" applies
also to the other Puerto Rican industries. All of these are essentially
broad classifications composed of several industries. All of them have
different industrial mixes in the New York Region, in the two states, and
in the country as a whole. [7] As a result, a comparison of the New York-
New Jersey earnings with the national data does not seem to be very useful for the purpose of providing support for Fleisher's hypothesis.

Finally, one should note that Fleisher aggregates all the data for the
Puerto Rican industries and then compares them with the national data.
This procedure further removes the results from any possible connection
with the impact of Puerto Ricans on the earnings of particular industries
in the New York Metropolitan Region. For then the comparison is really
between two aggregates that have very little in common with the earnings
of individual industrial classifications in the New York Region. [8]

Let me turn now to what Fleisher calls "Women's and Children's
Garments" industry. The ten industries that compose this classification
are concentrated in the New York Metropolitan Region. Therefore the
data for the two states may be considered as representing relatively well
the area under the influence of Puerto Rican migration. [9] But there is
still the important problem of the mix of the broad classification used by
Fleisher. As might be expected, the industrial mix of the broad category
of "Women's Garments" in the New York Region is quite different from
that of the nation. The component industries of this broad category
experienced in 1947-55 very different percentage rates of wage
increases -- rates varying from 28 per cent to 7 per cent. Inevitably
then, Fleisher's measures -- comparing earnings of two different industrial mixes -- cannot provide a reliable index of the course of relative
earnings in the individual industries of the New York Region that employed large numbers of Puerto Ricans. Thus industrial composition of
"Women's Garments" in the New York Region is heavily weighted with
two industries that in the 1947-55 period experienced the least percentage
increases in earnings -- "Women's dresses" and "Women's suits, coats

262

and skirts. "[10] This fact in itself probably constitutes a large part of the explanation why Fleisher's measures show a decline in the relative earnings of the New York Region during 1947-54. [11]

The available data on average hourly earnings of some component industries of "Women's and Children's Garments" do not indicate that the New York Region's relative earnings declined markedly during the years of large Puerto Rican migration and then remained stable during 1954-58. For example, the special B. L. S. computations made for my study show that during 1947-55 the New York Region's relative earnings declined in one industry ("Women's Dresses") by less than two percentage points and rose in another ("Women's Nightwear") by 1-1/2 per cent points. [12] Another relevant set of data is provided by the recent B. L. S. report on employment and wages in New York City during 1950-1960. [13] This report contains average hourly earnings of the two major three-digit industries and of one four-digit industry in the "Women's and Children's Garment" group. These data were used to compute relative earnings of the New York City industries in 1950, 1954, and 1958. [14] The results, shown in Table 1, do not provide any support for Fleisher's interpretation. The years 1950-54 were those of the largest Puerto Rican migration to New York City, with an average annual influx of 37,846. Yet the relative earnings in New York City -- the area presumably most intensively affected by the Puerto Rican workers -- actually increased. During the subsequent four years annual immigration fell to 25,521; the relative earnings of New York City garment industries still increased but somewhat less than in the previous period. Thus, as one might expect, more detailed data do not indicate that the course of relative earnings in the Region's garments was very responsive to the short-run fluctuations in the annual influx of Puerto Ricans.

Table 1. Relative Hourly Earnings of New York City's Women's and Children's Garment Industries in Three Selected Years (New York City's Earnings Expressed as Percentage of Earnings in the U.S.A.)

Industry	Years		
	1950	1954	1958
Women's Dresses	133%	136%	137%
Women's Outerwear	128	132	134
Women's Underwear and Nightwear	115	119	119

Sources: BLS Employment, Earnings and Wages in New York City, 1950-60, Table B-3; BLS Employment and Earnings, Annual Supplement Issues for 1955 and 1959.

Since Fleisher based his argument against the interpretation stressing inter-regional competition on the behavior of the New York Region's relative earnings, [15] it seemed appropriate to consider his statistics in some detail. Nevertheless the problem of statistical evidence is not the most troublesome aspect of Fleisher's paper. More critical is the fact that Fleisher's emphasis on the impact of Puerto Rican migration rather than on the influence of inter-regional competition implies a strange type of union policy in the garment industries of New York; and that it cannot provide a meaningful explanation of, and is indeed inconsistent with, some important changes in the inter-industry wage structure in the New York Region.

As noted before, during the 1947-55 period the industries of the garment complex in New York and particularly the women's garments, experienced significantly smaller percentage wage increases than the rest of industries in the area.[16] Thus in this period the median increase in the Region's construction and services was 49.6 per cent; in local market industries 52.4 per cent; in national market industries, except garment trades, 47.7 per cent. In the garment sector industries the median increase for the same period was only 17.8 per cent, with the women's and children's garment industries experiencing less than average gains.

Since the garment industries in the New York Region are completely unionized, this significant lag reflects in the main the policies evolved in collective bargaining. The garment unions have exercised very considerable restraint in their wage policies. For example, an 8 per cent increase in women's dress industry, negotiated in 1958, constituted the first general wage change in that industry in five years. Similarly, restraint was clearly exercised in the negotiation of piece rates, so critical in influencing unit labor costs, and also earnings of the garment workers. The reasons for union policies must be sought in the strong competitive pressures from outside production centers and in the need to preserve jobs and unionized shops in the New York Region. At the same time the collective bargaining policies were made feasible by the influx and hiring of Puerto Ricans. These factors explain why garment wages lagged so much behind those of the other sectors of the New York Region. Quite distinct from the basic phenomenon of wage restraint policies, the year-to-year changes in the Region's relative earnings must be explained by a host of other factors -- changes in locational distribution of firms outside of the Region (e.g., Cleveland vs. Wilkes-Barre); locational shifts within the Region (e.g., New York City vs. Paterson, N.J.); changes in the product mix and thus in the incentive earnings of the workers in particular industries, changes in the skill mix of the work force, etc.

In contrast to the above, Fleisher's interpretation seems to suggest (a) that the results of collective bargaining in the garments industry were influenced primarily by the fact that the in-coming Puerto Ricans were willing to work at relatively low wages; and (b) that during the lifetime of any long-term contract fluctuations in garment earnings responded -- presumably as a result of piece-rate setting -- to short-run changes in the yearly influx of Puerto Ricans. It is difficult to see why these phenomena should take place. The Puerto Rican migration has been adding annually only about 13,000-15,000 workers to the Region's labor force of 7 million. The New York garment industries, far from suffering marked unemployment, have been actually experiencing some difficulties in recruiting labor at the going wages. Moreover, an overwhelming majority of garment workers -- practically all of them unionized -- has not been Puerto Rican. Given these circumstances it does not appear reasonable to assume that the Puerto Rican migration could be the primary cause of a drastic change in the relative position of garments in the Region's wage structure; or that it led to union policies resulting in a decline in real earnings and a marked -- and politically embarrassing -- lag of garment wages behind those of the rest of the New York Region. The only factor that makes the union policies and the behavior of garment wages in the Region explicable is the compelling pressure of outside competition and the continuous threat of locational shifts of garment industries to lower -- and frequently non-union -- wage areas.[17]

An important piece of evidence pertaining to the issue at hand is provided by the comparison between the course of garment wages and that
of the other services and industries in the New York Region that employ
substantial numbers of Puerto Ricans. Several of these non-garment
industries are virtually free from inter-regional competition because of
transport costs, the need for immediate contact with customers, etc; in
others the impact of outside competition is not nearly as strong as in the
garment industries. During 1947-55 the median increase in all the non-
garment Puerto Rican industries and services for which data are available was 49. 5 per cent; in the local market industries and services --
i. e. , those free from outside competition -- the median increase was
slightly higher, over 50 per cent.[18] This is, of course, in sharp contrast with the median increase of 17. 8 per cent in the garment sector. It
should be also noted that the wage increases in the local market sector
employing Puerto Ricans (e. g. , hotels) were as large as or larger than
those in other metropolitan areas that in 1947-55 experienced little or no
Puerto Rican influx. And the over-all wage increases in the non-garment
Puerto Rican sector were about the same as in the Region's industries
employing no Puerto Ricans.[19]

The implication of the above cited data is, I think, fairly clear. If
the mere process of employing large numbers of Puerto Ricans had been
the critical factor influencing wage changes in particular industries of the
Region, one should not find the striking contrast between the course of
wages in garments and in the other Puerto Rican industries and services.
But, if the critical difference derives from the difference in the exposure
to the competitive forces emanating from outside of the Region, the con-
trasting wage behavior in garments and in the other industries becomes
fully explicable.

Finally, let me turn briefly to the problem of wage behavior in the
garment sector during the late fifties -- a period not covered by my study.
It is quite possible, and even likely, that wage series still less aggregated than those in Table 1 would show that in some garment industries the
Region's wages rose by larger percentages in 1954-58 than in the previous
period; or that they rose more than the wages of the same industries in
the rest of the country. But if this happened the reason lies not in the decrease of the annual influx of Puerto Ricans, as Fleisher's paper would
imply, but rather in the response of the Region's labor market to the
drastic and prolonged lag in the advances of garment wages. Because of
this lag New York employers have experienced difficulties in attracting
new workers; in various campaigns garment unions have been accused of
"neglecting" the interests of Puerto Ricans and Negroes; there has been,
undoubtedly, dissatisfaction among garment union members; there have
been open debates concerning the deteriorating status of New York earnings and the "exploitation" of newcomers to the Region. These factors,
reflecting basically the interdependence of wage movements in a local
labor market, produce inevitable pressures on employers and unions in
the garment sector. The consequence may well be that, at the cost of
further reduction in the employment of product lines most affected by
outside competition, the rate of increase of garment wages is being
brought closer to that of the rest of the Region. Fleisher's paper ignores
the post-war developments in the internal wage structure of the New York
Metropolitan Region. Because of this omission, and also because it does
not concern itself with the mechanics of the operation of the New York
labor market, the paper gives, in my view, an incorrect interpretation
of the Puerto Rican impact on the course of New York wages.

[1]Martin Segal, Wages in the Metropolis, (Cambridge: Harvard University Press, 1960).

[2]Segal, op. cit., pp. 101-04, 128-36, 160-63. Fleisher's statement of my interpretation is somewhat misleading since it implies that I do not attribute any influence to Puerto Rican migration. For example, the quote in footnote [21] of Fleisher's paper, taken in the context of his discussion, seems to imply that I was writing about the lag of New York's garment wages behind those of the rest of the nation. But the cited phrases refer to a different phenomenon, namely the lag of garment wages behind the wages of other industries in the Region that employ substantial numbers of Puerto Ricans.

[3]In the latter period the decline of New York's relative earnings is less marked.

[4]It is also puzzling that there should be a greater similarity between the course of garment wages and that of non-Puerto Rican industries than between garment wages and other Puerto Rican industries.

[5]Segal, op. cit., p. 130.

[6]The problems arising out of the differences between the industrial mix of the New York Region and of other areas or the country as a whole are treated exhaustively in R. Lichtenberg, One Tenth of a Nation, Harvard University Press, 1961. See also, Segal, op. cit., pp. 46-47. For a general discussion of the problem as it pertains to wage statistics, see F. A. Hanna, State Income Differentials, 1919-1954, (Duke University Press, 1959).

[7]For example, "Sporting and Athletic Goods" is, in terms of employment, an important component industry of the classification of "Toys and Sporting Goods" in the country as a whole. In the New York Region, the number of workers in that industry is insignificant.

[8]The non-Puerto Rican sector is an aggregation of an unspecified number of industries. In this case, the problem of mix is undoubtedly even more crucial.

[9]However, the relative importance of the Region has been diminishing considerably since 1947.

[10]The "Women's dresses" industry was particularly affected by inter-area competition -- particularly between low wage areas of Pennsylvania and New York -- and by rise of cheaper style lines. In the "Women's suits, coats and skirts" industry the course of earnings reflects to a significant degree the relative rise of employment in skirt shops. These shops employ much less skilled labor than "coats and suits" shops and pay lower wages. As a result, the over-all hourly earnings of the industry show very little increase during 1947-54. The same is, of course, true of average annual earnings of the workers.

[11]Another reason for the decline of relative earnings probably lies in the locational shifts of the garment industries within the New York Region and within the two states. During 1947-55 in several garment industries there was some redistribution of employment from New York City to the other areas of the New York Region -- areas in which wages were lower than in the city. This represented a shift from the area most affected by Puerto Ricans to areas in which there were few or no Puerto Rican migrants. In the measure such as that of Fleisher, this shift would reduce the rate of increase of New York's wages. See Segal, op. cit., pp. 136-142, and Appendixes B and C.

[12]Segal, op. cit., p. 89. My work was not carried beyond 1955.

[13]U.S., Bureau of Labor Statistics, Employment, Earnings and Wages in New York City, 1950-1960, New York, June, 1962.

[14]Year 1950 was the earliest in this publication. The other two years were chosen because these are the years used by Fleisher. Prior to 1950, average yearly Puerto Rican migration was actually less than in the later periods.

[15]See his statement toward the end of his paper: "If changes in technology and the structure of demand has been important in causing the 1947-54 decline in New York-New Jersey relative earnings and employment share, why did they not have the same effect in 1954-58." Actually, the share of employment in garments declined throughout the entire 1947-54 period.

[16]See Segal, op. cit., pp. 133-42 and Appendix B. The data in my study extend through 1955.

[17]As all the data, including most recent surveys, show, there has been in fact a very considerable shift of garments from the Region to lower wage areas.

[18]Segal, op. cit., p. 135, Table 29. In calculating the median changes in Puerto Rican industries, I excluded some industries in the table because they employ only Negroes (e.g., building trades).

[19]Segal, op. cit., Appendix B.

For Product Safety Concerns and Information please contact our EU
representative GPSR@taylorandfrancis.com
Taylor & Francis Verlag GmbH, Kaufingerstraße 24, 80331 München, Germany

www.ingramcontent.com/pod-product-compliance
Lightning Source LLC
Chambersburg PA
CBHW070611270326
41926CB00013B/2499